The Evidence for the Top Quark
Objectivity and Bias in Collaborative Experimentation

The Evidence for the Top Quark offers both an historical and a philosophical perspective on an important recent discovery in particle physics: the first evidence for the elementary particle known as the top quark. Drawing on published reports, oral histories, and internal documents from the large collaboration that performed the experiment, Kent Staley explores in detail the controversies and politics that surrounded this major scientific result.

At the same time, the book seeks to defend an objective theory of scientific evidence based on error probabilities. Such a theory provides an illuminating explication of the points of contention in the debate over the evidence for the top quark. Philosophers wishing to defend the objectivity of the results of scientific research must face unflinchingly the realities of scientific practice, and this book attempts to do precisely that.

This book will prove to be absorbing reading to a broad swathe of readers including philosophers, physicists, and historians of science.

Kent W. Staley is Assistant Professor of Philosophy at Saint Louis University.

The Evidence for the Top Quark

Objectivity and Bias in Collaborative Experimentation

KENT W. STALEY

Saint Louis University

PUBLISHED BY THE PRESS SYNDICATE OF THE UNIVERSITY OF CAMBRIDGE
The Pitt Building, Trumpington Street, Cambridge, United Kingdom

CAMBRIDGE UNIVERSITY PRESS
The Edinburgh Building, Cambridge CB2 2RU, UK
40 West 20th Street, New York, NY 10011-4211, USA
477 Williamstown Road, Port Melbourne, VIC 3207, Australia
Ruiz de Alarcón 13, 28014 Madrid, Spain
Dock House, The Waterfront, Cape Town 8001, South Africa

http://www.cambridge.org

First published 2004

Printed in the United Kingdom at the University Press, Cambridge

Typeface ITC New Baskerville 10/12 pt.   *System* LaTeX 2_ε [TB]

A catalog record for this book is available from the British Library.

Library of Congress Cataloging in Publication Data

Staley, Kent W. (Kent Wade), 1963–
The evidence for the top quark : objectivity and bias in collaborative
experimentation / Kent W. Staley.
p. cm.
Includes bibliographical references and index.
ISBN 0-521-82710-8
1. Particles (Nuclear physics) – Flavor. 2. Quarks. I. Title.
QC793.5.Q252S73   2004
539.7′2167–dc21   2003048465

ISBN 0 521 82710 8 hardback

For Dianne

Contents

Contents

Figures and Tables

TABLES

Preface and Acknowledgments

The origins of this project are located along a lonely stretch of Interstate 80 in Iowa in the early 1980s. Drasko Jovanovic, head of the Physics Department at Fermi National Accelerator Laboratory at the time, was in charge of Fermilab's "summer student" program. Students in the program had the opportunity to work in experimental groups at the lab and to learn about high energy particle physics. While driving along I-80 one day, Jovanovic saw a sign announcing "Grinnell College, next exit," which prompted him to note that Fermilab had not hosted a summer student from Grinnell in a while. I was the next Grinnell student to apply for the program. I spent the next two summers bending light guides and stringing cables under the imperturbable guidance of Fermilab physicist Mike Crisler, in the somewhat infamous "E-711" group.

More than ten years later, I was fishing around for a dissertation topic, and I called Drasko to ask whether anything interesting was happening at Fermilab. He dropped rather broad hints that something really big was in fact just about to happen: "I can't speak too freely on the phone, but . . ." He suggested that I call back in about a week. That was on April 19, 1994. Exactly one week later, the *New York Times* ran the headline, "Top Quark, Last Piece in Puzzle of Matter, Appears to Be in Place," on page 1. The historical and philosophical investigations presented here center on the events leading up to that headline.

Serendipity alone could not have brought about the completion of this project, however. For that I needed also the assistance of many people. Indeed, the nature of this project necessarily depends directly or indirectly on the efforts of a great multitude – indeed of more people than I can mention.

But I might start by referring you to oral histories listed in the references to this work. There you will find listed those who, through conversations, interviews, and e-mail messages, did their best to keep me out of the darkness of error. I want to thank especially those who took the time to read the dissertation that I wrote and to share their reactions with me. A few were especially generous with their time and deserve special mention. Henry Frisch played a special role in the early stages of my investigations, both encouraging me and putting me in touch with many other people. Henry also directed my attention to the inherent interest of the problem of bias. Tony Liss was especially generous in helping me to understand the process by which the Evidence paper was produced, and many of the details of the analysis described in that paper. Dave Gerdes and Bruce Barnett both served on my dissertation defense committee and provided me with very valuable guidance with respect to both matters of historical fact and philosophical argument. G. P. Yeh provided me with very detailed accounts of some of the debates within the collaboration over the Evidence paper results. I benefited tremendously from correspondence with Morris Binkley, Joe Incandela, Krys Sliwa, Paul Tipton, and John Yoh. I am grateful as well to Boaz Klima of the D-zero collaboration for providing me with his reactions to my project. To him and to the physicists of D-zero in general, I wish to express my regret that I could not give D-zero's search for the top the same kind of detailed attention that the CDF (the Collider Detector at Fermilab) collaboration receives in this work.

Lillian Hoddeson and Adrienne Kolb at Fermilab helped me by advising me about who to talk to and by helping me find my way through the Fermilab archives. Carol Picciolo, CDF's secretary, helped me to navigate around the collaboration and advised me on how to deal with the lab bureaucracy. While conducting research at Fermilab, I have at various times relied heavily on the apparently inexhaustible hospitality of my friends Chris McKeachie and Deb Saeger of Oak Park, Illinois.

Many times in the pursuit of this project I have sought correction and advice from physicists not directly involved in the search for the top quark, particularly in my work on Chapter 1. Both Jonathan Rosner and Serge Rudaz gave me helpful feedback on some of the ideas developed in that chapter. Roger Stuewer encouraged me to develop a shorter version of the first half of the chapter for his journal *Physics in Perspective*, and in the process gave me valuable editorial advice. An anonymous referee for that journal steered me toward some resources that I had missed. I especially wish to thank Sandip Pakvasa for his very helpful comments, as a result of which I steered clear of some serious errors, and for patiently answering many questions.

My work on this project spanned many years and several phases of my career. I wish to acknowledge the friends and colleagues who have helped me, beginning at the end.

This manuscript was very nearly complete by the time I began my present position at Saint Louis University. Nevertheless, I wish to thank the SLU Department of Philosophy for finding sufficient philosophical merit in my work to invite me to join them as a colleague, which I can only regard as a compliment.

I visited Boston University during the spring of 2001. I am grateful to the Philosophy Department there for providing a congenial work environment for me. I wish especially to thank the students in my course "The Uses of Experiment" for many stimulating discussions.

Much of this manuscript was written while I served on the faculty of Arkansas State University. The challenge of carrying out research in the midst of a busy teaching schedule and a paucity of good research materials may well have proved overwhelming. I was saved largely through the assistance of my department chair Chuck Carr in coping with the former, and by the friendly and efficient services of the interlibrary loan department at the Dean B. Ellis Library in overcoming the latter. I wish also to acknowledge the congenial intellectual environment created by my philosophical colleagues at Arkansas State, especially my friend Ron Endicott (now at North Carolina State), from whom I learned so much about philosophy.

In the summer of 1999, I participated in a National Endowment for the Humanities (NEH) seminar on induction and probability directed by Deborah Mayo at Virginia Tech. The discussions that took place during that seminar influenced my thinking in many ways. I gained a great deal from everyone in the seminar, but I especially want to thank Douglas Allchin, Prasanta Bandyopadhyay, Peter Lewis, Greg Mikkelson, Cassandra Pinnick, Dave Rudge, Dan Sloughter, and Susan Vineberg for many helpful and challenging discussions.

When I began this project as a graduate student at Johns Hopkins University, I had no idea what I was getting into. Many people provided me with valuable advice, particularly Bruce Barnett and Dave Gerdes in the Physics Department. Robert Smith gave me valuable advice on interviewing scientists. Rob Rynasiewicz served as a reader on my dissertation; his comments helped me to sharpen my thinking. More importantly, he showed me by example how to approach a historical episode in physics in a responsible and scholarly manner. Allan Franklin of the University of Colorado took an interest in my project early on and gave me helpful encouragement. He also provided me with helpful feedback as a reader for Cambridge University Press. I'm also grateful to another anonymous reader for Cambridge and an anonymous reader for Johns Hopkins University Press for their comments. I am glad for having gone through graduate school with Chuck Ward, the most amiable of philosophical companions and in many ways a kindred spirit.

I finally wish to thank two philosophers from whom I have learned much as mentors and friends. Their contributions to my thinking will be self-evident in the text that follows. When I found myself confused by the statistical concepts used in physics publications, I read some articles by Deborah

Mayo and found that some things suddenly made much more sense to me. Subsequent correspondence helped further, and Deborah saved me from much rewriting of my dissertation by giving me an advance copy of *Error and the Growth of Experimental Knowledge* while it was still in page proofs. Peter Achinstein served as my dissertation advisor. His encouragement and advice kept me clear of several dead ends. His remarkably sharp critical eye and demanding standards kept me on my toes early on. After I completed my Ph.D., he remained the vigilant advisor, constantly encouraging me to keep working. While I certainly hope that this book constitutes a contribution to philosophical understanding, it was also tremendously satisfying to be able to say "yes" when Peter asked one last time, "Is it finished yet?"

Financial assistance for my 1995 interviews with CDF and D-zero physicists came from a grant-in-aid from the Friends of the Center for History of Physics, American Institute of Physics (AIP). Recordings of my interviews are in the archives of the AIP's Niels Bohr Library, and I am grateful to Spencer Weart at the AIP for his help. Arkansas State University provided me with a faculty research assistance grant for a round of follow-up interviews in 1998. Completion of the manuscript was facilitated by a summer stipend in 2001 from the National Endowment for the Humanities. Aaron Cobb helped with the indexing.

A significant part of Chapter 1 was previously published as "Lost Origins of the Third Generation of Quarks: Theory, Philosophy, and Experiment," in *Physics in Perspective*, volume 3 (2001), 210–29. I am grateful to Birkhäuser Verlag for permission to reprint that material here. Parts of Chapter 7 have previously appeared in *Philosophy of Science*: "What Experiment Did We Just Do? Counterfactual Error Statistics and Uncertainties about the Reference Class," volume 69, 279–99, copyright 2002 by the Philosophy of Science Association, and "Novelty, Severity, and History in the Testing of Hypotheses: The Case of the Top Quark," volume 63 (Proceedings), S248–55, copyright 1996 by the Philosophy of Science Association. I am grateful to University of Chicago Press for permission to reprint material from those articles here.

Thanks to my friends and family, especially Caroline and Lisa Staley, for their unfailing encouragement and support through many transitions and uncertainties. Above all, I am thankful to and for Dianne Brain.

Introduction

I prefer you to practise by drawing things large. . . . In small drawings every large weakness is easily hidden; in the large the smallest weakness is easily seen.
                                                    Leon Battista Alberti ([1435] 1966, 94)

Big science presents a big opportunity for methodologists. With their constant meetings and exchanges of e-mail, collaboration scientists routinely put their reasoning on public display, long before they write up their results for publication in a journal. For philosophers who wish to understand scientific evidence and scientific reasoning, such thinking out loud can be a rich source of information. (And given the expense of mounting an experiment in a field such as high energy physics, why not get as much as possible out of it? For all the interesting physics that emerges, the episode may also yield results for philosophical, social, and historical studies. Think of it as scholarly recycling.)

So here is a study of big science: the search for the elementary particle known as the *top quark*, the last of the six quarks of modern physics' "standard model" to be experimentally confirmed, by the Collider Detector at Fermilab (CDF) collaboration. CDF is certainly *big*. By the time that they announced their "observation of the top quark," CDF had swelled to a size of approximately 450 physicists. But in virtue of what does CDF's enterprise constitute *science*?

Without here attempting to specify necessary and sufficient conditions for science, I do wish to suggest that a central aim of a scientific enterprise is to determine what general claims about the world can be supported with evidence. That is, scientific enterprises are evidence-oriented.

At first glance, such a claim may seem overly simplistic, suggesting that scientists engage exclusively in testing theories and evaluating the outcomes of such tests. The chapters to come should dispel any such notion. Theorizing on a speculative level, scheduling meetings, designing experimental apparatuses, raising funds, performing feasibility studies, developing review

1

and oversight procedures, arguing over the choice of persons to carry out such procedures, running computer-aided simulation of data, calibrating instruments, and gossiping about one's competitors – these are all part of the story that I am about to tell. Yet throughout this story, we will find that, even when it does not take center stage, evidence plays an important role in the unfolding of events. The exact nature of that role depends on the context. Questions about evidence, or the potential for evidence, enter in one way during the development of speculative physical theories (Chapter 1) but in quite a different way during the design of a new detector (Chapter 2) or when deciding on the best means for analyzing data (Chapter 3). Evidential considerations move into the spotlight when, with the theory, instrument, and analysis in place, scientists ask themselves whether they do indeed have evidence for a specific claim, or, to turn the coin to its opposite side, for what specific hypothesis they can say that they have evidence (Chapters 4 and 5).

I chose to study this particular episode because of the detailed publication on which it centers, the "Evidence for Top Quark Production" paper published in 1994 by the CDF collaboration, and because of the interesting methodological issues raised by the debate within CDF over their evidence claim. My choice is problematic on two counts.

First, by focusing on CDF's history and search for the top quark, I slight the D-zero collaboration, which joined CDF in announcing in 1995 the "Observation of the Top Quark." D-zero is also situated at Fermilab, and although they are a younger collaboration than CDF, they are just as large and could easily serve as the subject of a lengthy study such as this. However, neither of the 1995 "Observation of the Top Quark" papers involved the kind of methodologically significant controversies generated by CDF's "Evidence" paper. I wish to emphasize, however, that *this is not the history of the discovery of the top quark.* No work that gives such slight attention to the D-zero collaboration could claim that title. My decision to emphasize CDF's efforts should not be read as any kind of judgment regarding the relative worth or importance of the work done by the two collaborations.

Second, this episode is not in any way representative of a class of things such as "scientific experiments," or even "particle physics experiments." In many ways, the case I have chosen is extreme: extreme in its complexity, scale, and level of contentiousness as an experimental result. I do not propose here to reach philosophical conclusions by generalization from this single episode. My approach is much more piecemeal.

Instead, in the first five chapters, which are primarily historical in their emphasis, I seek to discover what *can* happen in the natural sciences, rather than what typically does happen, by examining some aspects of physicists' work on the top quark. Thus, this study serves as a source of suggestions for further historical study. Also, these historical chapters narrate the progress of a number of controversies that arose during CDF's search for the top quark and situate those controversies within the specific context of CDF. The fact that some methodological issues in experimental science are specific to particular contexts is relevant to the ensuing philosophical discussion.

That philosophical discussion occupies the last two chapters, in which I adopt a certain theoretical framework regarding evidence, the "error statistical" theory (Mayo 1996). According to the error statistical theory, an experimental test yields evidence for a hypothesis when (or insofar as) the result of the experiment not only fits the hypothesis but is also of a kind that would be very improbable if the hypothesis in question were not true. The latter requirement is the *severity* requirement. A convincing argument for a hypothesis must therefore convince the audience that the hypothesis has passed a severe test. In my concluding chapters, I seek to explicate the *concept* of evidence at the center of the error statistical theory, particularly with respect to two issues: (1) In what sense, if any, is that concept objective? (2) Does the error statistical concept of evidence relate in any useful way to belief? In addition, I pose two main questions for the error statistical *theory* of evidence: (1) Does it provide the resources for explaining the relevance of those considerations on which debate over CDF's claim to have evidence for the top quark turned? (2) Does it provide the resources for proposing potentially useful strategies for avoiding methodological difficulties of the sort encountered by CDF, and for coping with them once they have arisen? The ability to accomplish these two aims constitutes, I propose, a necessary condition of adequacy for any theory of evidence. My hope is that the historical details presented in the earlier chapters will provide the reader with sufficient information to evaluate my own answers to these questions.

Chapter 1 concerns the introduction into the Standard Model of the fifth ("bottom") and sixth ("top") quarks by Makoto Kobayashi and Toshihide Maskawa in 1973 (Kobayashi and Maskawa 1973). In particular, I discuss the largely forgotten background to Kobayashi and Maskawa's widely acknowledged work. Their work had its roots in the school of theoretical physics at Nagoya University over which Shoichi Sakata presided. Sakata and several other Nagoya physicists were committed to dialectical materialism, which served as the basis for a methodology for advancing physical theory. Sakata and colleagues invoked that methodology in turn in proposing a theory of the structure of matter – the "Nagoya Model" – that anticipated more famous later developments. The Nagoya model then came to serve as the framework in which a number of Japanese physicists interpreted the results of cosmic ray experiments suggesting the existence of a new fundamental particle, and it was this apparent extension of the class of elementary particles that prompted Kobayashi and Maskawa toward the theoretical studies that led to their introduction of an entirely new "generation" of matter, including what would later come to be called the top quark. In restoring part of the history of the developments that led up to the search for the top quark, I also discuss the role of philosophical commitments in physical theorizing.

Chapter 2 chronicles the origins of CDF and the design and construction of their remarkably complex detector. The story told here details how the group that became CDF chose a detector design from among many different possibilities, against a background of constraints that were evolving even as the members of the collaboration were attempting to finalize the blueprint for their detector, and while they attempted to recruit both talented physicists and funding to support a rapidly growing enterprise. I argue that during the ten years spent planning and building the detector before gathering any data at all, CDF's aims were quite general. They created a resource for an extended experimental program, capable of supporting tests of a number of general types of physical hypotheses, without locking themselves into the pursuit of specified theoretical claims. They expected the detector to accommodate a variety of experimental pursuits, in virtue of the detector's ability to measure certain quantities, the interest of which was largely independent of the particular theories to be explored. Allan Franklin has described how experimenters exhibit "instrumental loyalty," shaping their experimental programs to utilize already operable, well-understood instruments (Franklin 1997). The early years of CDF show the other face of instrumental loyalty, where experimentalists set out to build a device capable of rewarding such loyalty, even though the precise course of experimentation remains to be determined.

Crucial to CDF's top quark evidence claim was an upgrade of the detector resulting in better discrimination between particle decay events involving top quarks and other particle events yielding similar results. The improvement required CDF to insert into the heart of their detector a new component based on technology never before tried in such a setting. I explain in Chapter 3 how a group of Italian physicists interested in applying this new technology won over their colleagues by using persistence and ad hoc arguments and by deferring to physicists at another institution to solve a recalcitrant technical problem. Also helpful was the fact that CDF's "detector philosophy" was compatible with the proposed detector component. Peter Galison has discussed how such detector philosophies shape experimental programs (Galison 1997b), and I contrast my interpretation of that phenomenon with his. This episode shows how the development of new techniques of detection and instrumentation technologies exhibits a certain degree of autonomy from specific experimental and theoretical aims.

Also important to Chapter 3 is the struggle within CDF as physicists developed strategies for finding the top quark. Political intrigues and social forces figured prominently, but not simply by making some approaches look convincing and others not convincing, as a strictly "social constructivist" reading might suggest (see, e.g., Pickering 1984). Rather, alienation from colleagues made it difficult for some working on complicated methods to get others to study their proposals. Their analyses may or may not have worked, but it is possible that these methods were not adopted because not enough of the collaboration devoted the effort needed to evaluate them. (Others in the collaboration insist that, beyond the "political" difficulties of some of these proposals, the proposed methods simply did not work in a reliable way, even when one did take the time to study them carefully.) Although one cannot deny that the internal social dynamics of the CDF collaboration influenced the careers of these proposals, such social forces served not to *determine* or *constitute* the validity or invalidity of the techniques proposed, but rather to encourage or set obstacles to the study needed for assessing their validity.

Based on data collected during a period known as "run Ia," CDF submitted to the journal *Physical Review D* a lengthy account of the analysis that led to their evidence claim (the "Evidence" paper). I describe in Chapter 4 the months that CDF spent writing that article, and the many disagreements and controversies that boiled within the collaboration. Many CDF members criticized the interpretation of the data favored by group members directing the writing of the Evidence paper. I argue that part of the function of the published research report is to facilitate just such criticism of one's own work by those best situated to discover certain types of errors. If evidence is produced by exposing hypotheses to severe tests during an experiment, then writing a paper after the experiment can become the pursuit of experiment by other means. While writing, experimenters can test the various assumptions on which they rely in arriving at an evidence claim, often by reanalyzing existing data in different ways. Such a conception of the aim of writing a research report accounts, in part, for the fact, decried by some, that scientific papers typically provide misleading, cleansed narratives of the research process (Medawar 1964). My argument also adds a new dimension to Thomas Nickles's observation that in writing a research report "scientists are not writing *about* science; they are *doing* it" (Nickles 1992, 96).

I turn to the aftermath of the Evidence paper in Chapter 5. After a brief shutdown, CDF resumed taking data during "run Ib," and after another year, both CDF and D-zero announced that they had "observed" the top quark. In the meantime, some in CDF used new data and new methods of analysis developed *after* the Evidence paper to cast doubts on some aspects of the earlier results. Some CDF members believed that the earlier tests were biased, resulting in a misleading assessment of the statistical significance of their data. Subsequent developments, they said, yielded two sources of support for their suspicions: (1) reanalysis of existing run Ia data using newly developed algorithms, and (2) application of similar algorithms to larger bodies of data from both run Ia and run Ib. I argue that (1) constitutes an example of a "robustness analysis" (Campbell and Fiske 1959; Wimsatt 1981; Culp 1995) used to test the reliability of a result, while (2) reflects an awareness that, as data accumulate, genuine effects should yield more impressive statistical significance calculations. Both kinds of testing show that evaluating evidence may require further empirical (in some cases diachronic) investigations, as one would expect if evidence claims are indeed empirical (Achinstein 1995; 2001) and subject to additional testing.

With the historical narrative complete, Chapters 6 and 7 are more strictly philosophical. In Chapter 6 I discuss the error statistical theory in more detail. First, I develop an outline of an error statistical model of CDF's experiment as reported in the Evidence paper, utilizing a hierarchy of models connecting data to high-level theoretical claims via a (partial) model of the experiment itself. I further argue that the error statistical theory supports "local" comparative assessments of evidential strength (comparisons of different outcomes of a given experiment as evidence for a given hypothesis, or comparisons of a given experimental outcome as evidence for different members of a family of hypotheses), and that it also supports a classificatory concept of evidence, enabling one to make judgments that certain results do or do not constitute evidence for a particular hypothesis. However, the error statistical theory does not enable the kind of "global" comparisons (comparing the strength of e as evidence for h, and e' as evidence for h', for any e, e' and h, h' whatsoever) that would be required for a quantitative measure of evidential strength. Nevertheless, both the comparative and classificatory concepts of evidence described by the error statistical theory are fully objective in the sense that whether an experimental outcome is error statistical evidence for a hypothesis, or stronger evidence for one hypothesis than for another, is independent of whether any person believes it to be such. Furthermore, if a result is error statistical evidence for a hypothesis in the classificatory sense, then it constitutes a reason to believe that hypothesis independently of whether any person believes it to be such, and independently (in an epistemic sense) of the epistemic state of the experimenter or anyone else (see Achinstein 2001).

Finally, Chapter 7 addresses problems of bias in experimental testing and revisits the question of objectivity. A recurring point of controversy within CDF was the problem of biasing a test toward a favored outcome by changing one's data selection criteria in the face of knowledge of the data to which they will be applied, a phenomenon known to particle physicists as "tuning on the signal." Precisely in order to avoid such problems, experimenters in many contexts insist on "predesignating" the features of their testing procedure prior to acquiring any knowledge of any data. I compare the predesignation requirement to the requirement of "novel predictions" that has long preoccupied philosophers of science. I argue that if we take the experimenter's concern to be, not providing novelty as such, but ensuring that the test being administered is sufficiently severe, then we can better understand the ways in which failure to predesignate did and did not matter to CDF's evidence claim.

In Chapter 7 I also discuss a method for evaluating the strength of evidence when a test is suspected of being biased by tuning on the signal and argue that reliable experimental inference requires an adequate model of the error probabilities of the test. Such a model might require taking into account statistically relevant aspects of the testing context involving the behavior and dispositions of experimenters. However, when such social-psychological factors become relevant, an adequate model may be difficult for experimenters to generate. Many rules of experimental method are thus designed to make such factors statistically irrelevant. Hence, evidential relationships do not exist on a plane free of considerations of human sensibilities and preferences.

On the surface, such an acknowledgment may seem to pose a threat to the alleged objectivity of the error statistical concept of evidence. Certainly it has been taken as such by critics of classical statistical testing techniques and by some who have rejected the evidential significance of predesignation. The objection posed is roughly that intentional and epistemic factors cannot be relevant to an objective evaluation of the evidential significance of an experimental outcome. I respond by drawing a distinction between the causal and epistemic relevance of such states. Evaluating experimental evidence in an error statistical manner may require consideration of the causal relevance of experimental agents' intentional or epistemic states. Nonetheless, error statistical evidence is entirely objective in that such states are not epistemically relevant to its evaluation.

The experimenters discussed here worked in a world filled with physical and social complexities but sought to isolate a single and in some sense simple fact in the midst of that tangle. The experimenter interacts with her colleagues and competitors, but also – as an agent with her own beliefs, preferences, and weaknesses – with the objects on which the experiment is performed. This account seeks to elucidate simultaneously the thicket in which the experimenter works and the potential emergence from that thicket of an objective reason for believing some important truth about the world. That such a thing could happen at all is in itself wondrous and seemingly magical, but in truth hard work, careful planning, and fortuitous circumstances make it possible. In any case, that is how things happened in the story that follows.

1

Origins of the Third Generation of Matter

In the *standard model* (see Figure 1.1) of the elementary particles and forces
of nature that is endorsed by contemporary physics, the fundamental basis of matter is constituted by the *quarks*, which bear fractional electrical
charges and respond to the *strong force* that binds together atomic nuclei,
and the *leptons*, which are neutral or have integral electrical charges and
are immune to the strong force. The standard model groups these matter
particles into three *generations*, where each generation includes two quarks
and two leptons. The *up* (u) and *down* (d) quarks, the *electron* (e), and the
electron neutrino (ν_e) constitute the first generation. The second generation
includes the *charm* (c) and *strange* (s) quarks, the *muon* (μ), and the *muon
neutrino* (ν_μ). Finally, the *top* (t) and *bottom* (b) quarks join the *tau* (τ) and
the *tau neutrino* (ν_τ) in making up the third generation (see Kane 1993 for
an introduction).

Another class of particles known as gauge bosons serve as propagators
of the elementary forces. These include the massless *photon* (γ, carrier of
the electromagnetic force), the *gluon* (g, carrier of the strong force), and
the *intermediate vector bosons* W^+, W^-, and Z^0 (carriers of the weak force
responsible for various decay processes).

By the early 1990s, all these particles had been experimentally confirmed,
except for the top quark. So when the Collider Detector at Fermilab (CDF)
collaboration announced in 1994 that they had found the first evidence of
the top quark, it was a major scientific development. CDF presented their
results in a long, detailed article submitted to *Physical Review D*, entitled
"Evidence for Top Quark Production in $\bar{p}p$ Collisions at $\sqrt{s} = 1.8$ TeV" (Abe,
Amidei, et al. 1994b). Prominent among the many works that CDF cites in
this 60-page article is a 1973 article by Makoto Kobayashi and Toshihide
Maskawa (Kobayashi and Maskawa 1973; Figures 1.2 and 1.3).

Citations in a scientific paper perform a variety of tasks, such as referring the reader to other work that supports particular claims being made
in the present article and helping the reader to locate articles expressing

		first generation	second generation	third generation		
fermions	**quarks**	u	c	t	γ	**gauge bosons**
		d	s	b	g	
	leptons	ν_e	ν_μ	ν_τ	Z	
		e	μ	τ	$W^\pm$	

Hadrons	
Mesons	**Baryons**
$\langle q\bar{q}\rangle$	$\langle qqq\rangle$
e.g., $K^+ = \langle u\bar{s}\rangle$, $J/\psi = \langle c\bar{c}\rangle$	e.g., $p = \langle uud\rangle$, $\Omega^- = \langle sss\rangle$

FIGURE 1.1. The particles of the standard model.

FIGURE 1.2. Toshihide Maskawa. Courtesy of Makoto Kobayashi.

competing viewpoints. Citations may help to validate the competence of the researcher with respect to the subject matter at hand (Kitcher 1995). Many citations, however, are simply meant to acknowledge the contribution that a particular work has made in developing the ideas under discussion. CDF's citation of Kobayashi and Maskawa's article is honorific in this sense. It reflects physicists' consensus that Kobayashi and Maskawa, in their 1973 article, were the first to suggest that something like the top quark might exist.

The citation of the Kobayashi and Maskawa (KM) paper appears in the first sentence of CDF's article, which reads, "The standard model has enjoyed outstanding success in particle physics for two decades, yet one of its key constituents, the top quark, has remained unobserved" (Abe, Amidei,

FIGURE 1.3. Makoto Kobayashi. Courtesy of Robert Palmer, Brookhaven National Laboratory.

et al. 1994b, 2968). Two reference numbers follow the words "The standard model." The first directs the reader to citations of articles by Sheldon Glashow, Steven Weinberg, and Abdus Salam (Glashow 1961; Weinberg 1967; Salam 1968). These citations all point to what are regarded as the seminal works in the development of the standard model of electroweak interactions, which unifies the weak force and the electromagnetic force. The second number appearing after "the standard model" refers to citations of a 1970 article by Glashow, John Iliopoulos, and Luciano Maiani, and to Kobayashi and Maskawa's 1973 paper. The Glashow-Iliopoulos-Maiani (GIM) paper (Glashow, Iliopoulos, and Maiani, 1970) proposed a model of weak interactions based on a quartet of fermions. All the papers that CDF cites in connection with "the standard model" are among the most cited articles in the literature of particle physics.[1]

The KM paper is entitled, "*CP*-Violation in the Renormalizable Theory of Weak Interaction." The abstract reads as follows:

In a framework of the renormalizable theory of weak interaction, problems of *CP*-violation are studied. It is concluded that no realistic models of *CP*-violation exist in the quartet scheme without introducing any other new fields. Some possible models of *CP*-violation are also discussed. (Kobayashi and Maskawa 1973, 652)

The reader may reasonably be wondering what any of this has to do with the top quark.

The actual nature of Kobayashi and Maskawa's accomplishment is more complicated than simply having predicted the existence of an entity for which, 21 years later, CDF found evidence. In fact, Kobayashi and Maskawa are given credit for predicting the entire third generation of matter: both the top and bottom quarks and the tau lepton and its neutrino.[2] This assessment is certainly just. Kobayashi and Maskawa did suggest that there might be a third generation of matter, and they did so well in advance of the experimental discovery of any members of that group. In fact, they contemplated a fifth and sixth quark prior to the discovery of the *fourth* quark (although this statement will have to be qualified, as discussed below).

Yet the relationship between Kobayashi and Maskawa's theoretical work and the top quark discovery is more complex than the pattern of citations would make it seem. Although Kobayashi and Maskawa receive credit for predicting the entire third generation of matter, the top quark appears to be the only member of that generation that was discovered by experimenters consciously pursuing the speculations of Kobayashi and Maskawa's paper. For two other members of the top quark's generation (the bottom quark and the tau lepton[3]), the work of the two theorists seems to have had no direct influence on the experiments that revealed the existence of these particles. Few physicists noticed the KM paper when it was first published.

I wish to start, then, with the story of Kobayashi and Maskawa's paper. What theoretical and experimental issues motivated them to write their paper? Who read it and who did not read it when it was published? How did it come to receive the widespread attention that later came to it? The answers to these questions will require consideration of the context out of which the KM paper emerged, including experimental, theoretical, and philosophical aspects of the physics scene in Japan, especially the part of it that centered on the University of Nagoya and the theoretician Shoichi Sakata.

1. KAONS AND *CP* VIOLATION

As their title indicates, Kobayashi and Maskawa's theoretical speculations were in response to, and an attempt to accommodate, a surprising experimental discovery.[4] In 1964, Jim Christenson, Jim Cronin, Val Fitch, and René Turlay found that the meson known as K_2^0 (a member of the "kaon" family) decays occasionally to a final state of two pions (pions are also a kind of meson) (Christenson, Cronin, et al. 1964). This decay was unexpected because it violated a symmetry principle known as *CP symmetry*.

Prior to Cronin, Fitch, Christenson, and Turlay's experiment, it had been thought that weak interactions were symmetric with respect to the combination of spatial reflection and charge conjugation. Symmetry with respect to spatial reflection alone is known as parity conservation. The space reflection operator P inverts all spatial coordinates of the state on which it acts. A state

$|\psi\rangle$ is an *eigenstate* of P just in case $P|\psi\rangle = a|\psi\rangle$, where a is a scalar quantity called an *eigenvalue* of P. The *parity* of a state refers to its eigenvalue with respect to the spatial reflection operator, and can take the value of either $+1$ or -1. The charge conjugation operator (C) transforms a particle into its antiparticle. So the thesis that the weak interaction is symmetric with respect to the transformation CP amounts roughly to the claim that, in any weak interaction, the CP eigenvalue of the state of the system prior to the interaction (i.e., the scalar value a such that, if the initial state is ψ', then $CP|\psi'\rangle = a\,|\psi'\rangle$) will equal that of the state of the system after the interaction (the final state). Christenson, Cronin, Fitch, and Turlay discovered a weak particle decay in which the CP eigenvalue of the initial state differed from that of the final state, but their route to this discovery was by no means direct.

In an experiment at the Brookhaven Alternating Gradient Synchrotron, Christenson, Cronin, Fitch, and Turlay set out to examine the decays of long-lived strange neutral kaons (K_2^0). Although neutral K mesons had been discovered in 1947, the K_2^0 had only been recognized rather recently.

This new particle had been introduced by Murray Gell-Mann and Abraham Pais in response to some peculiarities of the K (which physicists at the time called the θ) (Gell-Mann and Pais 1955). The K^0 has a quantum number, strangeness, which is conserved in strong interactions. The K^0, however, can be interconverted with its antiparticle $\bar{K}^0$ through weak interactions ($K^0 \rightarrow \pi^+\pi^- \rightarrow \bar{K}^0$ or $\bar{K}^0 \rightarrow \pi^+\pi^- \rightarrow K^0$). In such an interaction, strangeness is not conserved. Gell-Mann and Pais suggested that, even though the K^0 and $\bar{K}^0$ are eigenstates of the strong interaction, they are not eigenstates of the weak interaction, and hence should not be regarded as true particles. In other words, while the K^0 and $\bar{K}^0$ can be satisfactorily regarded as real particles for purposes of describing their production through strong force interactions, this description is not completely general because it does not work for weak interactions. Gell-Mann and Pais proposed that, for purposes of describing the weak-decay interactions in which strangeness is not conserved, K^0 and $\bar{K}^0$ instead be regarded as superpositions of two new particles, K_1^0 and K_2^0, where

$$|K_1^0\rangle = (|K^0\rangle + |\bar{K}^0\rangle)/\sqrt{2},$$
$$|K_2^0\rangle = (|K^0\rangle - |\bar{K}^0\rangle)/\sqrt{2}.$$

Consequently, on this scheme, the K_1^0 and K_2^0 are superpositions of the K^0 and $\bar{K}^0$, and vice versa.

Gell-Mann and Pais predicted that the hypothesized K_1^0 and K_2^0 could be identified through a difference in their decays. The K_1^0 would have a short lifetime and would sometimes decay into two pions, whereas the K_2^0 would have a longer lifetime and would not decay into two-pion final states.

This prediction was at least partly born out in 1956 in a bubble chamber experiment at the Brookhaven Cosmotron. An experiment there found evidence that was "consistent with a K^0-type particle undergoing three-body decay" with a long lifetime. However, this group observed that "long lifetime and 'anomalous' decay mode are not sufficient to identify the observed particle with $[K_2^0]$" (Lande, Boothe, et al. 1956, 1903).

There was, however, another phenomenon that would, if detected, constitute much stronger evidence for the Gell-Mann and Pais model. Suppose that one starts with a beam of K^0s and allows this beam to travel far enough so that the short-lived K_1^0 components of the initial superposed states have been almost entirely depleted through weak decays. The remaining K_2^0 states are, of course, themselves a superposition of K^0 and $\bar{K}^0$ states. If the K_1^0-depleted beam is now passed through an absorber, the two components of the K_2^0 states will be attenuated by different amounts. (Strong interactions between the $\bar{K}^0$ and protons and neutrons in the absorber will tend to deplete the $\bar{K}^0$ component of the K_2^0, whereas strangeness conservation in strong interactions forbids such interactions with the K^0 component.) This means that the beam that exits the absorber will consist mostly of K^0s. Because the K^0 is a superposition of K_1^0 and K_2^0 states, the emergent beam will have a significant K_1^0 component that can be detected by its prompt decay into two-pion final states. The short-lived K_1^0 decays immediately to two pions. Abraham Pais and Oreste Piccione had predicted this phenomenon and described the experiment to detect it (Pais and Piccioni 1955). A group led by Francis Muller observed the phenomenon in 1960 (Muller, Birge, et al. 1960).

But then, in an experiment to study the regeneration phenomenon, R. K. Adair and W. Chinowski's group found an anomalously large number of events resulting in two pions. In fact, the peak at forward angles in the angular distribution of such events was too large by a factor of ten to twenty (Leipuner, Chinowski, et al. 1963).

This result turned out to be spurious (Franklin 1983, 216), though Cronin later noted that it "would have been a major discovery" if confirmed (Cronin 1997). Cronin and Fitch's group proposed to "check these results with a precision far transcending that attained in the previous experiment," as well as to obtain "a new and much better limit" on the CP-violating decay $K_2^0 \rightarrow \pi^+ + \pi^-$ (ibid., 135). The anomalous regeneration did not reappear. What the experiment revealed instead was that the long-lived neutral kaon K_2^0 decays in such a way that, while most of the time CP is conserved (three pions are produced), in a small fraction of instances CP is violated (two pions are produced).

This phenomenon was a major part of Kobayashi and Maskawa's motivation in writing their paper. They were explicitly looking for ways to accommodate CP violation within a theory of weak interactions of the Weinberg-Salam-Glashow-Iliopoulos-Maiani type.

2. THE STANDARD HISTORY OF THE STANDARD MODEL

So far, I have been recounting a history that is generally well known. Fitch and Cronin received the Nobel Prize for their work, and its importance for Kobayashi and Maskawa's paper is too obvious to require any special comment. The rest of the story to be told here is not uniformly well known among physicists, however. In particular, it is conspicuously missing from what one might think of as the "standard history" of the standard model. In order to appreciate the relevance of the experimental, theoretical, and philosophical background I am about to relate, it may therefore be helpful to review that standard history.

Among the many inhabitants of the particle "zoo" known to physicists in the 1960s were two classes of particles subject to the strong force, the baryons and the mesons. In addition, there were three known leptons – the electron, muon, and neutrino. In 1964, Gell-Mann and George Zweig independently introduced what Gell-Mann called quarks as a way of bringing order to the many baryons and mesons (Gell-Mann 1964; Zweig 1964a; 1964b). According to the quark theory, the baryons and mesons could be represented as combinations of quarks or anti-quarks.

The 1964 quark model featured three quarks: up, down, and strange (u, d, and s). This basic triplet fit nicely with symmetries that had been found earlier by Gell-Mann and Yuval Ne'eman among the various baryons, a symmetry that Gell-Mann dubbed the "eightfold way" but that exemplified a more general type of symmetry known as $SU(3)$ (Gell-Mann 1961; Ne'eman 1961). In the quark model, baryons are composed of three quarks bound together (for example, a proton is composed of two up quarks and one down quark: $\langle uud \rangle$), while mesons are composed of quark-antiquark pairs (the positive pion π^+, for example, is the pair $\langle u\bar{d} \rangle$).

Also in 1964, J. D. Bjorken and Sheldon Glashow proposed that there might be four fundamental strongly interacting particles, giving the name "charm" to this new theoretical entity, although none of the known baryons or mesons seemed to represent charm (Bjorken and Glashow 1964).

Abdus Salam and Steven Weinberg, working independently of one another, came up with a way of unifying the electromagnetic and weak interactions in 1967 (Weinberg 1967; Salam 1968), and their idea was improved upon and extended to a quartet scenario by Glashow, Iliopoulos, and Maiani in 1970 (Glashow et al. 1970).

The problem Kobayashi and Maskawa tackled in their 1973 paper, then, was to see whether the phenomenon of *CP* violation could be accommodated in theories such as that emerging from the Weinberg-Salam-Glashow-Iliopoulos-Maiani school of thought, employing a quartet of strongly interacting elementary particles. Kobayashi and Maskawa showed that, to account for *CP*-violating decays while respecting certain plausible constraints, additional strongly interacting fields were needed beyond the four that theorists were entertaining at the time. In particular, a six-field theory seemed to allow for *CP* violation in a fairly natural way. Thus, they introduced the possibility of a theory with six quarks, and the newly proposed particles came to be called "top" and "bottom."

Kobayashi and Maskawa were far ahead of their time, for as of 1973 no one had yet discovered any charmed particles experimentally. The charmed J/ψ particle was not discovered until November 1974, when groups led by Samuel Ting and Burton Richter discovered the new particle more or less simultaneously (Aubert, Becker, et al. 1974; Augustin, Boyarski, et al. 1974). In 1976, a group led by Martin Perl working at the Stanford Linear Accelerator Laboratory (SLAC) discovered the τ lepton (Perl, Abrams, et al. 1975), and Leon Lederman and a group at Fermilab found the bottom quark in 1977 (Herb, Hom, et al. 1977). When CDF and D-zero found the top quark in 1995 (Abe, Amidei, et al. 1995a; Abachi, Abbott, et al. 1995b), they completed in experiment the picture sketched in theory by Kobayashi and Maskawa more than 20 years earlier.

The standard history just related is not inaccurate, though it is of course incomplete in its brevity. My aim here is not to condemn this standard history but to ask what is missing from it – not in details, but in dimensions – for understanding how the top quark, and the entire third generation of matter, came to be part of the standard model. My claim is that what is missing is an understanding of the context of Japanese physics in the mid-twentieth century from which Kobayashi and Maskawa's work emerged.

3. THE SAKATA MODEL AND THE NAGOYA MODEL

One of the most important physicists in Japan in the 1950s and 1960s was Shoichi Sakata of Nagoya University. Sakata proposed, in 1956, a model of the fundamental particles that came to be called the "Sakata model" (Sakata 1956).[5] According to the Sakata model, the proton, neutron, and lambda were "fundamental" baryons. This raised a problem for the other baryons such as the Σ and Ξ, however. Why was the Λ fundamental, but the Σ and Ξ were not? Gell-Mann and Ne'eman had proposed, in their "eight-fold way" approach, that the Λ, p, and n be treated on a par with the other spin-$\frac{1}{2}$ low-mass baryons, the Σ^0, Σ^+, Σ^-, Ξ^0, and Ξ^-. Both the Sakata model and the eightfold way represented the fundamental baryons as expressions of a fundamental symmetry known as $SU(3)$; however, the two models represented that symmetry in different ways. The idea behind $SU(3)$ was that symmetry considerations would lead to predictions about how families of particles could be grouped together and assigned specific quantum numbers for strangeness and isospin. However, one can express $SU(3)$ symmetry with groups of various sizes: 1, 3, 6, 8, and 10 are all possibilities. The Sakata model grouped the fundamental spin-$\frac{1}{2}$ baryons as a triplet (p, n, Λ), and assigned the Σ and Ξ baryons to other multiplets. It predicted that the Ξ

baryons would have spin $\frac{3}{2}$, for example. The scheme proposed by Gell-Mann and Ne'eman represented the spin-$\frac{1}{2}$ baryons as an octet (including the Σ and Ξ baryons), and the spin-$\frac{3}{2}$ baryons as a decuplet, including an entirely new particle with strangeness –3, the Ω^-.

Interestingly, the Nagoya group itself had proposed an $SU(3)$ representation for the spin-0, negative parity mesons that more closely resembled the eightfold way than it did the Sakata model (Ikeda, Ogawa, and Ohnuki 1959), and this assignment led to the successful prediction of the η meson (Pevsner, Kraemer, et al. 1961). As Ne'eman pointed out later, "nobody knew the ins and outs of $SU(3)$ in 1960 as well as the Nagoya group. Still, they did not observe the seemingly obvious better fit provided for the baryons by that same octet representation that they were using for mesons!" (Ne'eman 1974, 12). The superiority of the eightfold way scheme for baryons became undeniable in 1964, when the Ω^- was discovered (Barnes, Connolly, et al. 1964).

But by that time, the Sakata model had already been supplanted by a still more ambitious theory that came to be called the "Nagoya model." This new model appeared in its earliest form in a 1960 paper by Sakata and his Nagoya University colleagues Ziro Maki, Masami Nakagawa, and Yoshio Ohnuki, published in *Progress of Theoretical Physics*. Titled "A Unified Model for Elementary Particles," this paper presented the core elements of the Nagoya model (Maki, Nakagawa, et al. 1960). As in the Sakata model, the fundamental hadrons were again proposed to be the proton, neutron, and lambda. But the Nagoya group here intended to present a model of the hadrons and leptons simultaneously. Thus, each of the fundamental hadrons was represented as a composite of a lepton (neutrino, electron, or muon) with something called a B^+, referred to as "B-matter."

The "Kiev Symmetry" figured as a major motivation for this introduction of an entirely new kind of matter. According to the Kiev Symmetry (or "baryon-lepton symmetry"), the Hamiltonian describing the weak interaction is invariant when one performs a simultaneous baryon-lepton transformation replacing proton fields with neutrino, neutron with electron, and lambda with muon.[6] Maki et al. noted that "[t]he existence of the Kiev symmetry tells us that, apart from the strong and electromagnetic interactions, proton, neutron, and Λ-particle have the same properties as those of neutrino, electron and muon respectively" (ibid., 1175). They concluded that "if we imagine that P, N, and Λ are the compound system of positively charged matter B^+ with ν, e^+ and μ^- respectively, all weak interactions will be reduced to those of leptons, and the strong interaction will be regarded as the properties of B^+" (ibid., 1175).

In 1962, experiments at Brookhaven National Laboratory began to yield evidence of a second neutrino (Danby, Gaillard, et al. 1962). That same year, two papers extending the Nagoya model to incorporate two neutrinos appeared: one by Maki, Nakagawa, and Sakata (1962), and the other by physicists at the University of Kyoto (Katayama, Matumoto, et al. 1962). The Nagoya group focused on an extension of the model in which one of the neutrinos (ν_2) does not bind to B-matter. According to that proposal, there would be four fundamental leptons but only three fundamental baryons – hence they sacrificed the baryon-lepton symmetry that was central to the original Nagoya model, but retained the $SU(3)$ structure of the baryon family. Maki, Nakagawa, and Sakata relegated to a footnote the comment that, "Alternatively, we can assume that $\langle B^+\nu_2\rangle$ corresponds to a new kind of baryon with a very large mass" (Maki et al. 1962, 872).

In the other paper, Kyoto University physicists Yasuhisa Katayama, Ken-iti Matumoto, Sho Tanaka, and Eiji Yamada considered a similar extension of the Nagoya model, discussing at length both three- and four-baryon models. They based their four-baryon model on the following four lepton–B-matter composites:

$$p = \langle \nu_1 B^+\rangle, \quad n = \langle e^- B^+\rangle, \quad \Lambda = \langle \mu^- B^+\rangle, \quad V = \langle \nu_2 B^+\rangle.$$

Here V is a new baryon with positive charge and a new quantum number (Katayama, Matumoto, et al. 1962, 680). Both the Nagoya and Kyoto groups treated the neutrinos ν_1 and ν_2 as superpositions of the observed electron and muon neutrinos (ν_e and ν_μ) (Katayama, Matumoto, et al. 1962, 679; Maki et al. 1962, 874).

A further development in the Nagoya model was proposed by Ziro Maki and Yoshio Ohnuki in 1964. They presented the core of their version of the model in two letters to the editor of *Progress of Theoretical Physics* by Maki (Maki 1964a; 1964b), and developed their model in a longer paper shortly afterward (Maki and Ohnuki 1964). Rather than taking the proton, neutron, and lambda as the most fundamental baryonic states, Maki and Ohnuki hypothesized what they called "urbaryons." In their "revised Nagoya model," the urbaryons were represented as a triplet $\chi = (\chi_1, \chi_2, \chi_3)$, and a fourth singlet urbaryon χ_0. Maki and Ohnuki retained the B-matter concept, and represented the χ_1, χ_2, χ_3 urbaryons, like the p, n, and Λ baryons of the original Nagoya model, as lepton–B-matter composites:

$$\chi_1 = \langle B^+\nu_1\rangle, \quad \chi_2 = \langle B^+ e^-\rangle, \quad \chi_3 = \langle B^+\mu^-\rangle.$$

They argued that the fourth urbaryon χ_0 had to be treated differently. In the 1962 two-neutrino versions of the Nagoya model, the combination of B^+ with the second neutrino state ν_2 was postulated either to correspond to no baryon or to a newly introduced baryon V, which would have quite different properties from the other baryons. But by 1964, Maki and Ohnuki were finding both of these proposals "hard to accept physically," since both required that "one and the same B^+ should behave in a quite different way as soon as it couples to ν_2" (Maki and Ohnuki 1964, 147). So Maki and Ohnuki introduced a "neutral counterpart" B^0 to B^+. They could then write the fourth baryon as

$$\chi_0 = \langle B^0 \nu_0 \rangle.$$

In this model, the strong interaction displayed a symmetry of the type $U(1) \times U(3)$. As it stood, such a model implied that in strong interactions the number of each type of baryon would be strictly conserved (i.e., $\Delta N_{\chi_i} = 0$, for $i = 0, 1, 2, 3$). This was implausible in light of experimental results, however. For example, such a strict conservation law forbade the subsequent decay of the χ_1 into a nucleon-meson pair, when a χ_1 and $\overline{\chi}_1$ were pair-produced.[7] That would mean the existence of massive, charged, stable χ_1 baryons, which had somehow never been observed in any experiments.

To avoid this implausible conclusion, Maki and Ohnuki introduced a new interaction violating strict baryon number conservation. This "moderately strong interaction," as they termed it, was to be characterized by the transition $\chi_3 \leftrightarrow \chi_0$. Taking the moderately strong interaction into account led in turn to a weakening of the baryon number conservation rule, which became

$$\Delta N_{\chi_1} = \Delta N_{\chi_2} = 0 \quad \text{and} \quad \Delta(N_{\chi_0} + N_{\chi_3}) = 0.$$

Maki and Ohnuki related this new class of unobserved urbaryons to observed particles by means of a two-stage construction. First, they suggested two triplets of new mesons $\theta = (\theta_1, \theta_2, \theta_3)$, and $\overline{\theta} = (\overline{\theta}_1, \overline{\theta}_2, \overline{\theta}_3)$, where $\theta_i = \chi_i \overline{\chi}_0$ and $\overline{\theta}_i = \chi_0 \overline{\chi}_i$ (ibid., 151–5). Maki and Ohnuki tentatively identified some already detected mesons as mixtures of these meson states. For example, they defined the two following states:

$$V_1 = \frac{\theta_3 - \overline{\theta}_3}{\sqrt{2}} \quad \text{and} \quad V_2 = \frac{\theta_3 + \overline{\theta}_3}{\sqrt{2}i}.$$

Maki and Ohnuki then suggested identifying the V_1 with the ω meson, which had been discovered in 1961 by Luis Alvarez's bubble-chamber group (Maglic, Alvarez, et al. 1961). They noted that since the dominant decay mode of the heavier V_2 was expected to result in the just-discovered V_1 (ω meson) and a photon, it was reasonable for it to have not yet been observed (ibid., 154). Maki and Ohnuki then proposed that baryons be regarded as bound states of θ mesons with a third urbaryon, for example the proton, $p = [\overline{\theta}_3, \chi_1]$, and the neutron $n = [\overline{\theta}_3, \chi_2]$. They noted finally that their model was a "natural generalization of the Sakata model for hadrons and of the Nagoya model" (ibid., 157).

Noteworthy are two aspects of the development of the Nagoya model that have a bearing on the standard history of the standard model:

1. *The Origin of Charm.* In light of the extensions of the original Nagoya model by the Kyoto and Nagoya groups in 1962, we know that when J. D. Bjorken and Sheldon Glashow published the proposal that there might be "charmed" hadrons carrying a new quantum number

(Bjorken and Glashow 1964), some proponents of the Nagoya model had already been working with quartet schemes for two years. Indeed, several authors appear to have introduced models with four fundamental baryon-type fields during the period from 1962 to 1964. Pekka Tarjanne and Vigdor Teplitz of Berkeley's Lawrence Radiation Laboratory published such a model in 1963, based on an $SU(4)$ rather than $SU(3)$ symmetry (Tarjanne and Teplitz 1963). Four CERN physicists explored various $SU(4)$ models in a 1964 paper (Amati, Bacry, et al. 1964), and Caltech's Yasuo Hara proposed a four-baryon model early in 1964 (Hara 1964). Bjorken and Glashow cited the papers by Hara and by Tarjanne and Teplitz, but none of the papers on the Nagoya model.[8] Maki and Ohnuki apparently exchanged papers with Hara prior to publication because these two papers cited each other as preprints. Only Maki and Ohnuki cited the 1962 papers on the extended Nagoya model. In spite of all this, it is common for Bjorken and Glashow to be given sole credit for the charm proposal.

2. *Origin of Quarklike Particles.* Second, in the urbaryon proposal of 1964, the Nagoya physicists had their own version of the idea that the known baryons were composite states of more fundamental particles, quite independently from the quark proposal of Gell-Mann and Zweig (Gell-Mann 1964; Zweig 1964a; 1964b). (In fact, Maki had already articulated the central claims of the urbaryon theory in 1963, in two letters to the editor of *Progress of Theoretical Physics* that did not appear in print until the following year (Maki 1964a; 1964b).) The two proposals even share certain features. They both classify the baryons as an eight-dimensional representation of an $SU(3)$ symmetry (although differently), and both represent mesons as particle-antiparticle pairs, while baryons are treated as composed of three of the more fundamental particles (although for the Gell-Mann/Zweig proposal this is true of all baryons, whereas in Maki and Ohnuki's scheme, other "observable" baryon states will consist of single urbaryons, or five urbaryons; likewise some mesons might be composed of four urbaryons).

4. DIALECTICAL MATERIALISM AND PHYSICAL THEORY

Shoichi Sakata himself, as well as a small number of his Nagoya colleagues, advocated the dialectical materialism of Engels, Marx, and Lenin.[9] In 1969, Sakata wrote that Engels's *Dialektik der Natur* "has been continuously sending invaluable light into my studies of about forty years as a precious stone" (Sakata 1971a, 1). In that work, Engels embraced the view, which he traced to the ancient Greeks, that

the whole of nature, from the smallest element to the greatest, from grains of sand to suns, from protista to men, has its existence in eternal coming into being and passing away, in ceaseless flux, in unresting motion and change. Only with the essential difference that what in the case of the Greeks was a brilliant intuition, is in our case the result of strictly scientific research in accordance with experience, and hence also it emerges in a much more definite and clear form. (Engels 1940, 14)

In particular, Sakata regarded his own theorizing as rooted in a doctrine of Lenin's dialectical materialism known as inexhaustibility.[10] According to the doctrine of inexhaustibility, each allegedly fundamental particle would ultimately reveal a lower-level structure, and the elements of that structure would in turn themselves display further microstructure, ad infinitum. A distinct set of laws would govern each level, or stratum, and each stratum would continually change.

The doctrine of inexhaustibility entailed a rejection of the idea that the elementary particles postulated at any given stage in the development of physical theory could be point particles without structure or extension. Sakata also drew the conclusion that quantum mechanics would eventually prove to be incomplete. In 1959 he wrote, "Elementary particles are not the ultimate of Matter. Quantum mechanics is, likewise, not the ultimate theory. . . . Elementary particles are not points. Now is the time to say good-bye to the point model and to, in some way, form a non-local theory" (Sakata 1959, quoted in Maki 1989, 87).

The dialectical materialist philosophy of the Nagoya group included also a three-stage epistemology attributed to Mituo Taketani. Sakata and Taketani had both been students of Hideki Yukawa when the latter was a lecturer at Kyoto University and collaborated with him when Yukawa took a position at Osaka Imperial University, contributing to his work on meson theory. Taketani also became involved in a leftist magazine, *World Culture*, both as an editor and as a writer (Brown and Rechenberg 1996, 142–4). According to Taketani's model, understanding of nature progresses through *phenomenological, substantial,* and *essentialist* stages: "The phenomenological stage is the one in which one observes and describes natural phenomena as they are. In the substantialistic stage, one investigates structure of the object. Finally, one finds physical rule governing the object in the essentialistic stage" (Sakata 1971a, 6). Sakata regarded his own model of the "fundamental" particles as an advance from the phenomenological to the substantial stage.[11]

Although only a few of the Nagoya physicists were full-fledged dialectical materialists, the associated three-stage epistemology was more widely influential. According to Ziro Maki, who studied and worked with Sakata beginning in 1955, "There was a tendency for particle theorists, who grew up after the War in Japan, to accept the three-stage theory without question." Maki noted how he himself used it "as a guide for his own research" (Maki 1989, 90). For Maki, the three-stage epistemology's usefulness lay in the fact that it "encourages the elucidation of the role of 'substance' as a theoretical moment which mediates between a phenomena [sic] and its essence. As a result one begins to ask questions as to the nature of that 'essence', and by doing so, the substance, which is tied to the essence and phenomena, gradually becomes specified" (ibid., 90). That is, the theorist, thinking in terms of these three stages, begins at the level of phenomena but avoids getting stuck at the level of mere correlations among phenomena or leaping too quickly to some postulated true nature of those phenomena. Thinking in terms of a substance or thing that underlies those phenomena, and then inquiring into the properties such a substance must have in order for those phenomena to be the result (its essence), provides a heuristic for the theoretical move from phenomena to underlying cause. Sakata used the slogan "from a logic of form to a logic of matter" to refer to this process.

Sakata seemed more committed to this general philosophical standpoint than to the fate of any particular theory of matter he proposed. In remarks made at Hiroshima University in 1963, Sakata spoke without apparent anxiety about the prospects of the experimental confirmation of the eightfold way proposed by Gell-Mann and Ne'eman, which conflicted with his own three-dimensional representation of the fundamental baryon $SU(3)$ symmetry. If the eightfold way proved experimentally sound (as it would be the following year), then

we must accept first the eight baryons to belong to the same level, and proceed to assume the existence of ur-proton, ur-neutron, and ur-Λ-hyperon behind them, – this is what I proposed. The viewpoint that there must underlie inevitably a 'logic of matter' beneath the symmetry, is characteristic to the methodology implied by my proposal of the composite model. (Sakata 1971c, 209)

In other words, if the specific particles (the observed proton, neutron, and lambda) turned out not to be fundamental relative to the other baryons, as Sakata had proposed ("fundamental" being a relative term in his view), then theorists must immediately propose some other "ur" particles to take their place. The models proposed by Gell-Mann and Zweig (quarks) and by Maki and Ohnuki (urbaryons) provided just the kind of theory that Sakata's philosophy called for, at least in very general terms.

In the case of Maki and Ohnuki's proposal, the fit with Sakata's philosophy was, according to its authors, no coincidence. In a paper presented at a 1965 conference in Kyoto, Maki, Ohnuki, and Sakata claimed a very strong methodological link between the Nagoya model and dialectical materialism. By this time, the Nagoya model had been developed to its revised form in terms of urbaryons, and Gell-Mann and Zweig had both made public their quark theory. Maki and Ohnuki, unlike Gell-Mann and Zweig, could contemplate a realist interpretation of either the urbaryon or quark theory without discomfort. The peculiarities of the quark model, which caused even its authors to hesitate giving it a realistic interpretation, were accepted by Maki, Ohnuki, and Sakata in 1965 as "no surprise to us," for they simply concluded that the quarks "might perhaps belong to the elements of the subquantal level. . . . our methodology will never stop at any particular level, and ever seeks for the deeper levels" (Maki, Ohnuki, and Sakata 1966, 112).

Maki, Ohnuki, and Sakata went further than simply embracing the physical possibility of a new stratum of nature, though. They positively denounced any instrumentalist interpretation of the new particles, any suggestion that quarks or urbaryons were "merely a mathematical trick to find out the symmetry." Such an approach was "positivism," and they predicted that such positivist thinking would lead to dire consequences for physical theorizing. Positivist theorists would get stuck at the level of phenomena, simply recording whatever symmetries could be found among the hadrons. They declared, "For these positivists, symmetries are considered to be the first principle of physics given by the *Providence of God*, and hence, the scientific thinking which goes beyond the hadronic level will actually be forbidden" (ibid., 112, emphasis in original).

Maki, Ohnuki, and Sakata saw their own efforts as an antidote to such positivistically mandated theoretical stagnation. They had "proposed in 1959 *the Nagoya Model* which belongs to the next level hidden behind" the baryonic level.[12] Thus, Maki, Ohnuki, and Sakata saw the Nagoya model as an attempt to delve one level deeper into the strata of nature than others had gone, and in so doing to explain, in terms of "the logic of matter," the symmetries that were apparent at the higher level, and described in terms of "the logic of form." Thus, they proclaimed, "the development of the composite model to the Nagoya model was performed by the faithful application of our methodology – to arrive at 'the logic of matter' by starting with 'the logic of form'" (ibid., 112).

In reflections on the Sakata and Nagoya models published in 1989, Maki still described the importance of the Nagoya model largely in terms of the three-stage epistemology. But his claims respecting the three-stage epistemology were more modest than in 1965. Maki noted that "[i]n *B*-matter, the basis for its own existence – form – is not given," so that the Nagoya model as a whole should perhaps be classified as "pre-physics" rather than "physics" (Maki 1989, 92). Indeed, the precise nature of *B*-matter was largely left unspecified in the Nagoya model. The Nagoya group had considered various ways of thinking about the B^+, including both as a "particle" and as a "fluid." One intriguing idea that they had presented was a model in which leptons act as "vessels" that can be either empty (ordinary lepton states), or "filled" with *B*-matter (baryon states) (Maki, Nakagawa, et al. 1960, 1179). But these proposals were merely offered as "possible forms of construction" and were presented "in accordance with our intuitive pictures" (ibid., 1179).[13] Maki consequently described the Nagoya model as an "imaginary construct" or "Vorstellung," the significance of which was that it "provided us with moral support, convincing us of the importance of baryon-lepton symmetry."

According to Maki, then, the Nagoya model served the kind of valuable purpose for which one often introduces a "good construct": it provided "a 'moment' by which one can proceed to substantialistic cognition" (Maki

1989, 92). In other words, by postulating a potentially imaginary "substance" (*B*-matter), with yet-to-be-determined properties beyond those necessary for saving the symmetry relations at the level of known baryons and leptons, the Nagoya model set the stage for proceeding to an investigation of the "essential" features of matter underlying those symmetries. Hence, even if this imaginary construct turned out to be completely wrong, for Maki it facilitated theorizing regarding a deeper stratum of nature. Hence, in Maki's mature view, although the initial value of the *B*-matter concept was precisely that of a good *instrument*, the aim promoted by that concept was not an instrumentalist aim, but ultimately the discovery of some new reality at a previously undisclosed level of nature.

5. THE X-PARTICLE AND QUARTETS

Japanese physicists intensified their scrutiny of the Nagoya model in 1971 when a group led by Kiyoshi Niu of the University of Tokyo seemed to find evidence for a new, short-lived particle with a mass of approximately 2 GeV/c^2 (Niu, Mikumo, and Maeda 1971).[14] Recording cosmic ray events with nuclear emulsions flown on an airplane, Niu's group found an extremely energetic event that could be reconstructed as a massive new particle decaying into a neutral pion and another charged hadron. They calculated the estimated mass of the unknown particle *X* according to two different plausible decay modes. Assuming the *X* had decayed to a neutral pion–charged pion pair, they estimated the mass of *X* to be 1.78 GeV/c^2. For decay into a neutral pion-proton final state, they estimated a mass of 2.95 GeV/c^2. This particle later came to be referred to as the *X*-particle.

This single cosmic-ray event initiated a flurry of articles appearing in *Progress of Theoretical Physics* (*PTP*). As Shuzo Ogawa pointed out (Ogawa 1985), *PTP* published fifteen articles in the years 1971–3 that were directly related to the *X*-particle event. In the United States, on the other hand, the *X*-particle seems to have received very little attention. During 1972, there were no citations of Niu et al.'s 1971 paper in either *Physical Review Letters* (*PRL*) or *Physical Review D* (*PRD*), the two most prominent journals in the United States that publish papers on elementary particle physics.

The fifteen articles in *PTP* can be divided into two general categories: those that examined the *X*-particle specifically in order to identify it from the standpoint of various theories, and those that considered quartet models of fundamental particles more generally, citing the *X*-particle as a motivation for doing so. In both categories, the Nagoya model was of central importance.

Analyses of the X-*Particle.* At Hiroshima University, Takemi Hayashi, Y. Koide, and Shuzo Ogawa had produced a modified version of the extended Nagoya model (in their version, the baryons *p*, *n*, λ, and the hypothetical heavy *p'* were formed from B^+ and lepton-antilepton pairs – conceived to be analogous to mesons – rather than single lepton states) (Hayashi, Koide, and Ogawa 1968). Hayashi and Ogawa, along with Ei-ichiro Kawai, Masahisa Matsuda, and Shinsei Shige-eda, immediately produced an interpretation of the event in terms of the Hiroshima modification of the extended Nagoya model (Hayashi, Kawai, et al. 1972a). Arguing that the event could not be produced either by a strange particle or a weak boson, the Hiroshima group interpreted it as a particle carrying the fourth baryon. Other articles also sought to provide interpretations of the new event, either in terms of the Hiroshima group's version of the Nagoya model (Hayashi, Kobayashi, et al. 1971; Hayashi, Kawai, et al. 1972b) or the original version of the extended Nagoya model by the Hiroshima group themselves (Hayashi, Matsuda, and Ogawa 1973) or another theory known as $O(4)$ Symmetry, for which one needed to interpret the event as a boson carrying strangeness $S = 3$, which then decayed to two pions, resulting in a change of strangeness $|\Delta S| = 3$ (Sato and Nakamura 1972). Shigeki Tasaka and Yoshiaki Yamamoto of Konan University suggested that among cosmic ray events recorded in a balloon-borne emulsion chamber, eight could be identified that could be of the same type as that recorded by Niu et al. (Tasaka and Yamamoto 1973).

General Evaluations of Quartet Hadron Schemes. Each one of the papers in this category, except for one, cited Niu et al.'s paper as motivation for an examination of quartet models. Toshihide Maskawa and Ziro Maki were among the authors on three such papers, which sought to follow up on Maki and Ohnuki's urbaryon revision of the extended Nagoya model (Maki and Maskawa 1971; Kondo, Maki, and Maskawa 1972; Maki, Maskawa, and Umemura 1972). Kondo et al.'s paper is titled "A Note on the Leptonic Decays of Charmed Mesons," indicating that the authors considered the X-particle likely to be a charmed particle, and that for them "charm" referred to the fourth fundamental hadron predicted by the revised Nagoya model. For some Japanese physicists, the origin of charm lay not in Bjorken and Glashow's 1964 paper but in the extended Nagoya model of 1962. (This is indicated also by some comments made years later by Kobayashi, which I discuss later.)

Hiroshima's Hayashi and Ogawa appeared as coauthors on two more theoretical papers, citing Niu et al.'s paper as a motivating factor (Hayashi, Karino, et al. 1973; Hayashi, Nakagawa, et al. 1973). Finally, Makoto Kobayashi, along with Hidemitsu Nitto and Masami Nakagawa, published a paper on yet another descendant of the Nagoya model, this time with fractional charges assigned to the quartet of fundamental hadrons (Kobayashi, Nakagawa, and Nitto 1972).

Following up on the 1971 paper in which Hayashi, Kobayashi, Nakagawa, and Nitto gave their interpretation of the X-particle, Nakagawa and Nitto elaborated on their version of the four-baryon Nagoya model and attempted to show that "various basic assumptions made in the hadron physics of the triplet model (p, n, λ) could be tested experimentally in a very sharp form for the hadron physics of the quartet model (p, n, λ, p'), and at the same time to point out that some predictions are derived for phenomena of super hadrons [hadrons carrying the quantum number of the fourth baryon p']" (Nakagawa and Nitto 1973, 1322–3). They then went on to refer to the X-particle as an example of such a "super hadron" (ibid., 1328).

In Shuzo Ogawa's 1985 retrospective, all fourteen of the papers just mentioned are included, along with the KM paper, in a list of "papers relating to the charmed particle which have been published in PTP before the finding of J/ψ after Niu's discovery" (Ogawa 1985, 58). The 14 papers that I have mentioned all explicitly cite the Niu et al. paper.

The one exception on Ogawa's list is the KM paper itself. It does not include an explicit citation of the X-particle discovery. However, there is ample evidence that Kobayashi and Maskawa were prompted to take up the problem of CP-violation in quartet theories as a response to the X-particle discovery in the context of a serious interest in the Nagoya model.

First, we know that both Kobayashi and Maskawa were well acquainted with the authors of the Nagoya model and with the theory itself. Both Kobayashi and Maskawa received their Ph.D.'s from Nagoya University. At the time of the X-particle discovery, Kobayashi was still a graduate student there, and Maskawa had just left a postdoc at Nagoya to take a position as research associate at Kyoto University. Kobayashi then took a position at Kyoto upon completing his Ph.D. in 1972 (Kobayashi 1997, 138). Kobayashi had collaborated with Hayashi, Nakagawa, and Nitto on one of the first papers to interpret the X-particle in terms of the Hiroshima variant on the Nagoya model (Hayashi, Kobayashi, et al. 1971). Furthermore, both Kobayashi and Maskawa had published other papers that cited Niu et al. (1971) explicitly as a motivation for considering quartet schemes (Maki and Maskawa 1971; Kobayashi, Nakagawa, and Nitto 1972; Kondo, Maki, and Maskawa 1972; Maki, Maskawa, and Umemura 1972). Thus, both Kobayashi and Maskawa were writing already about four-hadron versions of the Nagoya model prior to their 1973 paper, with the X-particle serving as an important motivation for looking at such theories.

Kobayashi's own later testimony establishes conclusively the connection between the X-particle, the Nagoya model, and the 1973 KM paper. He noted that their work on CP violation was done in the midst of speculation that Niu's team had in fact already discovered a fourth fundamental particle subject to the strong force. Kobayashi recalled that Shuzo Ogawa had first suggested that the X-particle might be the fourth baryon of the extended Nagoya model, and that

[f]ollowing Ogawa's suggestion, a few Japanese groups, including myself, started the investigation of the cosmic ray events based on the four-quark models, so that, more

or less, we were familiar with the structure of the weak interactions in the four-quark scheme. (Kobayashi 1998, 16)

As a result of these studies,

[w]e accepted [the Glashow-Weinberg-Salam theory of the weak interaction's] extension to the hadron based on the GIM [Glashow-Iliopoulos-Maiani] scheme as a quite realistic possibility, because the fourth quark already existed for us in a sense. Sometimes it is said that our *CP* paper was written before the discovery of charm. In this sense, however, our paper came after the charm. (Kobayashi 1997, 138)

All of this preceded the spectacular events that took place in 1974 in the United States. In November of that year, a group at SLAC led by Burton Richter and a group at Brookhaven led by Samuel Ting discovered a new, quite narrow resonance at 3.1 GeV. This new particle came to be known as the J/ψ (Ting's group had dubbed it the J, while Richter's group called it the ψ). These discoveries, along with a quick confirmation from a group at the ADONE[15] ring at Frascati led by Giorgio Belletini (later a CDF spokesperson), were taken to establish the existence of charmed particles, earning Nobel Prizes for Richter and Ting in 1976 (Aubert, Becker, et al. 1974; Augustin, Boyarski, et al. 1974; Bacci, Balbibi Celio, et al. 1974).[16]

Niu's lonely cosmic ray event, which had garnered so much attention in Japan, and had even been called charm by a number of Japanese physicists, was quickly overshadowed. In the wave of papers on charm and the J/ψ that appeared after the "November revolution" – as it came to be called – the *X*-particle and the four-baryon Nagoya model were almost never mentioned. An exception is a review of the charm issue, published in 1975, by Mary Gaillard, Benjamin Lee, and Jonathan Rosner (1975). The main body of this article was written before the Ting and Richter groups had made their discoveries, and the authors added a "note in proof," nearly as long as the original article, discussing the new findings. In the portion of the paper written before the November revolution, they did mention Niu et al.'s single event among the "few candidates" for already-observed charmed particles (ibid., 297). They also mentioned Maki and Ohnuki's 1964 urbaryon paper among theories including a charmed quark (but not the earlier Nagoya model papers) (ibid., 277).

This, however, is an isolated instance.[17] Although in the aftermath of the November revolution many articles that discussed the *X*-particle and the revised Nagoya model appeared in *PTP*, neither received attention in the *Physical Review*, and the developments in Japan that led up to the introduction of a third generation of matter have been omitted from most historical accounts.[18]

The many papers on the *X*-particle that appeared at the time in *PTP*, along with Kobayashi's remarks, clearly indicate that the mostly forgotten cosmic ray event found by Niu's group helped to stimulate an idea far ahead of the charm proposal. Furthermore, these publications indicate that a pre-occupation with the extended Nagoya model and its various modifications was largely responsible for the great interest taken in the *X*-particle in Japan.

Having already considered the existence of a fourth fundamental strongly interacting particle to be a very realistic possibility, Kobayashi and Maskawa turned to exploring the potential weaknesses of the quartet model. This led them to suggest that a third generation of fundamental hadrons would be a plausible way to accommodate *CP* violation. This suggestion eventually received renown, but first lay for several years in obscurity.

6. THE KOBAYASHI AND MASKAWA PAPER

Noteworthy is what Kobayashi and Maskawa did *not* discuss in their paper. Although they are generally said to have predicted the existence of the third generation of quarks, their paper did not explicitly employ the quark model. They considered different possibilities for the number of fundamental hadrons, but "hadron" is in this context a generic term for whatever particles are subject to the strong force. Kobayashi and Maskawa denoted the four fundamental fields in the quartet model as p, n, λ, and ξ, to which they assigned the indeterminate charges Q, $Q - 1$, $Q - 1$, and Q, respectively (Kobayashi and Maskawa 1973, 652). These charges were compatible with both the Nagoya model's assignment of 1, 0, 0, 1 and the assignments given in the quark scheme. In their 1970 paper based on a quartet model of quarks, Glashow, Iliopoulos, and Maiani discussed both a fractional charge assignment ($\frac{2}{3}$, $-\frac{1}{3}$, $-\frac{1}{3}$, $\frac{2}{3}$) and an integral charge assignment $(0, -1, -1, 0)$ (1970, 1290).

Furthermore, the term "predict" suggests a stronger assertion than one actually finds in the KM paper regarding the "6-plet" model they discussed. Kobayashi and Maskawa's central claim was that no "realistic" quartet model allows for *CP*-violating weak interactions: "in the case of the . . . quartet model, we cannot make a *CP*-violating interaction without introducing any other new fields," or somehow violating at least one of two additional conditions: (1) that the mass of the fourth particle ξ be "sufficiently large" (since a light fourth hadron would presumably have been detected previously), and (2) that "the model should be consistent with our well-established knowledge of the semi-leptonic processes" (Kobayashi and Maskawa 1973, 652). I. I. Bigi and A. I. Sanda have described this as "a prime example for deducing the existence of New Physics indirectly" (Bigi and Sanda 2000, 128). I would emphasize that this was the deduction of *some kind* of new physics, encompassing several possibilities.

Kobayashi and Maskawa divided their argument into several cases, and it is the first case that is most significant for the present discussion. First they noted that in one version of a four hadron (two "generation") model, the description of the *weak charged current* (weak particle decay processes with a net change in charge) involves a unitary matrix U,

$$U = \begin{pmatrix} \cos\theta_c & \sin\theta_c \\ -\sin\theta_c & \cos\theta_c \end{pmatrix},$$

known as the Cabibbo matrix, where θ_c is the "Cabibbo angle."

Specifically, the weak current in which the charge of a hadronic field is lowered is described by $J_\mu^- = \bar{q}_u\gamma_\mu U(1-\gamma_5)q_d$. If there are n families of quarks, $\bar{q}_u$ is the row of n antiparticles of the charge $\frac{2}{3}$ quarks, and q_d is the column of n charge $-\frac{1}{3}$ quarks. (The terms γ_μ and γ_5, which refer to Dirac matrices, need not concern us here.) Thus, for the kind of theory Kobayashi and Maskawa were considering, with the quark fields regarded (for purposes of the weak interaction) as forming two doublets, the charge-lowering weak current can be written as

$$J_\mu^- = (\bar{u} \quad \bar{c})\gamma_\mu U(1-\gamma_5)\begin{pmatrix} d \\ s \end{pmatrix}.$$

For *CP*-violating interactions to be allowed, there must be terms in U with a complex phase (an additional factor of the form $e^{i\delta}$, where $i = \sqrt{-1}$). Only if such complex phases are present can there be weak decays violating T conservation, the principle that interactions must be invariant under transformations by the time reversal operator T. This is significant because it is a basic theorem of quantum field theory that *CPT* symmetry holds (in keeping with certain well-confirmed relationships between particles and antiparticles). If both *CPT* and T conservation hold, however, then so must *CP* conservation. Therefore, for this representation of the quartet field, *CP* violation cannot occur.[19]

In this first case under consideration, the weak charged currents involve only *left-handed* components of the quark fields. A particle state is *left-handed* (its *helicity* $= -\frac{1}{2}$) if its spin is antiparallel to its momentum (i.e., the spin and momentum vectors point in opposite directions), whereas it is *right-handed* (*helicity* $= +\frac{1}{2}$) if its spin and momentum vectors are parallel. In the present context, though, the "left-handed" and "right-handed" components of the quark fields (and fermion fields generally) correspond to helicity $= -\frac{1}{2}$ and $+\frac{1}{2}$ states respectively only for states with zero mass.[20] The correspondence holds approximately in the "ultrarelativistic" limit, where the energy of the particle is much greater than its rest mass. In this modified sense of the term "left-handed," only the left-handed components of the quark fields are involved in the interaction described by the formulas given earlier for the weak charged current. That is to say, such a model includes only "left-handed currents."

Kobayashi and Maskawa considered three more possible representations of the quartet scheme, all involving both left- and right-handed currents. Two of these schemes do allow for *CP*-violating decays, but only at the expense of violating one of the two conditions on "realistic" quartet models: either the fourth hadron turns out to be so light that it should already have been discovered, or there is a conflict with well-established experimental knowledge about the semileptonic decay of baryons. A final case falls back into the difficulties of the first case (Kobayashi and Maskawa 1973, 654–6).

Kobayashi and Maskawa then took up three different possibilities in which new fields are introduced to allow for *CP*-violating weak decays. First, they begin within the framework of the standard GIM theory, including the reliance on the *scalar* doublet *Higgs* field ϕ.[21] To this model, they suggest adding an entirely new scalar doublet field ψ, with which the hadrons interact through an entirely new force. Then it would be possible for the interaction between the quartet of fundamental hadrons and this new field to be described via an operator from which complex phases could not be eliminated, thus allowing for *CP* violation. This possibility would require new fields and a new force, but not a third generation of hadrons (ibid., 656). The second *CP*-violating possibility involved the introduction of a new scalar field *S* mediating the strong interaction. Such a field would then interact both with the fundamental hadrons and with the Higgs field ϕ. In this case, the matrix describing the interaction between the new field *S* and ϕ contains complex phases, and *CP* violation is allowed (ibid., 656).

Only after these (forgotten) proposals were described did Kobayashi and Maskawa take up the "6-plet" model, which they described as "another interesting model of *CP* violation" (ibid., 657). Here they pointed out that if a third generation of fundamental fermions is introduced, then in the description of the weak interaction, the 2×2 Cabibbo matrix must be replaced by a 3×3 unitary matrix. With such a modification, the weak charge-lowering current would be written

$$J_\mu^- = (\bar{u} \quad \bar{c} \quad \bar{t}) \gamma_\mu V (1 - \gamma_5) \begin{pmatrix} d \\ s \\ b \end{pmatrix},$$

where *V* is the new 3×3 matrix. In order to retain unitarity, this matrix (now called the Cabibbo-Kobayashi-Maskawa, or CKM, matrix) must contain elements with a complex phase. Inclusion of a complex phase results in a theory that allows processes violating *CP* invariance: "Then, we have *CP*-violating effects through the interference among these different current components" (ibid., 657), and this could be done with only left-handed currents.

Kobayashi and Maskawa suggested a third generation of fundamental hadrons as one possibility among others that could accommodate *CP* violation. They did not discuss the top quark specifically. Indeed, they did not even present their discussion in terms of the quark model. They admitted that other kinds of schemes less closely modeled on that proposed by Weinberg (1967) could also accommodate *CP* violation, particularly that

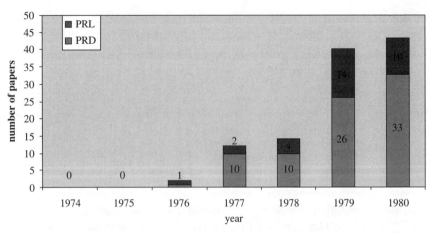

FIGURE 1.4. Citations of Kobayashi and Maskawa 1973 in the *Physical Review*, 1974–80.

proposed in Georgi and Glashow (1972). For their reserve in presenting a revolutionary idea,[22] they were rewarded with several years of neglect.

7. THE RECEPTION OF KOBAYASHI AND MASKAWA'S PAPER

During the first few years following its publication, the KM paper seems to have received little attention, although experimental developments gave impetus to the extension of the four-quark model toward a three-generation theory of the sort postulated by KM. The record of citations of the KM paper in the *Physical Review* is telling. The number of papers citing the KM paper in the *Physical Review* for each year following its 1973 publication through 1980 is as follows: 1974, 0; 1975, 0; 1976, 2; 1977, 12; 1978, 14; 1979, 40; 1980, 43 (see Figure 1.4).[23,24] In this section, I will examine how the KM paper came to be cited and used by other physicists, and what events might have been responsible for the increase in the attention that it received.

7.1. The Discovery of the Tau Lepton

In 1975, a group led by Martin Perl was looking for, among other things, new and heavier leptons than the four known members of that group (e, ν_e, μ, ν_μ). Perl was particularly motivated by Paul Tsai's "sequential leptons" proposal (Tsai 1971; Perl 1997).[25] Kobayashi and Maskawa's third genera-tion of fermions does not seem to have been a motivation for Perl. Using the SLAC-LBL (Stanford Linear Accelerator Center – Lawrence Berkeley Laboratory) detector at SLAC's SPEAR (Stanford Positron-Electron Asym-metric Ring) accelerator, Perl and his colleagues found an excess of events resulting in electron-muon pairs at around 4.8 GeV. The meaning of this excess was not immediately clear. They published the result in 1975, noting that the excess "cannot be explained . . . by the production and decay of any presently known particles," and concluding that the existence of a previously unknown particle would be a "possible explanation" of the events (Perl, Abrams, et al. 1975, 1492).

Perl later recalled that the publication of the 1975 paper was "followed by several years of uncertainty about the validity of our data and its interpretation" (Perl 1997, 90). It was not until 1978 that a consensus was reached in the particle physics community that this was indeed a new lepton, the τ. That one of the expected decay modes of the τ, in which τ decays to a τ neutrino (ν_τ) and a negative pion (π^-), proved mysteriously elusive contributed to the uncertainty. In the immediate aftermath of the Perl group's initial findings, several experiments failed to find this decay mode at levels anywhere near theoretical predictions. Until this decay mode was experimentally confirmed at approximately the predicted rate, physicists had reservations about identifying it as a heavy lepton. Years later, Martin Perl admitted that he could still not explain why this particular decay channel had been difficult to see, but by the middle of 1978, all of the various experiments looking for τ decays reported branching fractions for $\pi^- \nu_\tau$ final states at close to theoretical predictions, and physicists confidently concluded that the anomalous events were in fact τ decays (Perl 1997, 93–5; see also Treitel 1987).

Although the possible existence of a third generation of leptons was strongly suggestive of a third generation of quarks as well (recall the Nagoya physicists' emphasis on "baryon-lepton symmetry"), Perl's discovery does not seem to have directly motivated physicists to look at the KM paper (with one exception, which is discussed in Section 7.3).

7.2. Early Models with More Than Four Quarks

Nevertheless physicists were discussing, in 1975 and 1976, models with more than four quarks. In fact, such models proliferated, apparently without any prompting from the considerations raised by the KM paper. In a paper published in January 1975 in *PRL*, R. M. Barnett of Harvard proposed a theory with three charmed quarks and right-handed weak currents. His model involved two triplets of quarks, where the second triplet (p', n', λ') would have all the same quantum numbers as the first (p, n, λ), but with charm $+1$ (Barnett 1975).

Haim Harari also proposed a model with two triplets, but of a different sort (Harari 1975). He retained the usual $SU(3)$ triplet (u, d, s),[26] but added another $SU(3)$ antitriplet (t, b, r), thus introducing the "top" and "bottom" terminology for the new quarks. Harari proposed a new quantum number *heaviness*, and assigned heaviness $= +1$ for the heavy t, b, and r quarks. The other three quarks were to have heaviness zero. On this scheme the newly discovered ψ could be interpreted as a mixture of heavy quark-antiquark bound states: $\psi = (\frac{1}{\sqrt{3}})(t\,\bar{t} + b\bar{b} + r\bar{r})$. Had this been correct, the discovery made by Ting and Richter would have meant the discovery all at once of three new quarks, an accomplishment that in fact would take 20 years.

In 1975, three groups published papers that presented variations on Harari's proposal of six quarks with fractional charges. Called "vector" or "vectorlike" models, these approaches added new, right-handed weak charged currents to the purely left-handed charged currents in the Weinberg-Salam-Glashow-Iliopoulos-Maiani-type theories.

1. A. DeRújula, Howard Georgi, and Sheldon Glashow initially published a modification of the GIM quartet model itself featuring the addition of right-handed currents (DeRújula, Georgi, and Glashow 1975a). When they refined and elaborated the model in a longer article in *PRD*, they proposed expanding from a quartet to a six-quark model, citing both Barnett's and Harari's papers, as well as unpublished work by F. A. Wilczek (DeRújula, Georgi, and Glashow 1975b).

2. Similarly Sandip Pakvasa, W. A. Simmons, and S. F. Tuan (Pakvasa, Simmons, and Tuan 1975) adopted the "heaviness" scheme proposed by Harari, yielding a two-triplet model, but amended Harari's representation of charge-changing weak decays in keeping with the DeRújula-Georgi-Glashow right-handed currents proposal, including citations of both Harari's paper and the early *PRL* paper by DeRújula, Georgi, and Glashow.

3. In an article appearing in *Physics Letters* in November 1975, Harald Fritzsch, Murray Gell-Mann, and P. Minkowski presented a "vectorlike gauge theory" with six quark flavors, again explicitly following Harari's proposal (as well as adding a new lepton – Perl had first talked about the data that would form the basis of the τ discovery in June of 1975, although the first paper on the topic did not appear until December). Fritzsch, Gell-Mann, and Minkowski noted that their theory was a "minimal" one, and that more complex schemes with additional fundamental fermions were possible as well (Fritzsch, Gell-Mann, and Minkowski 1975).

A less well-known proposal came from Yukio Tomozawa and Suk Koo Yun in a paper appearing in *PRD* in May of the same year. This model, drawing on a suggestion by Moo-Young Han and Yoichiro Nambu (Han and Nambu 1965), proposed three triplets of quarks *plus* three charmed quarks, making 12 integer-charged quarks in all. The authors considered different symmetry schemes and also discussed a fractional-charge variation on their model, but in every case only left-handed currents were invoked. Like Kobayashi and Maskawa, Tomozawa and Yun were motivated by an attempt to incorporate *CP*-violating interactions by means of introducing irreducible complex phases into the matrix used to calculate the weak currents (Tomozawa and Yun 1975).

None of the papers I have just mentioned cited the KM paper. Neither did other papers concerned with models featuring more than four quarks (Wilczek, Zee, et al. 1975; Barnett 1976; Gürsey and Sikivie 1976).

7.3. Early Citations of KM

The earliest citations of KM appeared in four papers published within a few months of one another. The first of these to be received by the journal in which it appeared was a paper titled "*CP* Violation in the Six-quark Model" by Sandip Pakvasa and Hirotaka Sugawara, at the University of Hawaii at Manoa (Pakvasa and Sugawara 1976). Sugawara brought the KM paper to Pakvasa's attention sometime after they began work on this article in July 1975. Pakvasa and Sugawara noted that "A few years ago Kobayashi and Maskawa pointed out that *CP* violation can be incorporated into the standard *V–A* Weinberg-Salam model if we increase the number of quarks from four to six" (305). They then went on to summarize the central argument of the KM paper.[27] This paper, submitted to *PRL* in August 1975, contained the first reference to the KM paper in the *Physical Review*. It appeared in the "Comments and Addenda" section at the back of the July 1976 issue of *PRD*, after some difficulty getting past *PRL*'s referees (Pakvasa, personal communication).

The second paper to draw attention to the KM paper was by Luciano Maiani, titled "*CP* Violation in Purely Lefthanded Weak Interactions." It was published in *Physics Letters* in May 1976. The original draft was received on November 3, 1975, and the revised draft was received on February 18 of the following year. The opening sentence declared Maiani's motivation: "The possible evidence recently found at SPEAR of a new charged lepton suggests that new quarks exist, beyond the usual p, n, and λ quarks, and the charmed p' quark" (Maiani 1976). Maiani noted the possibility of a simple extension of the quartet model to six quarks, citing Harari's paper. He then mentioned, in a footnote, that Kobayashi and Maskawa had already devised a similar scheme for *CP* violation and noted that "After completing the paper, I learned that some of the arguments presented here have been independently discussed by S. Pakvasa and H. Sugawara" (183n).

In *PRL*, Steven Weinberg, noting the two papers just mentioned, cited KM not to praise it but to bury it. Weinberg sought to establish that "[r]enormalizable gauge theories of the weak and electromagnetic interactions provide a mechanism which could violate *CP* conservation with about the right strength: the Higgs boson" (Weinberg 1976, 657). Citing the KM paper as well as the work of Maiani and Pakvasa and Sugawara, Weinberg accepted the argument that in four-quark models, the weak current matrix could always be written without complex phases by choosing the appropriate phases for the quark fields, but argued that for "a wide class of theories" involving more Higgs fields than required in the usual standard model formulation, *CP* would not be conserved in the exchange of a Higgs boson. In his footnote acknowledging Kobayashi and Maskawa as the originators of the argument based on four-quark fields, Weinberg noted that "[f]rom the standpoint of the present paper, it is hoped that there are *not* more than four quarks, to insure that *CP* violation arises only from Higgs exchange" (661).[28]

The most generous and possibly the most noticed citation of KM amongst these papers appeared in "Left-handed Currents and *CP* Violation," by John Ellis, Mary K. Gaillard, and Dimitrius V. Nanopoulos, three Conseil Européen pour la Recherche Nucléaire (CERN) theorists, which appeared in *Nuclear Physics* (Ellis, Gaillard, and Nanopoulos 1976). Ellis, Gaillard, and Nanopoulos highlighted the importance of Kobayashi and Maskawa's work in the first sentence of their abstract: "An analysis is presented of a model for *CP* violation due to Kobayashi and Maskawa, which has six quarks and purely left-handed currents" (213). And in the opening paragraphs of the article itself, the authors noted that "A possible reason to introduce six quarks was pointed out some time ago by Kobayashi and Maskawa," and that "The Kobayashi-Maskawa model has recently been revived by Pakvasa and Sugawara, and independently by Maiani" (214). Ellis, Gaillard, and Nanopoulos went on in this paper to give detailed analyses of possible ways in which *CP* violation could occur on the KM model. The first paragraph of their conclusions is worth quoting at length:

We like the KM model because it introduces *CP* violation in a natural way, as a result of weak mixing between the quarks analogous to the Cabibbo angle in the GIM model. Thus two puzzles are reduced to one. Looked at another way, it gives a *raison d'être* to the fifth and sixth quarks, needed in the simplest Weinberg-Salam theory to cancel anomalies and restore renormalizability if there is a heavy lepton. Unfortunately, it is a very well concealed theory, at least as long as the top and bottom degrees of freedom are hidden, and naked top and bottom states are not produced in e^+e^- annihilation or elsewhere. (239)

"Left-handed Currents and *CP* Violation" was received at *Nuclear Physics* in February 1976 and published in June of that year. A year later, the prospects for confirming the KM proposal would appear significantly improved.

7.4. The Discovery of the Upsilon

Although these scattered citations of the KM paper began to appear in major physics journals in the United States and Europe in 1976, the first of the new "heavy" quarks, the bottom quark, was unearthed at Fermilab by people at least some of whom were completely unaware of this theoretical development. At Fermilab, physicists from Columbia University had joined a group of the accelerator lab's own scientists to collect data using a two-arm mass spectrometer, which looked at particles emerging downstream from a fixed target. The detector was designed to allow measurements of the mass of initial states based on the energy of leptons produced in decays. The group, which was led by Leon Lederman, was collecting events with two-electron final states from 1974 to 1976, after which they were joined by a group from the State University of New York (SUNY) at Stony Brook. They had reported on an apparent peak in the distribution of their data at around 6 GeV. They took this peak to be the first indication of a new particle and named the putative particle upsilon (Υ) (Hom, Lederman, et al. 1976). According to the experiment coordinator John Yoh, group member Walter Innes had pointed out that the name "upsilon" could be turned into a "Leon-type joke" if the result turned out to be erroneous (which it soon did). The particle could then be renamed the "Oops-Leon" (Yoh 1998b, 35). The name stuck, and the pun-loving Lederman has used it in his self-effacing recollections ever since (Lederman 1997, 105).

Although the peak at 6 GeV eventually disappeared, John Yoh noticed in November of 1976 a very slight excess (10 events in a range of 300 MeV, where the expected background was about 1.75 events) at around 9.5 GeV based on all of the dielectron data and a small number of dimuon events. He placed a bottle of French champagne labeled "Υ 9.5" in the refrigerator, presumably to be opened should this excess turn into a great discovery.[29] According to Lederman, "Everyone thought he was kidding" (Lederman 1997, 106).

The collaboration upgraded their detector to run in "dimuon" mode, looking for events in which the initial particle decayed to a pair of muons. They increased the sensitivity of their detector, and achieved a good mass resolution with muons. The new version of the detector also allowed for a number of redundant measurements to increase confidence in the identification of particles as muons (Lederman 1997, 107; Yoh 1998b, 36). As a result, the little cluster of 10 events at around 9.5 GeV grew to a peak of about 600 events with a significance equivalent to greater than eight standard deviations (Lederman 1997, 106; Yoh 1998b, 38). The announcement of the true upsilon was made at Fermilab at the end of June 1977, and the paper was published shortly thereafter (Herb, Hom, et al. 1977). (Leon Lederman went on to become the director of Fermilab. John Yoh became a member of CDF.)

Although the discovery of the Υ eventually came to be recognized as the discovery of the bottom quark, Kobayashi and Maskawa's paper does not seem to have had any direct influence on this experiment. In their

PRL article announcing the discovery, the Columbia-Fermilab-Stony Brook (CFS) collaboration does not include a citation of the KM paper. John Yoh, in later recollections, wrote:

> The discovery of Υ (bottomonium) was actually more unexpected than that of the J/ψ (charmonium). The Kobayashi-Maskawa paper speculating on six quarks, though published in 1973, was totally unknown in the U.S., having been published in the obscure Japanese journal *Progress of Theoretical Physics*. The preliminary evidence for the τ from Mark I in 1975 was weak, and not established for a long time, becoming believable only after more data were collected by PLUTO and Mark I. (Yoh 1998b, 38)

So while the KM paper may have been emerging from obscurity among some theorists, it remained unknown to at least some of the experimentalists who were generating supporting evidence for part of its six-quark model.

    After the CFS collaboration's discovery, Ellis, Gaillard, and Nanopoulos, along with Serge Rudaz, declared it to be "likely that the -onium of one or more new quarks has been discovered," and noted that the "simplest model which could incorporate such quarks is a six-quark generalization of the standard Glashow-Iliopoulos-Maiani (GIM)-Weinberg-Salam-Ward $SU(2) \times U(1)$ model which was first discussed by Kobayashi and Maskawa (KM)" (Ellis, Gaillard, et al. 1977, 285–6). This perfectly prefigures the citation that CDF gave years later for "the standard model," with which this chapter began.

7.5. No Weak Right-Handers

The "standard Glashow-Iliopoulos-Maiani(GIM)-Weinberg-Salam-Ward $SU(2) \times U(1)$ model" was by no means the only game in town. However "simple" and "natural" Kobayashi and Maskawa's six-quark extension of that model to accommodate *CP* violation may have seemed to Ellis, Gaillard, Nanopoulos, and Rudaz, there were other models to contemplate for those with different conceptions of simplicity and naturalness, or who were indifferent to such theoretical virtues. In particular, numerous models included left- and right-handed weak currents of equal strength ("symmetric models"). The previously mentioned "vector models" proposed by DeRújula-Georgi-Glashow, Pakvasa-Simmons-Tuan, and Fritzsch-Gell-Mann-Minkowski were all symmetric models, for example.

    In 1978, though, experimental results from a group working at SLAC eliminated such "ambidextrous" theories from further consideration. One difference between symmetric models and the purely left-handed models of the "standard GIM-Weinberg-Salam-Ward" type was that according to the symmetric models, there should not be any "neutral" (not charge-changing) weak interactions in which parity is violated.[30] But the group at SLAC, led by Charles Prescott, found just such parity violations in weak neutral currents, thus ruling out the symmetric models (Prescott, Atwood, et al. 1978). However, those results had not ruled out an asymmetric "hybrid" model (Cahn and Gilman 1978). In the hybrid model, the left-handed components of the quarks were assigned to weak isospin doublets, while their right-handed components were assigned to singlets, just as in the purely left-handed models. However, the right-handed component of the electron was assigned to a weak isospin doublet with a hypothetical heavy lepton. Prescott's group, having collected additional data, eliminated the hybrid model as well in a 1979 paper (Prescott, Atwood, et al. 1979).[31]

Thus, by the end of the 1970s, Kobayashi and Maskawa's proposal could be seen as an anticipation of a very promising line of inquiry. One member of a third generation of quarks had been experimentally confirmed, as had a third-generation lepton. Furthermore, left-handed weak currents alone sufficed in fact to accommodate *CP* violation. Still, the KM six-quark model was not guaranteed to succeed. A 1979 publication, for example, pointed out the possibility of a five-quark model, in which the *b* quark is treated as a left-handed weak isospin singlet (Barger and Pakvasa 1979). In any case, no experimentalist would rest content with the claim that six quarks were required for theoretical reasons. To vindicate the standard model required establishing the existence of the top by a direct experimental search. After the exciting years of the middle 1970s, which had yielded neutral currents, two new quarks, and a new lepton, particle physicists were in for a long wait.

8. DISCUSSION: "DOGMATISM" AND THEMATIC COMMITMENT

The doctrine of inexhaustibility and the methodology of the three-stage epistemology constituted the core of the dialectical materialism that motivated much of the theorizing of the Nagoya school. Inexhaustibility makes the substantive claim that there is no truly fundamental level of analysis in nature, and each level of the material constitution of the universe is governed by a distinct set of laws. The three-stage epistemology makes the methodological recommendation that the physical theorist advance the understanding of nature by postulating a substance governed by laws at a lower level than those currently known, thus pursuing the course of inquiry from the level of phenomena to those essential lower-level material facts that provide the proper explanation for the known higher-level phenomena.

In one of the few discussions of the Nagoya model published outside of Japan, the Israeli physicist Yuval Ne'eman described the adherence of the group surrounding Shoichi Sakata to dialectical materialism as "an extreme orthodoxy which can only be compared with a fundamentalist's attachment to the biblical story of Genesis" (Ne'eman 1974, 15). While Ne'eman acknowledged that "interest in a materialist lower stratum as predicted by Dialectical Materialism had indeed been useful in triggering the introduction of the $U(3)$ group," he claimed that "the motivation was so overwhelming that it overshadowed the experimental facts" (11). Ne'eman further charged the Nagoya group with dogmatically attacking other theorists whose approaches disagreed with their own, as when Yoichi Fujimoto criticized Gell-Mann for entertaining the possibility of treating elementary particles as points (unacceptable on dialectical materialist grounds because such an attitude allegedly poses an obstacle to discovery of another stratum of matter beneath the current level). Ne'eman remarked, "It is . . . a sad development when a scientist falls back into the medieval way of preferring dogmas to actual physical theory" (21).[32]

I wish to explore another way of understanding the role that dialectical materialism played in the development of the Nagoya model. In their embrace of the doctrine of inexhaustibility, the Nagoya physicists were adopting, not a dogma, but what Gerald Holton calls a *thematic hypothesis*, while the three-stage epistemology constituted what he terms a *methodological thema*. Holton maintains that understanding how scientists make discoveries and accept or reject ideas requires analysis of the role of *themata*, which he describes as "those fundamental preconceptions of a stable and widely diffused kind that are not resolvable into or derivable from observation and analytic ratiocination" (Holton 1988, 13–14). Examples of thematic hypotheses that have been prominent in science include the claims that certain properties are strictly conserved, that certain processes are symmetric in various ways (an important thema for the developments surveyed here), that natural processes are directed toward some purpose, and so on.[33]

According to Maki, Ohnuki, and Sakata, as noted earlier, the search for some level of structure underlying the known baryons led to the proposal that there were more fundamental "urbaryons" of which these particles were composed. Ne'eman rightly pointed out that the "bootstrap" model, which completely abandoned the search for underlying structures beneath the hadrons, and declared a "nuclear particle democracy" was no less useful for promoting phenomenological studies of hadrons.[34] Two opposing themata can contribute to the advance of a scientific discipline, even simultaneously. However, when the adherents to a thema allow their commitments to "overshadow experimental facts," or oppose other theories because they run afoul of the thema, according to Ne'eman they are being dogmatic.

Perhaps, but it is difficult to see how commitment to a thematic hypothesis could serve its positive function of guiding theorizing without at the same time causing a theorist to judge negatively theories that are inconsistent with that thema. One can find thematic underpinnings in various episodes in physics in which one theorist has found another's ideas to be unsatisfactory in some way not directly linked to empirical inadequacy (arguably, for example, Einstein's opposition to Bohr's espousal of complementarity, or Galileo's failure to recognize the importance of Kepler's elliptical planetary orbits). While some of the Nagoya physicists may have been more explicit, or even strident, in declaring allegiance to dialectical materialism than other particle physicists typically have been in the deployment of such beliefs, their reliance on a thematic commitment was not a violation of proper scientific method, but an important part of at least one phase of scientific advance.

Of course, not all themata are equally fruitful, and one might object that the thema employed by the Nagoya school yielded only limited benefits for its adherents in terms of good theoretical ideas. This certainly was the position taken by Ne'eman in his paper, presented in 1971. Yet it could be countered that Ne'eman judged the Nagoya school prematurely, since he spoke two years prior to the Kobayashi-Maskawa paper, arguably a product of the Nagoya line of thought.

Here we run up against a difficulty in assessing the "fruitfulness" of a thematic hypothesis. It is true that the Nagoya school was influenced by dialectical materialism. It is also true that Kobayashi and Maskawa were part of a movement to update the Nagoya model and to interpret the X-particle within that framework. Finally, it is true that, in pursuing this task, they arrived at the idea of a third generation of fundamental, strongly interacting particles. All of this indicates a historical link between Sakata's dialectical materialism and the third generation of quarks. This does not mean, of course, that Kobayashi and Maskawa themselves subscribed to a dialectical materialist philosophy. Nothing in their writings indicates that they did. It would therefore be rash to regard the third generations of quarks as the direct fruit of the thema of dialectical materialism or to conclude that the KM paper stands as a vindication of the dialectical materialism of Sakata and his cohorts.

The KM paper and the dialectical materialist philosophy of the Nagoya school are connected via the Nagoya model, with the X-particle acting as catalyst. And while it may be difficult to assess the precise extent of dialectical materialism's influence on the development of Japanese physics in the period from the mid-1950s to the mid-1970s, the impact of the Nagoya model and of the X-particle is undeniable. The many articles appearing in *Progress of Theoretical Physics* that were concerned with these two subjects attest to the influence of the Nagoya model and the X-particle on the work of quite a few Japanese physicists, including Kobayashi and Maskawa. Yet these events received very little attention in the United States.

In fact, some Japanese physicists felt strongly that Western physicists, especially in the United States, systematically ignored their work. Some remarks made by Taketani at a conference held in Kyoto in 1965 express this belief:

Some English speaking people talk ten words when we talk one word, and do hardly take care of one word which we talk. This point is one of our complains [sic] in any international conference. We should like to ask English speaking people to hear about our talks with the special care. Otherwise our attendance to the international conferences would lose its true meaning, and we are led to consider that we are not welcomed as a matter of fact. (Taketani 1966, 179–80)

This feeling of having been slighted had historical roots. In the same comments, Taketani charged that the proposal by Sakata and Takesi Inoue (Sakata and Inoue 1946) to distinguish the muon and pion as distinct mesons (the "two meson theory") had been "intentionally neglected by some of the foreign physicists," who gave credit instead to Robert E. Marshak and Hans Bethe, who published a similar proposal somewhat later (Marshak and Bethe 1947). Taketani lamented,

> It is regrettable for us to find some workers in the major country insisting that the article which they did not read could have no contribution to the progress of science. We know that the works done by Sakata and his co-workers made the important contribution at least to the progress of physics in Japan. No one will deny that the achievements made by Japanese workers played an important role in the international developments of the meson theory. (Taketani 1966, 179)

In reply, Marshak acknowledged that "without question" Sakata and Inoue had priority on the two-meson theory but noted that "due to the war their paper did not reach to U.S. [sic] until 1948, which was at least 6 months after I presented my theory" (Tanikawa 1966, 180). In fact, Sakata had proposed the theory several years earlier, in 1942, but during the war Japanese physicists were not in communication with the West, and *Progress of Theoretical Physics* only came into existence in 1946, with Sakata and Inoue's paper in the first issue (Crease and Mann 1986, 167–68). Marshak added that both the Sakata-Inoue and Marshak-Bethe proposals were only partly correct. Sakata and Inoue underestimated the π lifetime, and Marshak and Bethe assigned incorrect spins to the two particles (Tanikawa 1966, 180–81). (This does not necessarily mean that the two papers were in fact on a par, however, since it is arguable that the quantum numbers predicted by Sakata and Inoue are of more fundamental importance than the lifetime.)

In light of this contentious background, Ne'eman's indictment of the Nagoya school's dialectical materialism raises a troubling question. If themata are to be judged according to their fruitfulness, then to *whose* fruits should one look? In the preceding comments, Taketani insisted on the significance of the work of Sakata and his colleagues for its importance to the "progress of physics in Japan." At the time that Ne'eman pronounced the Nagoya philosophy to have been a stone around the neck of Japanese physics, Japanese physicists had established a kind of cottage industry based on investigations of the Nagoya model, the results of which were published prominently in *PTP*, and interest in the theory continued unabated through the rest of the 1970s. The seminal papers that formed the core of the Nagoya model continued to be cited with great regularity in each issue of *PTP* during that decade. While it seems apparent at the present time that this approach became in some sense a dead end, it is difficult to see how one could declare it to have been barren as of 1971, unless one were unaware of the proliferation of papers on the subject appearing in Japan (or regarded those papers as themselves having no value). Even knowing that the Nagoya model ultimately did not pan out, we do know at least that it provoked a considerable amount of speculation in Japan, leading not least to the Kobayashi-Maskawa paper itself.[35] Most of this work remains little known, or at least little appreciated, in the West.

Did scientists outside of Japan, and particularly in the United States, systematically ignore their Japanese colleagues? Sakata and some of his colleagues certainly believed that this was the case, but this is not a question that can be easily settled by looking at one slice of history. Making such a claim requires a certain amount of highly problematic second-guessing regarding scientific developments. It is not enough to show that certain articles appearing in *PTP* went uncited in *Physical Review*. One would also need to show that they *ought to have been* cited. But how does one substantiate such a claim?

Leaving that vexing question aside, let us consider the apparent difference in perceptions between the United States and Japan as to the importance of the journal *PTP*. For the physicists of the Nagoya model, the journal, founded by Hideki Yukawa during the hardships of 1946, was the primary outlet for their ideas. Certainly, these articles did not receive nearly as much attention in the pages of the *Physical Review* (recall that the CFS experiment coordinator John Yoh described *PTP* as an "obscure Japanese journal" – although this may also reflect a difference between the attention given it by experimentalists like Yoh and attention it receives from theorists, since *PTP* has historically published little experimental work in high energy physics). It is not easy, however, to distinguish two possibilities: (1) U.S. physicists paid little attention to the journal in general, or (2) U.S. physicists did not think much of the Nagoya model specifically.

The period during which the KM paper was ignored in the United States, followed by more widespread recognition of its importance, might suggest that the problem really is as Yoh's description suggests: the paper was not noticed because the journal was not read. If U.S. physicists had simply been aware of the KM paper, they would have recognized its significance. This interpretation runs into the following difficulty, however: it is not obvious that the KM paper received any more attention in the pages of *Progress of Theoretical Physics* itself than it did in *Physical Review*. The number of papers citing KM in *PTP* for each year between 1974 and 1980 is as follows: 1974, 0; 1975, 0; 1976, 1; 1977, 10; 1978, 3; 1979, 3; 1980, 14 (see Figure 1.5).

It is difficult to know how to compare these numbers to those for the *Physical Review* (see Figure 1.4). *PTP* is a much smaller journal, which covers more subjects in fewer pages than *PRD*, and is published monthly, rather than semimonthly (like *PRD*) or weekly (like *PRL*). Nevertheless, the general pattern is the same for both: until the very end of the 1970s, physicists in both

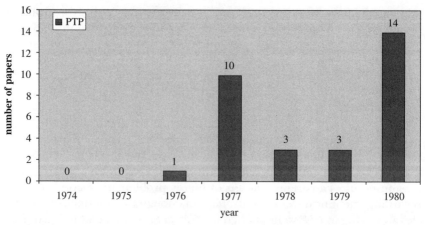

FIGURE 1.5. Citations of Kobayashi and Maskawa 1973 in *Progress of Theoretical Physics*, 1974–80.

the United States and Japan simply failed to notice this unassuming paper on *CP* violation. As David Hull has pointed out, to get one's ideas noticed requires more than merely publishing them. They need to be publicized as well.[36] Only after the KM paper had been quite prominently mentioned (and not merely cited) as an important contribution by well-known theorists at CERN in papers appearing in 1976 and 1977, after the existence of the fifth quark was well established, and after the resort to right-handed currents to account for *CP* violation had been ruled out, did the citations of KM increase significantly.

The fact that a paper is widely cited may not reflect a material impact on the course of inquiry, however. The theoretical physics community in the mid-1970s was apparently moving inexorably along to a six-quark theory of the type proposed by KM, without being aware at all of the paper's existence. So it is not clear that the paper made a great difference to subsequent events, although it did correctly anticipate them. Having the right idea was sufficient to get KM cited once physicists became aware of the paper's existence, thus suggesting that physicists have a sense that credit must be given to the originators of an important idea. On the other hand, having the right idea was not enough by itself to get the paper noticed in the first place.

David Hull and Robert Merton both argue that, while science as an institution operates to enforce a norm of disinterestedness, this does not translate, nor could it, into individual selflessness on the part of scientists. On the contrary, the fact that scientists are often aggressive in their drive to have the originality and priority of their work recognized contributes in important ways to the advancement of scientific thought, even if it brings with it the risk that vigorous, successful self-promotion might sometimes result in community enthusiasm for ideas destined to fail (Merton 1973, e.g., ch. 15; Hull

1988, esp. ch. 10). Hull has gone so far as to assert that "on my view, truly unappreciated precursors are the most extreme form of failure. In the annals of science, they are villains, possibly victims, but hardly heroes" (Hull 1988, 362).

Here the term "villain," suggestive as it is of a kind of moral failure, is certainly out of place. Perhaps we can state Hull's point in a form less objectionable (though less memorably provocative): since the pursuit of a particular science is a social activity, contribution to that activity requires social skills in addition to intellectual skills, including the ability to persuade others to notice and appreciate the importance of one's work. Yet the present case suggests the importance to science of social *obligations* as well. Kobayashi and Maskawa, as of 1975, might appear to have been on their way to becoming unappreciated precursors, uncited even in papers that appeared to be moving in the direction to which their arguments pointed. But a few high-profile citations appear to have averted such a fate. As Merton sees it, these cases in which "friends or other more detached scientists see the assignment of priority as a moral issue not to be scanted" indicate that scientists recognize, and strive to protect and enforce, a norm that requires them to acknowledge their intellectual debts (Merton 1973, 340). Such a norm indicates that the "assigning of all credit due is a functional requirement of science itself" (ibid., 340; see also 291–2).

The present case illustrates this point well. Neither Pakvasa and Sugawara nor Ellis, Gaillard, and Nanopoulos appear to have had anything obvious to gain by making prominent mention of the priority of Kobayashi and Maskawa. Indeed, they would have fulfilled even the obligations of collegiality merely by citing the KM paper, without incorporating more explicit and enthusiastic acknowledgments into the text of their papers. That they chose to acknowledge so prominently a previously neglected work suggests strongly that they attached value directly to giving credit for a good idea. As Merton notes, "Deeply felt praise for work well done . . . exalts donor and recipient alike; it joins them both in symbolizing the common enterprise." That scientists do sometimes find it rewarding to acknowledge the work of others in this way expresses, according to Merton, "the character of competitive cooperation in science" (Merton 1973, 339).

In exploring the literature surrounding the Nagoya group's philosophical commitments and theoretical contributions, one encounters the unpleasant reality of physicists accusing one another of various forms of cultural bias. Worse, I have found myself unable to resolve the worry that these accusations raise. One cannot draw conclusions from citation statistics without making problematic assumptions about the reasons behind such patterns of citation.

The discomfort of an unresolved and troubling historical question is alleviated, however, by the pleasure of bringing into the light some historical connections that have otherwise been consigned to obscurity. This chapter has focused on just such a connection between three major claims about the structure of the world: the doctrine of inexhaustibility, the Nagoya model, and the KM six-quark account of *CP* violation. The first of these was ignored (at least for purposes of developing theories of fundamental particles) by all but a handful of Sakata's cohorts. The second was regarded as highly important within Japan, but received little attention elsewhere. The reasons for this remain unclear. As for the KM paper itself, in time, after receiving prominent mention in a few important articles, and after some competing models were ruled out, it became one of the most-cited articles in particle physics.[37] Still, to be fully vindicated, the KM proposal had to be supported by experimental evidence for each of the new particles it proposed, including the top quark, the weak isospin partner to the bottom quark. How one collaboration sought to provide just such evidence is the subject of the story to come, which will continue to illustrate Merton's concept of "competitive cooperation in science."

2

Building a Detector and a Collaboration to Run It

The humanity of science is revealed in a humbler manner by the consideration of our instruments. They illustrate that science was not created only by our minds, but to a far larger extent than is usually supposed, by our hands.

George Sarton (1937, 28)

1. INTRODUCTION

Although the standard model requires a top quark, it does not specify the mass of this particle. This silence regarding the top's mass proved a major source of frustration for the CDF collaboration because the mass turned out to be rather high relative to people's expectations. In their 1977 paper in the wake of the upsilon discovery, Ellis, Gaillard, Nanopoulos, and Rudaz estimated the lower and upper "extreme values" for the mass of the top to be 5 and 65 GeV, respectively (Ellis, Gaillard, et al. 1977, 296). More definite predictions regarding the mass tended to rely on little more than what University of Chicago physicist Mel Shochet called "numerology," but theorists had little else to go on (Shochet 1995).

Some physicists suggested a "rule of three," for example, since the strange, charm, and bottom quarks had masses of "approximately" $\frac{1}{2}$, $1\frac{1}{2}$, and $4\frac{1}{2}$ GeV, respectively.[1] Based on this hunch, some expected the top quark to have a mass of about 15 GeV. Other estimates relied on less arbitrary, but not necessarily better-tested, theoretical assumptions. Imposing an additional symmetry constraint led one Japanese theorist to an estimate of 16 to 19 GeV (Yanagida 1979). Another model led to a prediction of $m_{\text{top}} = 148$ GeV (Terazawa 1980), which was considerably closer to the values that CDF and D-zero would eventually produce.

Although such estimates did not carry much weight among experimentalists, the prospects for finding the top in the near future depended heavily on this unknown quantity. The higher the top quark's mass, the lower

45

its *production cross section* (a parameter characterizing the probability that a process will result in the production of a particular particle), and hence the more slowly evidence of its existence would emerge. Consequently, the search for the top quark initially resembled feeling around for the light switch in an unfamiliar, darkened room. If there is a switch to be found, you will encounter it eventually, but aside from some quite vague, possibly unreliable, guidelines, you have nothing to indicate when you are about to discover it.

Because a top quark with large mass would be much rarer than a lighter top, a very massive top quark could not be found at accelerators unable to produce very high energy collisions at a rather high rate. Two experiments at electron-positron colliders searched for the top quark at low masses at the close of the 1970s and in the early 1980s – one at the Positron-Electron Tandem Ring Accelerator (PETRA) located at the German accelerator facility Deutsches Elektronen Synchrotron (DESY), and another at the Positron Electron Project (PEP) at the Stanford Linear Accelerator Center. These experiments looked in the 10–20 GeV range, where they failed to find the top quark (Brandelik, Braunschweig, et al. 1979; Adeva, Barber, et al. 1983). As it turned out, at the time that CDF and D-zero finally established that they had evidence for the top quark, they were the only collaborations in the world that had any chance of finding the massive particle.

Yet physicists laying the foundations for CDF in the mid-1970s did not give much thought to finding the top quark. Rather, the physicists who initiated both the collaboration and the detector around which that collaboration formed sought to make the best use of the impending improvements to the Fermilab accelerator, as well as other recent technological advances, in order to have the best general-purpose detector possible for investigating whatever interesting physics might come along.

When they did contemplate specific physics projects, the founding members of CDF focused on intermediate vector bosons. In the standard model, these particles, known as the W and the Z, play the role of carriers of the weak force, just as photons carry electromagnetic forces, and confirming their existence was an important step in solidifying the status of the standard model. One of the main themes of this chapter, however, is that CDF physicists were more committed to building a detector suitable for a general program of research, and a collaboration of physicists capable of carrying that program out, than they were to pursuing a particular experimental success such as finding the top quark or intermediate vector bosons. Such successes become the focus of historical interest retrospectively, when we come to regard them as the *point* of what experimenters were doing. But before a collaboration can even think of holding a press conference, they need an apparatus, and in the kind of high energy physics done at institutions such as Fermilab and CERN, the apparatus may serve as a resource for hundreds of physicists for many years. Consequently, the detector has to be a general-purpose tool. Therefore, CDF focused on building a detector defined not so much by the new particles it could find but by properties it could measure. And because the quantities to be measured depend on the physics projects that the collaboration pursues, CDF framed their project in general terms: What are the relevant measurements to make in order to identify various effects predicted by the standard model and its competitors? Which measurements are technologically feasible? Which effects will CDF be particularly well-positioned to find using Fermilab's accelerator?

The founders of CDF believed that they could be uniquely situated to look for particles with higher masses than those found before. Justifying this belief were some powerful technological advances. These new technologies, combined with the enthusiasm generated by the theoretical and experimental developments of the 1970s discussed in the previous chapter, motivated a handful of physicists to undertake a long and demanding project that would not yield the gratification of new physics results for years to come.

2. THE ORIGINS OF CDF AND COLLIDING BEAMS

Many scientific instruments are made for use in a wide variety of settings. An ordinary telescope can be used nearly anywhere. But CDF built their detector to work in one place alone: at Fermilab's Tevatron Collider, where protons and antiprotons, bearing opposite charges, travel in opposite directions inside a subterranean ring of superconducting magnets, colliding headlong at designated locations. The CDF detector surrounds one such collision point and measures the products of these proton-antiproton catastrophes. Indeed, the Tevatron and the CDF detector evolved together. The two machines – accelerator and detector – were built to accommodate each other, and the CDF collaboration could not have done the kind of experiments it did without an accelerator delivering high-energy colliding beams. Building that accelerator required in turn the solution of a number of technological, political, and financial difficulties.

Two developments in the early 1970s helped pave the way technologically for the colliding beams on which CDF's experiments depended: at the instigation of its director Robert Wilson, Fermilab began a program to develop superconducting magnets for a new ring, and two proposals for "beam cooling," one from Gersh Budker and the other from Simon van der Meer, proved effective. These crucial achievements made possible two requirements for high-energy colliding beams: an efficient means of accelerating protons to energies of about 1 TeV (one trillion electron volts), and the ability to accumulate sufficient numbers of protons and antiprotons to achieve a high rate of collisions between two counterrotating beams. In this section I will briefly describe these developments.

Fermilab had been operating its Main Ring – a ring of magnets in an underground tunnel 4 miles in circumference – with conventional magnets since 1972. Director Robert Wilson proposed operating the accelerator with magnets utilizing superconducting materials, which conduct electricity with very low resistance (so that large currents can be maintained using less power), provided they are cooled to nearly 0 K. Superconducting magnets offered two advantages. First, Fermilab could install more powerful magnets, and thus accelerate protons to an energy of 1,000 GeV (1 TeV), instead of the 500 GeV then possible using conventional magnets.[2] Second, the electricity bill for the accelerator would actually decrease because much less energy would be lost owing to resistance in the magnet coils. With rapidly increasing energy costs, when the President of the United States turned down the thermostat in the White House and addressed the nation wearing a cardigan, superconductivity was the right technology for the time in more than one way. Robert Wilson called superconductivity "a magic potion, an elixir to rejuvenate old accelerators and open new vistas for the future" (Wilson 1977, 23; see Hoddeson 1987).

When Fermilab opened in 1968 (as the National Accelerator Laboratory – it was later renamed Fermi National Accelerator Laboratory), Wilson was already considering the possibility of colliding particles from two different rings. The Main Ring was built for stationary target experiments, in which particles drawn from a single accelerating ring strike stationary targets. Experimenters then measure the downstream products of those collisions. But Wilson had mandated that enough space be left beneath the Main Ring for a second ring in the same tunnel. In addition, construction of the Main Ring had left a $20 million surplus, which Wilson proposed to use to start work on a higher-energy superconducting accelerator (Tollestrup 1996, 505). Wilson's idea was to run the new ring, called the Energy Doubler or Energy Saver, as a 1-TeV fixed target accelerator, with the original Main Ring as a 100-GeV *injector*, a lower-energy accelerator that would feed particles into the Energy Doubler. The entire accelerator ("the machine" in Fermilab jargon), with protons reaching energies of 1 TeV, came to be known as the Tevatron.

Colliding 1-TeV protons with stationary targets, however, does not make 1-TeV available for particle production. Much energy is lost in the interaction because the energy released in a collision depends on the energies of both beam and target particle. More specifically, consider the square of the sum of the beam and target particle energy-momenta (denoted s). This quantity is Lorentz invariant, meaning that for a given system it is the same for any inertial frame of reference in which one evaluates it. Considered in the center of momentum (CM) frame, $s = (E_b + E_t)^2$, where E_b is the energy of the beam particle and E_t is that of the target particle. The total energy available for particle production is then $\sqrt{s}$. For a stationary target, however, with a beam particle energy of E_b in the lab frame of reference, this equation reduces to $s = m_b^2 + m_t^2 + 2m_t E_b$. At high energies, the mass terms in this expression become negligible (the mass of a proton is about 938 MeV). Consequently, the amount of energy available for particle production, $\sqrt{s}$, grows only as the square root of the beam energy in fixed-target experiments. Pursuing higher energies with fixed targets yields diminishing particle returns for increasing energy investments. No such loss occurs in the colliding beam scenario. In that case, $\sqrt{s}$ simply equals the sum of the energies of the two colliding particles and denotes the energy available for particle production in the case where both particles are annihilated in the collision.

Colliding-beam experiments pose a different problem, however. In a fixed-target experiment, a bunch of relatively loosely spaced particles srike a stationary, dense cluster of particles (targets are often a thin foil of a metal such as beryllium or a volume of a gas such as helium). Colliding-beam experiments, on the other hand, involve running two clusters of dispersed particles into each other (think of the difference between firing a shotgun blast at the side of a barn and firing a shotgun blast at another shotgun blast). When a collision occurs, the results can be spectacular, but this is of little use if such collisions occur too infrequently to yield a sample large enough to draw statistical conclusions.

Thus, using colliding beams to produce usable numbers of interesting events requires squeezing the beam particles into a more densely packed bunch. That is, the accelerator physicist needs to achieve a high beam *luminosity*. High luminosity requires large numbers of both protons and antiprotons. Protons, of course, are pretty much everywhere, but antiprotons must be created by artifice. One produces antiprotons by colliding protons with a fixed target, gathering the few antiprotons that result, and diverting them into a *storage ring*, where the antiprotons are kept in a holding pattern, while more and more antiprotons are added. When the storage ring holds enough antiprotons they are injected into the main accelerator, where they collide with protons.

While in a storage ring, however, particles tend to "heat up" – that is, they start to develop momentum components transverse to the beam line. If this process is not controlled, the stored particles will disperse too rapidly to accumulate in significant numbers. For the Fermilab collider to be of use, then, physicists there needed some means of *cooling* the beam, or reducing the transverse momentum of the stored particles.

The first practicable proposal for beam cooling, called electron cooling, came from Gersh Budker, of the Institute of Nuclear Physics at Novosibirsk. First described in 1966, the technique was demonstrated in 1974 at Novosibirsk (Budker 1967; see also Tollestrup 1996; Richter 1997, 278–9). Budker's technique involved injecting a beam of electrons in a straight section of beam pipe with the same velocity as the antiprotons. The transverse oscillations of the antiprotons are then damped by Coulomb interactions between the two beams. This cooling scheme never became widely implemented, but it demonstrated the possibility of solving the cooling problem.

In 1968, Simon van der Meer invented an alternative technique, called stochastic cooling, which was demonstrated in 1975 at the Intersecting Storage Rings (ISR) at CERN (van der Meer 1972; see also Tollestrup 1996, 505; Richter 1997, 279). In stochastic cooling, a sensor on one side of the storage ring measures deviations from the correct path of the beam and then transmits a signal to a "kicker" magnet on the other side of the ring, instructing it to apply a correcting field at the moment that the errant portion of the beam arrives. Whereas electron cooling becomes less effective as beam energy increases, the stochastic method cools at a rate independent of beam energy. The stochastic method would eventually prove very useful at CERN, and Fermilab would ultimately adopt this method for the Tevatron. Before one could consider building a proton-antiproton collider, though, one needed a reason to believe that one could accumulate significant numbers of antiprotons. Budker's electron cooling scheme provided physicists with that reason.

During this period of advances in the development of colliding beams, no one really knew when, or whether, a proton-antiproton collider would be built at Fermilab. The lab held a workshop on colliding beams in January of 1976, and Peter McIntyre of Texas A&M suggested colliding protons and antiprotons in the Main Ring. Because protons and antiprotons will travel in opposite directions in a given magnetic field, the proton-antiproton scheme had the advantage of requiring only one ring. Consequently, proton-antiproton collisions appeared to offer the quickest way to get colliding beams. This proposal won the enthusiasm of more senior physicists David Cline of Wisconsin and Carlo Rubbia of Harvard and CERN, who joined McIntyre in proposing a proton-antiproton experiment, to be run either at Fermilab or CERN (Rubbia, McIntyre, and Cline 1977; Tollestrup 1996, 505–6).

The physics motivating this proposal was not, however, the search for the top quark. The next big discoveries that physicists sought were the W and Z bosons, for which more reliable mass estimates were available than those for the top quark. Physicists expected the W bosons (the W^+ and W^-) to have masses close to 80 GeV and the Z to be approximately 90 GeV. Running protons and antiprotons in the Main Ring looked like a quick way to achieve the energies needed to find these particles. But Wilson was also eager to build the superconducting Energy Doubler, and building a "quick and dirty" proton-antiproton collider would divert funds and efforts from that project. Furthermore, Wilson had no construction funds from the U.S. Department of Energy (DOE), which funds Fermilab, for converting the Main Ring to a collider (Taubes 1986, 16–37).

Rather than killing the collider idea, though, Wilson let it drift along. In June 1976, Fermilab's Program Advisory Committee (PAC), which guides decisions at Fermilab about which experimental proposals to support, considered three colliding beam ideas. Two proposals involved colliding

200–250 GeV protons from the Main Ring with 1,000 GeV protons from the Energy Doubler. A third possibility was the Rubbia-Cline-McIntyre proposal to run protons and antiprotons in the same ring. Rather than singling out one of these approaches, Wilson allowed all three to proceed slowly (Taubes 1986, 23). A small proton storage ring was started at Fermilab in May 1976 to test electron cooling (Tollestrup 1996, 519). In 1977, Jim Cronin, on sabbatical at Fermilab from the University of Chicago, became head of the Colliding Beams Department at Fermilab. In addition, a "P-bar Source Department" was created within the Accelerator Division.[3,4]

If physicists at Fermilab were to do experiments using colliding beams, they would need to build a detector adapted to this new way of doing physics. The "Colliding Beams Department Working Group" began to discuss detector issues early in 1977. From the beginning, they faced the problem of what Jim Cronin called the "strong interaction" of plans for the detector and for the collider. In particular, he noted that if the detector was to use a magnet (a feature greatly facilitating particle identification), its design would need to take into account the magnets used to bring the beams into collision. Cronin urged the group to begin thinking systematically about possible detectors for the colliding beam facility (CDF meeting minutes, 2/15/77).[5] Detector issues were scheduled for discussion at a Summer Study planned for that year in Aspen, Colorado. In the spring of 1977, the group estimated that building a detector would cost $10 million, based on the goal of being able to do "*e*, μ, hadron detection in a diameter of 2.8m" (*CDF* 4/22/77).

The physicists who went to Aspen that summer for the meeting titled "Colliding Beams at Fermilab" examined five general problem areas: the means of bringing beams into interaction; how to store particles in the main ring; what kind of detector to build; how to produce, cool, and store antiprotons; and the kind of physics to be done with colliding beams (Walker 1977, vol. I, vii).

In 1977, expertise on colliding-beam physics was concentrated at the electron-positron collider at SLAC (Tollestrup, personal communication). A SLAC physicist, Dave Hitlin, organized the section of the meeting devoted to detector designs. The main feature of this portion of the conference was a lengthy report prepared by the "Detector Group," which included a number of future CDF members as well as Hitlin and other SLAC physicists.

In their report, the Detector Group assumed that the most important physics problem to be tackled would be the search for the *W* and the *Z*. Kobayashi and Maskawa's 1973 suggestion of a third generation of quarks was, as discussed in the previous chapter, just beginning to receive attention outside of a small circle of theorists. Likewise, the bottom-carrying upsilon was just emerging from obscurity in the CFS collaboration's data around the time that the Aspen meeting took place, and its significance was far from apparent at this early date. In the proceedings of this conference, one can find no mention of using the detector to look for heavy quarks (i.e., quarks beyond charm). A CERN physicist referred to the search for intermediate vector bosons as the "key present motivation" for building a hadron collider (Jacob 1977, 227).

The Detector Group was trying to design a detector for a colliding-beam accelerator that remained indeterminate not only in fine details but even in its basic conception. The manner in which beams might be made to collide remained "to be announced." The Detector Group did not know whether 250-GeV protons would be collided with 1,000-GeV protons (and if so, how this would be accomplished), or whether 1,000-GeV protons would be collided with 1,000-GeV antiprotons. In spite of this uncertainty, and in spite of changes to come in the specific physics problems to be pursued (especially the emergence of heavy quark studies), the Detector Group's report anticipated fairly accurately the features of the detector that CDF eventually built.

This is no accident. What mattered most to the physicists who built the CDF detector was the ability to make the measurements needed for solving the kind of experimental problems posed by the standard model and its competitors, by employing the best technologies available given the group's resources. This general goal was a more important factor in their decisions regarding the constitution of the detector than the desire to achieve any particular experimental goal, such as finding the W or the top quark. When the physics problems changed, radical detector changes were not needed because the detector had from the beginning reflected the group's interest in having a general-purpose machine appropriate for whatever problems turned out to be interesting and important.

3. THE CDF DETECTOR

Intelligibility requires that I begin to chronicle the building of the CDF detector at its end (more accurately, at *an* end). Thus, I will start with a description of the detector as it was at the end of the first phase of construction, when it was a more or less complete and functional particle detection device. A little knowledge of what the CDF detector came to be, how it functioned, and its different components will make the steps of the construction process more comprehensible.

Beginning with this picture of a complete detector might give this story an overly teleological cast, as though the complete detector existed virtually at the beginning, and the subsequent train of events followed a set of tracks already laid down, from nothingness to a complete device. Let me, therefore, state at the outset that, although CDF did develop a design plan for the detector before they built it, each feature of the completed detector resulted from a trade-off between competing factors. The CDF physicists had to decide what kinds of measurements mattered, which detector features were compatible with other desired features, how much money they could spend on various detector elements, what kinds of technology could be trusted, and so on. That the collaboration succeeded in building a detector at all required that they solve complex problems of funding, technology, and social cooperation. So while I begin by giving away the ending, I do this only so that I can tell intelligibly a story about how *that* ending was reached, as opposed to many other endings that might have been.

In fact, what I will describe is not really the *end* of the building of the detector at all, but only the state of the detector as it stood in 1987, at the beginning of CDF's serious efforts to find the top quark. I have two reasons for choosing this benchmark. First, 1987 marked an important turning point for CDF. In 1987, they first started "doing physics," as opposed to the previous ten years spent building the detector. After a short data-taking run in 1987, the collaboration published a series of articles in *Nuclear Instruments and Methods in Physics Research*, including an overview of the entire detector (Abe, Amidei, et al. 1988) and detailed descriptions of individual detector components, constituting a thorough survey of the detector at the first time that one could consider it to be, at least temporarily, "complete." Second, some of the changes to the detector subsequent to 1987 have interesting relationships with developments in CDF's strategies for finding the top quark. Some new detector components were in fact crucial to CDF's evidential argument for the existence of the top. Hence, these later changes are really part of the story of how the top search strategies evolved, the subject of a later chapter.[6]

The geometry of the barrel-shaped CDF detector (see Figure 2.1) reflected the intent that the detector surround the proton-antiproton collision area.

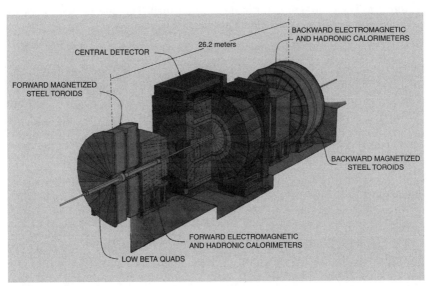

FIGURE 2.1. A perspective view of the CDF detector. Courtesy of Fermi National Accelerator Laboratory.

FIGURE 2.2. The central detector in the assembly hall. The calorimeter arches are retracted. Courtesy of Fermi National Accelerator Laboratory.

By surrounding the collision point, CDF sought to gather a variety of information about the products of each collision, with a special emphasis on collision products exiting at large angles relative to the beam axis, without allowing too many particles to escape undetected. Different parts of the detector served to measure different quantities, and facilitated the detection of different kinds of particles. Elements closer to the collision area typically produced information about the tracks of individual particles, while parts of the detector further from the beam line, or further up or down the line from the collision area, generally facilitated particle identification and measurement of energies of individual particles or particle "jets."

The detector consisted of three main sections: central detector, forward detector, and backward detector, each of which had in turn several components. I will first describe each main section in general terms, and then describe the functions of the various components.

The *central detector* consisted of tracking chambers (used to record tracks left by individual charged particles) immediately surrounding the collision region, enclosed within a superconducting solenoid generating a powerful (1.5-Tesla) magnetic field. Surrounding the solenoid were electromagnetic calorimeters (used to measure the energy of electrons and photons), which were in turn ringed with hadron calorimeters (used to measure the energy of hadrons). Constituting the calorimeters were 48 wedges projecting radially from the beam axis and stacked into four retractable semicircular arches of 12 wedges each (see Figure 2.2). Muon chambers then enclosed the calorimeters. All of this was contained within a steel yoke. The central detector weighed about 2,000 tons.

On either end of the central detector were the *forward and backward detectors*, which were the mirror-reflections of one another. These consisted of time-of-flight counters, electromagnetic shower counters, hadron calorimeters, and muon toroidal spectrometers. The forward and backward detectors picked up particles exiting the collision area at too small an angle to the beam axis to register in the central detector.

The detector as a whole, when collecting data, sat inside the accelerator tunnel, with beam passing through its center (see Figure 2.3). The area in which the detector sat was known as the collision hall, and was connected to the CDF assembly hall, where CDF workers put together many of the detector components. A thick shield door covered the opening between the assembly hall and the collision hall and would slide into place during runs to shield against the considerable radiation produced by particle collisions. The forward and backward detector segments remained in the collision hall at all times, while the central detector could be moved from the collision hall to the assembly hall and back (see Figure 2.4). The trip, a move of 31.4 m, took an entire day (Abe, Amidei, et al. 1988, 389).

Different detector elements provided different kinds of particle information (see Figure 2.5). Closest to the interaction point on the beam line, tracking chambers provided information about the identity of particles by means of the curvature of their trajectories in a magnetic field and helped to make sense of calorimeter readings by allowing an identification of the first section of a particle's trajectory. The *vertex time projection chamber* is an example of a "pictorial" drift chamber (Fernow 1986, 242). Drift chambers use sense wires to pick up ionization electrons produced by particles passing through a gas. In the vertex time projection chamber, the close placement of these wires yielded high-resolution tracking information. Further tracking information came from two more drift chambers surrounding the vertex time projection chamber: the *forward tracking chamber* and the *central tracking chamber*. One of the most important changes to the detector would be the 1992 replacement of the forward tracking chamber and vertex time projection chamber, based on the well-established drift chamber technology, with the very high-resolution *silicon vertex detector* (SVX), using a newer

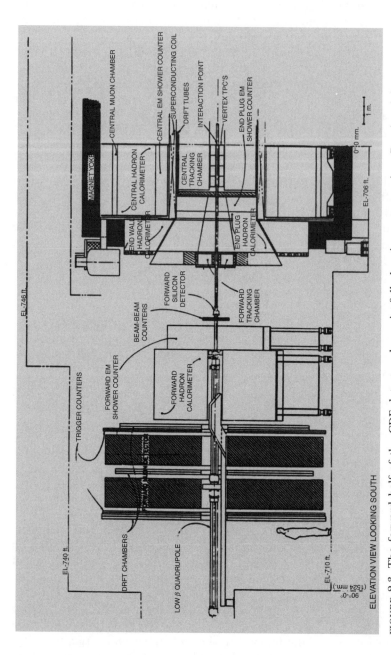

FIGURE 2.3. The forward half of the CDF detector shown in full elevation cross section. Reprinted from *Nuclear Instruments and Methods in Physics Research.* **A271**, Abe, Amidei, et al., "The CDF Detector: An Overview," 391, Copyright (1988), with permission of Elsevier Science.

CENTRAL MUON CHAMBER

CENTRAL EM SHOWER COUNTER

SUPERCONDUCTING COIL

DRIFT TUBES

INTERACTION POINT

VERTEX TPC'S

END PLUG EM SHOWER COUNTER

MAGNET YOKE

CENTRAL HADRON CALORIMETER

CENTRAL TRACKING CHAMBER

END WALL HADRON CALORIMETER

END PLUG HADRON CALORIMETER

FORWARD SILICON DETECTOR

BEAM-BEAM COUNTERS

FORWARD TRACKING CHAMBER

TRIGGER COUNTERS

FORWARD EM SHOWER COUNTER

FORWARD HADRON CALORIMETER

FORWARD MUON TOROIDS

DRIFT CHAMBERS

LOW β QUADRUPOLE

EL-746 ft.

EL-740 ft.

EL-710 ft.

EL-706 ft.

0-90 mm.

(1524 mm.)
0"-0-.90°

1 m.

ELEVATION VIEW LOOKING SOUTH

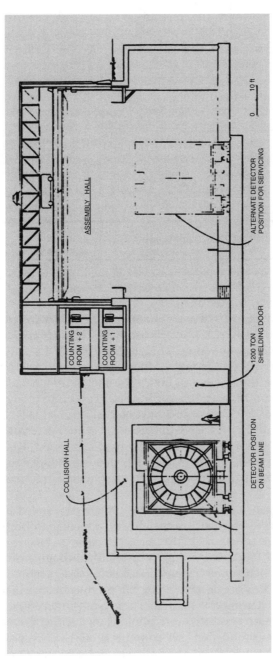

FIGURE 2.4. The B-zero collision hall and assembly hall. Reprinted from *Nuclear Instruments and Methods in Physics Research.* **A271**, Abe, Amidei, et al., "The CDF Detector: An Overview," 390, Copyright (1988), with permission of Elsevier Science.

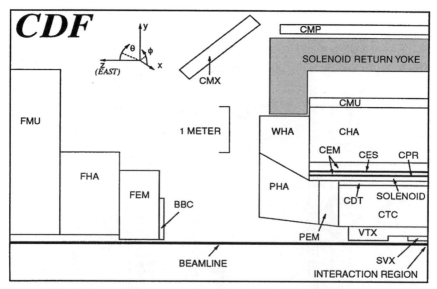

FIGURE 2.5. A one-quarter cross-sectional view of the CDF detector with upgrades for run Ia. Note the coordinate frame of reference (Abe, Amidei, et al. 1994b, 2970).

technology based on strips of semiconductor, and a *vertex drift chamber*, built to accommodate the new SVX and measure the primary interaction vertex with a resolution of 1 mm (compare Figures 2.6 and 2.7).

Outside the superconducting, 1.5-T magnet surrounding the tracking detectors sat, first, the electromagnetic calorimeters and, next, the hadron calorimeters. The CDF calorimeters consisted of towers pointing toward the interaction region. Each tower included both electromagnetic calorimetry and hadronic calorimetry. Each segment in the central detector covered an area 0.1 units of η by 15° in the ϕ direction.[7] The forward, backward, and plug calorimeters (the "plug" region immediately surrounded the beam axis on either end of the central detector) were segmented in 5° sections in the ϕ direction.

Electromagnetic calorimeters measure the energy carried by electrons (I follow the usage of particle physicists in referring to both electrons and positrons generically with the term "electron"). Electrons with energy greater than 100 MeV lose energy primarily through a process called *bremsstrahlung*, which produces photons. Photon interactions at energies greater than 100 MeV are dominated by pair production, in which a photon decays into an electron-positron pair. These pair-produced electrons and positrons will in turn produce more photons through *bremsstrahlung*. This process of "shower production" will continue until the energy of the resulting particles drops below 100 MeV. Hence, electromagnetic calorimeters that measure electron energies by promoting shower production are also called shower counters.

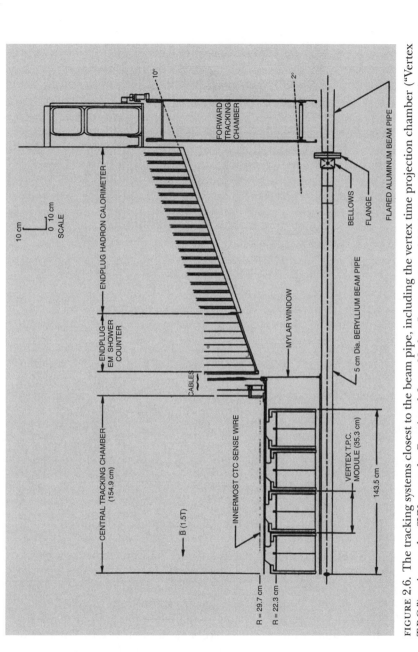

FIGURE 2.6. The tracking systems closest to the beam pipe, including the vertex time projection chamber ("Vertex T.P.C.") prior to the SVX upgrade. Reprinted from *Nuclear Instruments and Methods in Physics Research.* **A271**, Abe, Amidei, et al., "The CDF Detector: An Overview," 394, Copyright (1988), with permission of Elsevier Science.

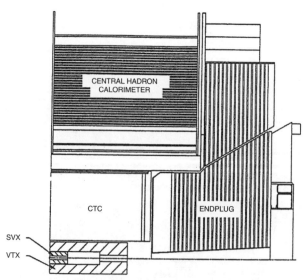

FIGURE 2.7. The tracking systems closest to the beam pipe with SVX/VTX modifications. Reprinted from *Nuclear Instruments and Methods in Physics Research.* **A350**, Amidei, Azzi, et al., "The Silicon Vertex Detector of the Collider Detector at Fermilab," 74, Copyright (1994), with permission of Elsevier Science.

CDF used two different electromagnetic calorimeter technologies. The CDF *central electromagnetic calorimeters* consisted of lead plates, to promote the production of charged particle showers, interspersed with scintillating plastic, to detect those showers. Scintillator materials produce photons in reaction to charged particles. The photons produced in the scintillator would be picked up by a phototube, which would in turn produce an electrical signal. In the *forward, backward, and plug shower counters,* CDF interspersed lead plates with gas-filled proportional chambers. A proportional chamber measures the electron-ion pairs formed when a charged particle passes through a gas in an electric field, by measuring the signal produced by the electrons attracted to the positive electrodes (Fernow 1986, 16).

Hadron calorimeters surrounded the electromagnetic calorimeters. Collisions of high energy hadrons produce narrow sprays of hadrons called *jets.* This happens because removing a quark from a hadron creates new quark-antiquark pairs. These newly generated quarks then bind to one another in a process called *hadronization.* The newly formed hadrons form a cluster of particles all moving along similar trajectories, constituting a jet. Because individual particles in jets are difficult to analyze owing to their dense trajectories, the energies of jet particles are measured en masse. The task of a hadron calorimeter is to measure the total energy of a jet, and, by the segmentation of the calorimeter into distinct "towers," to distinguish one jet from another.

Hadron calorimeters in the CDF detector used steel plates alternating with plastic scintillators in the central region, and alternating with gas proportional chambers in the plug, forward, and backward regions. CDF attached an additional scintillator-based hadron calorimeter to the endwall of the yoke on the central detector.

Finally, each tower in the central detector was capped by a segment of the *central muon detector*, four layers of drift chambers designed to detect muons. Drift chambers track charged particles by measuring the drift time of electrons ionized by the passage of the particle through a gas. They can also help physicists to identify particles by measuring energy loss in the chamber. Energy loss provides a clue to a particle's identity because, at relativistic velocities, energy loss varies with the mass of the particle. Muon chambers needed to be outside of the hadron calorimeters to ensure that they recorded the tracks of muons rather than hadrons. Hadrons, which are subject to the strong force, would be stopped by collisions with nuclei in the steel plates of the hadron calorimeter. Muons are not subject to the strong force and would simply pass through the many layers of steel in the hadron calorimeter and into the central muon detector.

In the forward and backward regions, muons were detected by means of a toroidal magnetic *muon spectrometer*. The spectrometer employed planes of drift chambers and scintillation counters to provide tracking information. From the curvature of the particle trajectory in the magnetic field, one could determine muon momentum.

In 1992, CDF added another layer of muon chambers outside of the steel yoke, the *central muon upgrade*, and a set of free-standing muon chambers covering the region $0.6 < |\eta| < 1.0$, the *central muon extension* (see Figure 2.5).

A plane of scintillation counters on the face of both forward and backward shower counters monitored the luminosity and provided a minimum-bias trigger. These *beam-beam counters* yielded the best measurement of the time of interaction between beams, as well as a crude measurement of the location of the interaction vertex.

Just as important to the collection of data as the detector itself was the *triggering* mechanism. Events occurred in the interaction region at a much faster rate than CDF could possibly record them. To avoid a flood of mostly useless information, CDF had to define in advance criteria that would indicate that an event was interesting enough to record. By means of triggering, CDF could filter out uninteresting events without using memory to store information about them offline. The CDF detector employed three levels of triggering. All three levels of triggering could be programmed as needed for the kind of events in which one was interested.

At level one, information was taken directly from the calorimeters on dedicated cables. Typically the level-one trigger required some minimum amount of total energy in a cluster of calorimeter towers, where the energy

in each tower in the cluster must also meet some threshold. Other requirements could also be included in the level one trigger.

An event that passed the level one trigger would face the level two trigger.
Here energies from clusters of towers were summed separately for hadronic
and electromagnetic calorimeters. A list of this information was sent to a
level two processor, which matched the clusters to tracks found in the central tracking chamber by a fast tracking processor. If a match was made –
a preliminary indication that both the calorimeter energy deposits and the
central track resulted from a single particle – a coarse momentum measurement was added to the list. Muons were matched to the track in the central
tracking chamber and added with their momentum to the list. The level
two trigger could be programmed to select according to some combination
of requirements for electrons, photons, jets, and the "missing" transverse
energy that indicates the passage of a neutrino. The level three trigger exe-
cuted algorithms that could include the reconstruction of incomplete central tracking chamber tracks. This made possible a more refined selection
on the features of the event. Only events that passed all three levels of the
trigger were written to tape for offline analysis.

4. A DETECTOR DESIGN FOR UNCERTAINTY

In 1977, the detector just described was only a conjecture. CDF chose the
various features that I have described from among numerous alternatives.
While I cannot here discuss each such decision individually, I will describe
how CDF chose the basic concept behind their detector and give an overview
of the assembly of this remarkable device.

Particle colliders created new challenges and opportunities for experimental high energy physicists. As Mel Shochet, a former CDF spokesperson, pointed out, fixed target experiments typically sought to detect one
particular kind of phenomenon. In fixed-target experiments, the products
of particle collisions projected forward from the point where the beam
struck the target. In such an experiment, uniform detection capabilities
over the entire solid angle surrounding the interaction were very difficult
to achieve. Consequently, physicists would build a detector to single out
a particular phenomenon of interest, and measure only the products of
that process. Investigating a different phenomenon would often require a
new apparatus. For colliding beams, however, one can much more feasibly
build a device that detects products at nearly every angle with fairly uniform
sensitivity, a detector suitable for a wide variety of experimental investigations (Shochet 1998). This is fortunate because it would be prohibitively
expensive and time-consuming to build such a detector unless one could
expect to benefit from it for many years. As we will see, the CDF detector was
just such an "allgemeine" detector, as John Yoh described it (Yoh, personal
communication).

Collider detectors were also part of a general development in the instrumentation of high energy physics, described elsewhere by Peter Galison (Galison 1997a; 1997b), in which two earlier traditions of particle detection came together to produce a "hybrid" detector. In what Galison has called the "image" tradition, physicists relied on detectors such as cloud chambers and bubble chambers that could produce highly detailed images of individual events. The "logic" tradition, on the other hand, selected particular types of events using detectors that yielded partial measurements, but not a complete "picture" of those events. In the logic tradition, large numbers of events could be collected and analyzed, but the data would include much less complete information about individual events.

The instruments in each tradition had complementary strengths and weaknesses. For example, logic-tradition instruments gave one the ability to trigger the detector on a particular kind of event, but the backgrounds for any given process were typically high relative to those found in image tradition experiments. Image detectors such as bubble chambers were notoriously difficult to trigger – one had to take whatever one got, and sort through it all by hand afterward – but the information contained in bubble chamber photographs of individual events was so detailed that backgrounds sometimes shrank to negligible levels, allowing one occasionally to argue for the existence of some new phenomenon based on a single event.[8]

Experimentalists in both traditions could not fail to notice that the other camp had coveted abilities that their own instruments lacked. Hence, it was natural for each tradition to try to incorporate what it could from the other. In particular, physicists and technicians designed electronic detectors such as drift chambers with ever-finer resolutions and devised ways of reading out the data from these detectors in "visual" displays so that a picture of each event could be constructed from the detectors' electronic signals. The results combined the strengths of both traditions. They provided detailed information, thus reducing backgrounds, but were also triggerable and produced data that could readily be analyzed using increasingly powerful computers and increasingly greater expertise in their use, both of which have been so important in the development of experimental high energy physics.

These broader developments provide the background to the design of CDF's detector. Deciding what a detector for a hadron collider should be like required creativity and innovative thinking, but it did not rest on hunches alone. The SLAC physicists who contributed to the 1977 detector group report had experience with the Mark I detector there. Built to gather information from electron-positron collisions, the Mark I can be seen as the first of the general-purpose collider "hybrid" detectors (Galison 1997a; 1997b, 515–45), and thus as a kind of ancestor to CDF's detector. Roy Schwitters, who was central to the building of the Mark I and later joined CDF as a cospokesperson during CDF's building phase, points out three criteria for a good collider detector: it must provide good solid angle coverage, it must respond uniformly over the extent of that coverage, and it must facilitate detailed, online analysis of data and of detector performance. As Schwitters puts it, "Collider runs are long, and the most interesting events are usually extremely rare; the detector must be live and well to not miss potential discoveries" (Schwitters 1997, 304). Additionally, a good collider detector must support a wide variety of investigations over time. This additional feature (which is partly dependent on those just mentioned) was important in shaping CDF's detector.

The 1977 detector group report discussed four different magnetic detector designs, each distinguished by the type of magnet to be used. The group discussed superconducting solenoid, superconducting dipole, iron toroid, and conventional solenoid designs. All of these shared certain features. They all emphasized the study of "high-angle" (high-θ) physics. They were all symmetric about the beam axis and about the center of the interaction region. They were all "large, heavy, complex and expensive" (Atac, Breidenbach, et al. 1977, vol. II, 3).

The detector group advocated the superconducting solenoid in order to achieve the "optimal" field strength of 15 kG (1.5 T) while saving on energy costs in the long run, and because the solenoid configuration yielded better momentum resolution for measurements centered around $\theta = 90°$. The group's emphasis on high-angle physics deserves notice. Although CDF sought to cover the entire interaction region, detecting particles at nearly every angle, they especially valued information about those "hard" collisions that produced particles perpendicular to the beam line, rather than the glancing blows of "soft" interactions. In part, this was a matter of capitalizing on the unique capability of a collider to produce hard collisions at very high energies. But such a capability was desirable in the first place because it gave physicists a powerful means of looking for very massive particles, owing to the tremendous energy made available for particle production in such hard collisions.

Summarizing the results of their study, the detector group listed the capabilities that any detector worthy of consideration should include:

1. Large solid angle coverage (2π in ϕ, as large as possible acceptance in [η]).
2. A large volume magnetic field containing tracking chambers with good rate capability and designed to facilitate pattern recognition in a high multiplicity, high background environment.
3. Electromagnetic shower detection with good segmentation and good energy resolution over a large solid angle.
4. Muon identification by passage through at least 1.5m of magnetized iron, over a large solid angle.

5. Hadron calorimetry with moderate segmentation and energy resolution over a large solid angle. This will most likely be integrated with the flux return steel of the magnet. (Atac, Breidenbach, et al. 1977, vol. II, 161)

To a surprising degree, this list accurately described the detector CDF eventually built. Some things changed, however. For example, the decision to pursue the top quark generated a need for greater segmentation in the hadron calorimeter. As Caltech physicists Geoffrey Fox and Bob Walker pointed out a couple of years after the 1977 meeting, one could not identify individual jets – crucial for identifying top decays – given the original, "moderate" segmentation design. Effective jet identification would require at least 600 calorimeter towers (Tollestrup 1995). Early detector plans also underestimated the importance of high-resolution tracking information in the interaction region for identifying the decays of hadrons containing *b* quarks (Tollestrup, personal communication). CDF would eventually satisfy that demand with the silicon vertex detector, but the importance of *b* identification for the top search would not be apparent to everyone until *after* the silicon vertex detector had proved essential for making a credible evidence claim for the top quark.

Participants in the 1977 conference were well aware that physicists at CERN sought the *W* and the *Z*, and the Aspen conferees, sensing the beginning of a race, clearly wished to move quickly. The detector group noted that the desired detector "can be constructed in the available time, but doing so requires the full-time commitment of a large number of physicists and engineers. . . . This commitment of people and financial resources should be made in the near future, if it is to be made at all" (Atac, Breidenbach, et al. 1977, vol. II, 163). Such a commitment would not be immediately forthcoming.

After the Aspen conference, Alvin Tollestrup and Tom Collins, both of Fermilab, raised the question "whether a magnet is necessary for the large detector we are planning," although all of the designs considered in Aspen had involved magnets. Tollestrup announced that he would "entertain detector development ideas . . . with some support (and money)" (*CDF* 11/4/77). One group formed to submit a proposal for a magnetic detector of the sort advocated by the detector group at the Aspen conference. David Cline headed this group. Another group led by Fermilab's Peter Limon pursued the idea of a purely calorimetric detector (somewhat like that later built by the D-zero collaboration). Both groups were constrained by a "cost ceiling of $4M" and a "physics goal of *W/Z* detection." The collaboration would choose between the two proposals in March. Everyone was invited to join, "but only if they are will[ing] to commit their major research efforts to it at some point in the next two years" (*CDF* 1/13/78).

As these competitors began their work in early 1978, the Colliding Beams Department began to refer to itself as the "Colliding Detector

FIGURE 2.8. Alvin Tollestrup, right, with Joe Incandela, in 1995. Courtesy of Fermi National Accelerator Laboratory.

Facility Department." Jim Cronin left his position as head of the colliding beam effort, a decision that some attributed to the Fermilab political environment, particularly a lack of leadership in establishing a direction for the lab (Tollestrup 1995; Frisch 1995). Henry Frisch of the University of Chicago recalled an "air of unreality" that pervaded the earliest days of CDF (Frisch 1995). Evidence of such an environment can be found in an early CDF schedule forecasting proton-antiproton collisions in 1979 (the first "engineering run" of the collider took place in 1985). With Cronin's departure, Alvin Tollestrup became head of the CDF department (Figure 2.8).

Cline's group, developing a magnetic detector design, had to choose a magnetic field configuration. Tollestrup reported on February 3 that they had rejected the dipole option in favor of a large, cheap, and thin solenoid (*CDF* 2/3/78). But the very next week group members discussed a possible toroidal design (*CDF* 2/10/78), and a subgroup led by Jim Walker pursued that option.

March came and went, and the decision on a detector design was postponed until after a "detector fest" scheduled for May in which no more than three detector designs would be discussed. Each proposal had to include drawings of the detector, an outline of the component parts, a "brief rationale for the choices made," and estimates of the cost and time needed for construction (*CDF* 4/7/78). Proposals for both magnetic and nonmagnetic detectors would be considered, although in late April, it already seemed to Tollestrup that "no one wants to build a non-magnetic detector." Still, the meeting would be important for deciding the magnetic field configuration and detector construction sequence. Optimistically – buoyed by what Tollestrup saw as a "favorable Congressional situation" – the group discussed the possibility that the lab might produce colliding beams as early as 1980. This gave special importance to sequencing because CDF could not expect to have a complete detector in place by that time but would want to take advantage somehow of beam collisions as soon as they were produced. They wanted to be able to collect useful data even in the intermediate stages of detector construction (*CDF* 4/28/78).

Meanwhile, funding problems were squeezing Fermilab's many projects. Funding new construction required congressional approval. Wilson, however, was funding work on the Energy Doubler and other construction as research and development projects, a different budgetary category in which he had more discretion but less money. Fermilab's projects were also competing for funding with high energy physics projects at other labs. Brookhaven National Laboratory in New York had proposed building a superconducting accelerator called Isabelle, which would collide 200-GeV protons on 200-GeV protons. Congress approved funds for the construction of Isabelle in 1977. In January 1978, the budget for Isabelle increased to $275 million to achieve 400 GeV on 400 GeV collisions. Isabelle had its own problems, though. Rumors indicated trouble in Isabelle's magnet program. Furthermore, physicists at Fermilab were developing proposals for projects much more ambitious than Isabelle, in spite of administrative ambivalence at Fermilab. Collider development took resources away from Wilson's efforts to complete the Energy Doubler. But he was unwilling to terminate definitively the pursuit of a collider. Wilson threatened to resign if the DOE did not provide more money for building the Energy Doubler. The DOE stood its ground, and Wilson carried out his threat to quit (Taubes 1986, 34–6; Hoddeson 1987).

When Wilson resigned as director of Fermilab, he left behind a lab lacking direction. The resulting uncertainties greatly complicated CDF's efforts to settle on a single detector design. On the one hand, proton-antiproton collisions using conventional magnets might be achieved quickly, long before the Energy Doubler was completed, possibly in late 1980 (*CDF* 4/28/78). This gave a reason to build a cheap detector very quickly. On the other hand, colliding beams might await completion of the Doubler, presumably years later, leaving time to build a much more sophisticated detector. But this latter scenario involved great uncertainties. If the lab chose to produce colliding beams using the Energy Doubler, the collision scheme was indeterminate, leaving unanswered such fundamental questions as the extent to which the detector ought to be designed for symmetrical events.

This struggle between different plans for the accelerator produced a complex political situation. As Tollestrup described matters to CDF members just prior to the May 1978 detector fest, three groups involved in colliding-beam efforts had three distinct agendas. Those working on the Energy Doubler wanted the lab to commit fully to completing the Doubler as quickly as possible, even if it meant suspending other projects, like colliding beams. Meanwhile, the group working on beam cooling was rapidly achieving successes, raising the exciting prospect of proton-antiproton collisions prior to the Doubler. Finally, CDF sought to put in place the best possible detector whenever colliding beams of whatever sort became available. Delaying colliding beams would give CDF time to build a more sophisticated detector, but even an incomplete detector might yield exciting results very soon if colliding beams were produced quickly. Tollestrup commented that "If all 3 [groups] can be kept together, which will be a difficult task, Fermilab can have an excellent CB [colliding beams] facility" (*CDF* 5/12/78). Peter Limon and Mel Shochet expressed the sentiment that

if everyone had their choice between the first (primitive?) detector and the later (elegant?) detector, they would all rather be on the first detector because that is where the physics results will come from for the first few years. No one would prefer to take the second look! (*CDF* 5/12/78)

After the detector fest, Tollestrup solicited comments from CDF members on the relative merits of the magnetic and nonmagnetic detector proposals. The responses varied. One attendee felt that the competition should be reiterated so that each of the groups would have an opportunity to respond to criticisms. Another felt that "to an extent both groups ... failed," while yet another indicated that the magnetic detector had "an overwhelming advantage." Most favored the magnetic detector, although concerns remained over unresolved technical issues. The magnetic detector, it was argued, could "react better to the possibilities of 'new' physics." Uncertainty over the future plans for the accelerator again caused concerns. One individual worried that "the asymmetry in pp vs. $\bar{p}p$ collisions ... was not addressed adequately" at the detector fest. And another noted that "we must keep options open since it is not clear whether pp or $\bar{p}p$ will appear first."

After all of the responses had been registered, Tollestrup concluded that "we should do the best we can so we *will* try for the magnetic detector." He noted that Wilson's resignation meant continued uncertainty "until a new

Director assumes control." Tollestrup reminded the group that, in selecting the magnetic detector, they retained the "option of falling back to the non-magnetic detector ... at any point we choose" (*CDF* 5/19/78).

Whatever specific weaknesses CDF members saw in the proposal for a magnetic detector, they clearly believed that having a magnetic field in the detector was advantageous. In this respect, comments made in favor of the magnetic detector anticipated the considered judgments of CDF members years later. The appeal of the magnetic detector lay in its flexibility. In response to any surprising "new physics" phenomena, a magnetic detector could be used to gather more information than a nonmagnetic detector. In 1995, Tollestrup indicated that the magnetic detector was chosen over a purely calorimetric one because of the physics advantage conferred by the magnetic detector's capacity for measuring momentum and distinguishing positive from negative charges (Tollestrup, personal communication).

Asked to reflect on the decision to build a magnetic detector, Mel Shochet noted that the collaboration believed that "a magnetic detector could do everything ... that a nonmagnetic detector could do, but it had additional handles that we thought were going to be crucial," such as the ability to measure the momentum of an individual electron track in the central tracking chamber. The magnetic detector would also give them more "handles" to eliminate "unknown backgrounds" for such things as the *W* search (Shochet 1995).

CDF submitted a design report for the magnetic detector to the lab administration on May 2, 1978. In their summary, they described the proposed detector as

a detector facility based on a superconducting solenoid of 15 kG, $R = 1.5$m and $L = 5$m. In addition to precisely measuring particle trajectories over ~98% of 4π solid angle, the inner drift chamber system will give some dE/dx information; shower counters and hadron calorimeters will also yield hadron energy information over most of 4π, as well as lepton identification. Although the detector was designed specifically to detect intermediate vector bosons [*W*s and *Z*s], it will clearly be useful in the study of a wide range of high energy processes initiated by both pp and $\bar{p}p$ colliding beams. The equipment cost of the initial system is estimated to be five million dollars. (Ankenbrandt, Atac, et al. 1978, title page)

This report detailed the constraints guiding the design process, including cost, suitability for both proton-proton and proton-antiproton collisions, and flexible physics capability. The authors noted, "we have focused on a definite physics goal: *The detection of intermediate vector bosons within the expectations of present theory*" (ibid., 1, emphasis in original). However, they considered these last two constraints – flexibility and the ability to detect intermediate vector bosons – to be complementary: "By designing the facility to have good sensitivity to both the hadronic and leptonic decay modes [of intermediate vector bosons], we obtain a general purpose device capable of a wide variety of physics" (ibid., 11). The report did not mention the top quark.

CDF submitted their design report in the midst of a very unclear political situation at Fermilab. A general-purpose detector such as they proposed had empirical advantages – they could respond in a flexible way to surprises that nature might have for them. But the general-purpose detector was also a wise choice in an uncertain administrative climate. Whether they had explicitly considered it or not, their emphasis on a general-purpose detector was well suited to the unpredictable course of events at the lab. Their specific experimental goal – discovering the intermediate vector bosons – was only plausible if the lab chose to pursue hadron collisions quickly, rather than emphasizing completion of the Energy Doubler first. It was unknown at the time whether the lab would take such a course, and in the end, things would go quite differently. But they would not need to alter the detector's design in any significant way as a result.

Carlo Rubbia had taken the idea of building a proton-antiproton collider using conventional magnets to CERN, and in July 1978 CERN decided to convert their Super Proton Synchrotron (SPS) to a Super proton-antiproton Synchrotron (Sp$\bar{\text{p}}$S), which would collide 270-GeV protons into 270-GeV antiprotons. Fermilab had to decide whether to compete directly with the European facility.

5. GETTING REALISTIC

In October 1978, Leon Lederman became director of Fermilab and immediately tried to set the lab on a definite course of action. On November 11 he held a "shootout," a meeting to decide the future of Fermilab. Faced with a tight budget and several projects being pursued with half measures, the Fermilab physics community had to decide what to pursue and what to abandon. They chose not to put themselves in a race with CERN to find the *W* and *Z* by building a proton-antiproton collider with conventional magnets (Rubbia and van der Meer would win the Nobel Prize in 1984 for finding the *W* and *Z* particles at CERN's UA1 detector, using the Sp$\bar{\text{p}}$S). Fermilab would finish the Energy Doubler first, and only then convert the Tevatron to a proton-antiproton collider capable of achieving a center-of-mass energy of 2 TeV. Effectively, the lab decided to surrender to CERN and Isabelle at the energy range of 540–800 GeV in the center of mass. Instead, Fermilab would focus on being the first to achieve a beam energy of 1 TeV for fixed target experiments and a center-of-mass energy of 2 TeV for colliding beams. The Program Advisory Committee endorsed this decision at its summer 1979 meeting (Groves 1979). Fermilab's physicists chose to concentrate on what could be done best, or only, at such high energies, and leave the just-around-the-corner discoveries to others.

Over the next few years, CDF engaged in ongoing debate over the problems that they could profitably tackle using 2-TeV colliding beams. Their primary interest in the closing months of the 1970s was to ensure that various types of processes could be detected, and above all to get construction of the detector started. CDF physicists were not thinking very much about the top quark at this time. That would change.

CDF members began 1979 by distributing amongst themselves a first draft of the Detector Design Report, a document that would take two and a half years to complete, and would serve as the blueprint for the detector (*CDF* 1/5/79), and by meeting with the new lab director Lederman to discuss the constraints facing CDF in terms of finances, manpower, and physics (*CDF* 1/19/79). Producing the design report, recruiting talented physicists, and building a home for the detector were the collaboration's primary initiatives during 1979. They planned to build an experimental area that would have a "two-sided," "push-pull" arrangement with assembly areas inside and outside the beam tunnel. A 12-foot-thick wall would shield the outside-tunnel area (later known as the assembly hall) from radiation produced in the collision area (the collision hall) (*CDF* 7/13/79).

In these early stages, CDF was badly understaffed. They could not entice physicists to join with prospects of impending discoveries because they would not be able to collect any data for at least a few years. Furthermore, although the Fermilab administration planned to convert the Tevatron into a collider after the Energy Doubler was completed, the DOE remained supportive of colliding beams at Isabelle, not Fermilab, casting doubt on CDF's prospects for executing their plan. The founding members of the collaboration had to convince physicists that joining them would be a good long-term investment. The problem was acute. To test prototypes of detector components, the collaboration had requested time at Fermilab's M5 test beam, and Tollestrup noted that "we need a vigorous recruiting effort to have enough hands to do the M5 testing (let alone the full detector)." He requested that people notify him of "good and available" physicists (*CDF* 7/27/79).

One way in which CDF attracted physicists was by bringing in collaborators from countries without comparable opportunities. Japanese and Italian groups were the first international collaborators to join.

In 1977, Tsukuba University physicist Kuni Kondo attended a workshop at the Brookhaven accelerator facility in New York. In a bar on-site at Brookhaven, Kondo met with Alvin Tollestrup and Shouroku Ohnuma, then a theorist in the accelerator division at Fermilab. Ohnuma told Tollestrup that Kondo might be in a position to provide both personnel and funding support to the CDF colliding-beam effort. Tollestrup suggested to Kondo, "Why don't you bring five million and five people" (Kondo 1995).

On this occasion CDF derived an advantage from taking a back seat to Isabelle. Japan sought opportunities for its young physicists to gain

72   *Building a Detector and a Collaboration*

experience in high energy physics so that they could apply their knowledge to Japan's own efforts, particularly the Tristan accelerator, an electron-positron collider with beam energies of 25 to 35 GeV. Kondo believed that his Japanese colleagues could more easily play an important role at CDF, by getting involved at the beginning, than at Isabelle, which had already attracted many people.

Furthermore, Kondo was optimistic about getting funds from the Japanese government for this project in light of the 1978 "Fukuda initiative," an agreement between the United States and Japan for developing energy resources that somehow included support for high energy physics. Looking back on the Fukuda initiative, Kuni Kondo speculated with a laugh that perhaps the United States had proposed including high energy physics research in the agreement, and Japanese Prime Minister Fukuda "didn't realize that it *consumes* high energy rather than *create* high energy" (Kondo 1995). However it happened, the DOE reached an agreement with Monbusho, the Japanese ministry of education, science, and culture. Physicists from Tsukuba University and KEK, the Japanese accelerator facility, joined CDF in April 1979.

By December, CDF members could examine a preliminary detector design report. But they still had no promise of funding from the DOE for either the detector or the antiproton source needed to convert the Energy Doubler into a collider (Tollestrup 1996, 518). Looking back on this period, Alvin Tollestrup cited the DOE's resistance as the major obstacle to the construction of the detector:

The DOE never supported the project for a long time. The early proposals that we made for colliding beams, and for a colliding beam detector, and things like that, went completely unheard by [the head of DOE]. He was unhappy with . . . even the Energy Doubler itself. . . . The fact that we were talking about colliding beams really bothered the DOE. . . . There was no money for a detector, there was no money for sources or anything like that. So that was the biggest obstacle. We had to get rid of Isabelle. (Tollestrup 1995)

At a December 14, 1979, meeting Tollestrup summarized his assessment of CDF's schedule. He noted that there would be no assembly area for the detector until early 1982, and that there would be no detector until the middle of 1983 at the earliest. An additional $20 million had been requested from the DOE to convert the detector from a research and design project to a construction project. Failure to win construction project status in 1981, he noted, would delay the project by an entire year (*CDF* 12/14/79).

Fermilab had appointed a Review Committee to oversee the CDF project. On May 14, 1980, CDF reported to the Review Committee on physics goals, detector design, and the current status of their project. The Review Committee reported in turn to the Program Advisory Committee. On June 27,

Tollestrup informed CDF that the "Review Committee report on our detector was generally favorable." He also announced that Roy Schwitters of Harvard was joining the collaboration, bringing a group of Harvard physicists with him. Schwitter's work on the Mark I and Mark II detectors at SLAC's SPEAR electron-positron collider made him a logical choice to join CDF as a cospokesperson for the group alongside Tollestrup. His first undertaking would be to review the detector design. The design was becoming increasingly complex, although Lederman had warned the group in late 1977 that apart from Mark I, "most complicated detectors have not met expectations" (*CDF* 12/16/77). With the increase in the proposed detector's complexity, its estimated cost had risen to a "completely unrealistic level" of $26 million. This estimate did not even include the cost of forward detectors, which CDF greatly desired in order to increase the detector's solid angle coverage. Tollestrup noted that "anything over $20 M is probably unrealistic" (*CDF* 6/27/80). CDF hoped that an outside expert who had experience with a complicated yet successful colliding-beam detector would help bring the design into the realm of "realistic" possibilities.

The $20 million spending limit was broken down, at a later meeting, into yearly suggested funding from the lab, with the detector "ready at the end of 1984." The plan included $5 million expected from Japan. Tollestrup noted that a design report would be needed by June 1981 (*CDF* 8/29/80).

Theoretical physicists occasionally met with CDF to discuss various physics problems. J. D. Bjorken gave the first talk that addressed the top quark on December 7, 1979. In a presentation titled "Physics at the Tevatron Pbar-P Collider," Bjorken described the top's "signature" and possible detection strategies, as well as phenomena predicted by the non–standard model theory called "technicolor" (*CDF* 12/7/79).

In the middle of 1980, CDF intensified their focus on the physics potential of their detector. In the midst of a stream of assessments and decisions regarding magnet materials, shower counter design, whether to use lead glass or lead scintillator for the electromagnetic calorimeter (they chose scintillator), configuration of electronics, and much more, the collaboration planned a physics workshop. They invited theorists to come to CDF and discuss the kinds of physics that could be done with the slowly emerging detector.

The November 1980 workshop covered such topics as "*W, Z* production," "Heavy quarks," and "new particles" (*CDF* 9/26/80). Perhaps in preparation for this workshop, Geoffrey Fox and Larry Romans, two theorists, prepared a CDF note titled "Longer Note on Top Quarks."[9] Fox and Romans noted that the search for top decays could look for either or both of two important top signatures. First, one could recognize top production events where the top decays to a *W* and a *b* quark by an excess of leptons (when the *W* decays leptonically and the *b* decays semileptonically). Second, if the top were to decay to a *W* and a *d* quark (with the *W* decaying hadronically to light *u* and *d* quarks), one could look for an excess in the number of jets produced. Romans and Fox pointed out that a similar final state would be produced by a top decaying to a hadronically decaying *W* and a *b*, provided that the mass of the *b* quark is small compared to that of the top. Because the second, jet-counting approach had great potential importance, Fox and Romans devoted much of their paper to the issue of developing an algorithm to identify jets (Fox and Romans 1980, 1–3). Jet identification was a difficult task, for which the original CDF detector design was inadequate, as Fox had already argued.

Jet identification would require both a good algorithm and good calorimeters. Jets are too messy to be identified with tracking detectors such as drift chambers. Calorimeters measure the total energy carried by a jet, and physicists individuate jets by dividing their calorimeters into segments, so that the energy carried by a jet will be recorded in only one part of the calorimeter at a time. In early 1980, calorimeters were very much on the minds of CDF physicists, as they accumulated raw materials for them and prepared to build and test prototypes. Late in 1980, the CDF meeting minutes reported that "Iron for calorimeter will be ready next week. Scintillator is almost ready to go in. ANL [Argonne National Laboratory, a collaborating institution] shower counter has scintillator being cut and wires strung" (*CDF* 12/5/80). CDF shipped the scintillator for the shower counters over from Japan. They placed prototypes of both the electromagnetic and the hadron calorimeter into test beams for appraisal during the first two months of 1981 (*CDF* 1/23/81, 2/13/81), and the Argonne shower counter went into the test beam on March 4 (*CDF* 3/6/81).

CDF's detector would be useless without colliding beams, but producing those beams was Fermilab's responsibility. The lab took an important step on October 16, 1980, when they successfully implemented Gersh Budker's electron cooling (*CDF* 10/17/80). This experience helped guide the design of the antiproton source, but a review panel formed by Lederman eventually rejected the initial design as inadequate for the Tevatron's expected capacities. Fermilab later replaced the electron cooling scheme with Simon van der Meer's stochastic cooling, which proved very effective at CERN (Tollestrup 1996, 505).

Around this time, Fermilab proposed construction of a second collision area, D-zero. (The tunnel ring at Fermilab is divided into six sections, A–F, with the center of each section designated location zero. The CDF detector is located at B-zero.) Lederman discussed the plans for the D-zero area with CDF at a February meeting following his approval of the B-zero experimental area. The lab reserved $1.7 million for the D-zero experimental area (*CDF* 2/13/81). This decision would have enormous consequences for the top search and for the frame of mind in which physicists at Fermilab sought that elusive particle.

In January 1981, cospokesperson Roy Schwitters presented proposals for the detector and the collision hall at a DOE review. When Schwitters joined CDF a few months earlier, the collaboration hoped that he would help them bring the cost of the detector down from the "unrealistic" level of $26 million. Yet by the time of Schwitters's presentation, the price tag (for both the experimental area and the detector itself) had grown to $49 million, which the DOE judged to be excessive. Schwitters proposed building the detector in stages, progressively matching its capabilities to the performance of the collider. "Staging" the detector in this way would not make it cheaper to build, but, by prolonging the building process, it would decrease the construction cost to the DOE in any given year. CDF discussed various strategies for staging the detector, but Schwitters also emphasized that approval of the project would require reducing the total cost to less than $40 million. He pointed out that CDF had to "prove to bureaucrats that you are not in a... mode where it's all or nothing as an infinite sink for money" (*CDF* 2/13/81).

In March of 1981, CDF issued a call for a "big push" to put together a more comprehensive design report (*CDF* 3/13/81). Collaboration members received a draft of this report on May 22, and subgroups spent the next week editing sections of the report (*CDF* 5/22/81). By the end of May, new plans for the experimental area were complete. Dennis Theriot of Fermilab, who coordinated construction of the experimental area and the detector, showed them to the collaboration, giving the revised cost estimate for the experimental area at $5.5 million. Construction was scheduled to begin in the fall.

CDF released its Design Report in August. Henry Frisch and Fermilab's Hans Jensen and John Yoh were heavily involved in editing the report, which provided the blueprint from which the detector was built (Tollestrup 1995). In Frisch's opinion, the 1981 Design Report correctly anticipated the physics to be done with the detector (Frisch, personal communication). The introduction to the report noted that the uniquely high energies to be achieved at the Fermilab collider would mean that

for a long time the collider will be the only place where collisions of the fundamental constituents of nature with c.m. [center of momentum] energies ranging up to 2 TeV may be observed. It is in this unexplored region that particles with masses in the range 150–500 GeV/c^2 and jets with momentum transfers of 100–400 GeV/c can be uniquely investigated. (CDF 1981, 1)

CDF devoted the report's first chapter to the various kinds of physics that they could use the detector to investigate, beginning with a discussion of

W and *Z* physics. The authors stressed the need to observe these particles directly and determine their masses, production cross sections, and decay modes. These decay modes could involve the decay of a *W* to a *t* and $\bar{b}$, and the decay of a *Z* to a *t* $\bar{t}$ pair (these decay modes for the intermediate vector bosons would be ruled out once the mass of the top was established to be greater than that of the *W* or *Z*) (ibid., 6). "Heavy flavors" was the second physics topic taken up in the report. The authors noted the still fairly recent discovery of the *c* and *b* quarks, and the need to measure their properties. "In most theoretical models," the report continued,

> a sixth quark (called "top") should exist. Data from PETRA have already ruled out *t* quarks with masses less than 18 GeV/c^2. If the mass of the *t* is less than 150 GeV/c^2, it can be produced with detectable rate at the Fermilab collider. (ibid., 8)

The report went on to describe the potential for investigations of exotic events such as "Centauros," tests of the non–standard model theory technicolor, searches for lepton and quark substructure, and other physics problems to be tackled with the proposed detector.

The 1981 design report represented a turning point for CDF. It set the agenda for construction and gave the collaboration a focus for its efforts. "Real construction," according to Hans Jensen, "began soon after" the design report. Jensen noted that it was at this time that the collaboration had reached a "critical mass" (Jensen 1995). In the opinion of Henry Frisch, adding Roy Schwitters and Dennis Theriot to the collaboration in leadership positions contributed greatly to the collaboration's newfound ability to progress with construction: "Until then we flailed around" (Frisch 1995). It was Theriot, John Yoh noted, who provided the most important leadership in getting the detector built (Yoh 1998a).

The design report settled many issues, but CDF members continued to disagree amongst themselves over a number of details of the detector's design. Calorimeter segmentation was a vexing issue. The more finely segmented the calorimeter was, the more detailed the information one could gather with it, facilitating jet identification. But finer segmentation meant greater complexity and cost. How finely segmented did the calorimeters need to be? On September 11, CDF members debated the segmentation of the central calorimeter (*CDF* 9/11/81). Two weeks later, and again the week after, they argued over the same issue with respect to the endwall calorimeters (*CDF* 9/25/81 and 10/2/81).

6. BUILDING THE DETECTOR – A CHRONICLE

The DOE reviewed the entire Tevatron I project (including the Energy Doubler, the antiproton source, as well as CDF's B-zero and the D-zero experimental areas) in March 1982. Fermilab could not solicit bids for the construction of the B-zero experimental area without the project passing this review. In a March meeting, Tollestrup reported that the DOE technical review "went very well, and we are waiting for things to percolate through Washington bureaucracy." The collaboration still hoped "to get B0 started this summer" (*CDF* 3/19/82), one year later than they had hoped for in 1980.

Meanwhile, five tons of scintillator for calorimeters accumulated at the Laboratori Nazionali di Frascati of the Istituto Nazionale di Fisica Nucleare (INFN) in Italy, where technicians would use a laser to cut it into the appropriate pieces (*CDF* 5/14/82).

The Italians were the other major international collaborators besides Japan in the early years at CDF. They got involved immediately in the design and construction of the hadron calorimeter. The Italian participation began when two Frascati physicists, Giorgio Bellettini and Paolo Giromini, visited the United States in the fall of 1979 to attend the Lepton-Photon Conference, held that year at Fermilab. Alvin Tollestrup and Hans Jensen approached Bellettini and Giromini, telling them of the great opportunities at CDF for talented physicists willing to make an early commitment to the project (Jensen 1995).

Bellettini had been trying to get an accelerator and detector project built in Italy, but, he recalled, "that was trailing along and never ending, and funds were always promised and never located." On the flight back to Italy, Bellettini and Giromini discussed accepting the CDF invitation, so they could "forget about fighting for impossible and . . . lower level prospects." Back in Italy, Bellettini called Tollestrup and expressed his interest in entering the collaboration, and later met with CDF member Bob Diebold of Argonne National Laboratory. In Bellettini's words, the Italian involvement originated "in the typical Italian style" as simply the initiative of two people, himself and Giromini. It did not result from the plan of a major organization or political body, "just two research people who decided to switch from here to there." At the time of the 1981 design report, four people from Frascati were in CDF, along with eleven from INFN at the University of Pisa (Bellettini 1995).

Those five tons of scintillator that the Italians provided for the hadron calorimeter were destined for Fermilab's Industrial Building Four, where Hans Jensen was in charge of assembling the central calorimeter. Jensen recalled, "We had parts coming from all over the world" (Jensen 1995).

By the end of May 1982, the DOE had given permission to open bids for the construction of the B-zero experimental area (*CDF* 5/28/82). Fermilab broke ground for B-zero on July 1 (*CDF* 6/26/82). Workers were pouring concrete as of September 24 (*CDF* 9/24/82). In November, the CDF review subcommittee reported favorably to the Program Advisory Committee on the progress at CDF (*CDF* 11/19/82). In December, Hans Jensen announced a series of upcoming "previews," intended to "discuss ideas before they are cast in concrete," and "reviews," intended to "see how the detectors perform relative to Design Report; how they perform relative to new physics from CERN; and voice concerns about technical difficulties" (*CDF* 12/13/82).

Accompanying this suggestion of an institutionalized review process, Roy Schwitters proposed that standing review committees be established. (These committees were originally called "godfather" committees. CDF later adopted the gender-neutral term "godparent" committees. This did not change the fact that most committee members, like most CDF members generally, were men.) Schwitters argued that such a system would be "a good sounding board for those of you actually building the equipment" and an "excellent mechanism for educating others in the collaboration about the details of various detector components." Above all, the aim of the system was to keep watch over "the physics performance and goals of the various detector systems" in a "cooperative rather than adversary relationship." Schwitters attached to this memo a list of fourteen suggested standing review committees and the three to four physicists who would sit on each (*CDF* 12/20/82). The role of godparent committees would eventually include preliminary review of physics results, thus taking on a role of inestimable importance within the collaboration.

The "new physics from CERN" to which Hans Jensen alluded in December 1982 was probably the discovery of the *W* boson, announced in January 1983 by the UA1 (Underground Area 1) and UA2 (Underground Area 2) collaborations (Arnison, Astbury, et al. 1983; Banner, Battiston, et al. 1983; see Watkins 1986 for an account by a member of UA1 and Krige 2001 for an intriguing historical discussion). Carlo Rubbia's UA1 collaboration announced the discovery of the *Z* boson in late May of the same year. These two discoveries were major victories for the standard model. CDF physicists had known that pursuing the full Tevatron project rather than the "quick-and-dirty" collider that Rubbia had urged meant letting CERN take the lead in the search for the *W* and *Z*. But for CDF physicists, the *W* and *Z* remained tremendously important, since CDF would ultimately be able to produce these particles in far greater numbers than UA1 and UA2 could ever hope to achieve. The masses, lifetimes, cross sections, and other properties of the intermediate vector bosons would have great significance for physics. Anyway, CDF still expected to have the best chance of any collaboration anywhere for finding the top, which appeared to be the next big discovery to be made.

In March of 1983, the door giving access to the collision area at B-zero was installed and painted "decorative colors" (*CDF* 3/25/83). In April the first version of the CDF detector computer simulation was distributed (*CDF* 4/15/83). The prototype of the vertex time projection chamber (VTPC) was completed in June (*CDF* 6/10/83). On June 24, Roy Schwitters reported on the meeting of the Program Advisory Committee, at which CDF presented its current status. The lab had scheduled the first run of the Tevatron in colliding-beam mode for May 1, 1985, and CDF hoped to have the central detector and the plug calorimeters in place by then (*CDF* 6/24/83).

The summer of 1983 was kind to Fermilab. In July of that year, the Tevatron accelerated protons to 512 GeV. The lifetime of the beam was about 7 hours (*CDF* 8/19/83). One month later, the lab accelerated protons to 700 GeV. That same summer saw the official demise of Isabelle (by this time renamed the Colliding Beam Accelerator), the funding of which had so frustrated Wilson in the late 1970s. It had been clear for some time that the SppS collider at CERN was yielding all the exciting discoveries in the energy range proposed for the Colliding Beam Accelerator, and that Brookhaven had taken too long to solve that project's problems. These facts, combined with the progress taking place on Fermilab's Tevatron prompted the High Energy Physics Advisory Panel, which represents high energy physicists to the DOE, to recommend terminating work on the Colliding Beam Accelerator (Hoddeson 1987).

At the end of September 1983, CDF took possession of the B-zero assembly hall, just in time to hold an October workshop there on the expected 1985 run (*CDF* 9/30/83). The aim of this workshop was to determine what kind of detector they would have in place for this run, what kind of physics they could do with that detector, what they needed to do to prepare the detector, and what they might learn about the detector during the run (*CDF* 9/19/83). CDF expected to begin taking data by June 1, 1985 (*CDF* 10/7/83).

In February 1984, the accelerator reached 900 GeV, the laboratory's goal (Tollestrup 1996, 515). By the end of March, the helium refrigerator needed for CDF's superconducting magnet was running (*CDF* 3/30/84). The magnet coil itself, which had been assembled in Japan, arrived at the B-zero site on July 20 (*CDF* 7/20/84). This magnet by itself constituted an innovation. Unlike the superconducting magnets used in the accelerator, it had to contain as little material as possible, because collision products would have to pass through it before entering the calorimetry. Too much interaction of these particles with magnetic material would degrade the quality of the information obtained from the calorimeters. There had been problems with a similar magnet at Berkeley, where the superconductor had been bonded to an aluminum backing. The Japanese physicists discussed the problem with the manufacturer Hitachi, and Hitachi responded by developing a new coil by extruding superconductor and aluminum together (Kondo 1995).

Assembly of major components of the CDF detector began in mid-1984. Parts that had been under construction at numerous collaborating institutions were slated to be brought to B-zero and put into place (*CDF* 3/30/84). Nineteen calorimeter wedges were complete as of June 1 (*CDF* 6/1/84).

The minutes to the CDF meeting on May 18, 1984, note some discussion of the latest results from UA1 and UA2, without specifying the nature of those results. The discussion may have concerned events leading up to Carlo

Rubbia's claim in June of that year to have discovered the top quark at UA1. Rubbia told the *New York Times* that the evidence for the top "looks really good," and sent out a press release claiming discovery of the top on July 4. By the time that the paper reached *Physics Letters* in October, the claim had been weakened to stating that the observed events were a "clear signal" that was "not consistent with expectations of known quark decays," but "in agreement with the process $W \rightarrow t\bar{b}$" (Arnison, Astbury, et al. 1984, 493). Although not a discovery claim, this remained a strong suggestion, since calling the effect a "clear signal" indicated that the observed events were not background, and hence not the result of any *known* processes. UA1 went so far as to infer limits on the top mass: "If this is indeed the case, then the mass of the top is bounded between 30 and 50 GeV/c^2," but softened their stance by concluding that "more statistics are needed to confirm these conclusions and the true nature of the effect observed" (ibid., 507). Over the next months, the UA1 consensus emerged that the "true nature of the effect" was that it was neither a top quark nor even a "clear signal" – UA1 had overlooked a significant source of background. Correcting that oversight, the result was consistent with the background estimate (Taubes 1986, 100n).

UA1's pseudodiscovery of the top quark made an impression on CDF physicists. Many felt that in his eagerness to claim a major discovery, Rubbia had put out results without sufficient scrutiny, and that his reputation suffered as a result. In my conversations with them, collaboration members referred to this incident as a cautionary tale, and as a reason for CDF to take an approach to new results that many CDF members considered to be very "conservative" (but that was not conservative enough, in the opinion of other group members). No one wanted to be known as the *second* physicist to go to the press over what was really just more background to the top search.

At the PAC meeting in June of 1984, the committee recommended that CDF "be given highest priority within the experimental program," even if keeping them on schedule entailed delaying other fixed-target experiments and D-zero "due to financial constraints" (*CDF* 6/29/84). In September, CDF held a workshop on offline data analysis to "arrive at a detailed scenario for the development of data handling, data reduction, and data analysis software to be ready by first data" (*CDF* 7/27/84). They now expected their first $\bar{p}p$ collisions in the fall of 1985. They installed the magnet coil in its yoke in November (*CDF* 11/9/84). Fermilab's accelerator operators extracted the first beam from the Energy Saver ring in December, and the lab commenced a fixed-target run at 800 GeV in January 1985 (*CDF* 12/17/84).

CDF spent most of 1985 in a flurry of activity in order to assemble as much of the detector as possible before the first $\bar{p}p$ collisions in the fall. During the bitter Illinois winter, they began to put pieces of the central detector in place, beginning with the endwalls and then the endplugs. By late January, they had stacked the first calorimeter wedges (*CDF* 1/25/85), and built all 48 of the wedges needed for the central calorimetry (as well as two spare wedges) (*CDF* 2/1/85). Cooldown of the superconducting magnet began on February 28 (*CDF* 3/1/85). They successfully operated the magnet at 1.5 T in late March, after CDF began to map the solenoid's magnetic field (*CDF* 3/29/85). The vertex time projection chamber and central tracking chamber prototypes were placed in the magnet, and physicists used them to observe tracks from a radiation source. Meanwhile, a team prepared the vertex time projection chamber itself for installation in the detector so that it would be in place when they moved the central detector into the collision area, which they planned to do in September (*CDF* 5/3/85). Cabling of the calorimeter arches began in late June (*CDF* 6/28/85), and "cabling and checkout" of the two east arches was complete by mid-August (*CDF* 8/16/85). They had assembled and installed the entire central calorimeter, with an "initial checkout of all electronics," by early September, just in time to move the detector into the collision area.

A magnet in the Tevatron had failed in late August, terminating the fixed target run a few days ahead of schedule. Fermilab continued to operate the Main Ring for work on the antiproton source (*CDF* 8/30/85) until September 9, when the accelerator was shut down completely, allowing CDF to open the door to the collision area. They moved the detector from the assembly hall to the collision area the next day (*CDF* 9/6/85). At the time the detector included all the calorimeter wedges. Central muon chambers were only partially installed. All eight modules of the vertex time projection chamber were installed, and nearly all were instrumented. The Level 1 trigger was "substantially operational" (*CDF* 9/13/85). The accelerator produced no colliding beams at first, but during brief periods when the accelerator was down, CDF tested the detector using a radiation source (*CDF* 9/27/85).

On October 13, 1985, Fermilab produced the first proton-antiproton collisions at the Tevatron, and CDF recorded them with their partially assembled detector. The accelerator achieved beam energies of 800 GeV, or 1.6 TeV, in the center of momentum, and the luminosity was a faint 10^{24} events/cm^2/sec, but collaboration spokesperson Tollestrup and lab director Lederman judged it a successful "engineering run" (Schwarzschild 1985). A mere 23 proton-antiproton events were detected on this momentous occasion, but their appearance allowed the collaboration to see some pieces of the detector in action. They were able to view, for example, a display of tracks from the vertex time projection chamber (*CDF* 10/18/85). Having recorded its meager 23 events, the gargantuan central detector began its slow, painstaking journey back out of the collision area on November 12 (*CDF* 11/1/85).

Reporting on the October 1985 engineering run at a conference held at Aachen, Germany, the following summer, Mark Eaton of Harvard noted that the detector components that were operational during this run were the "Central and Endwall Calorimeters, the Vertex Time Projection Chambers,

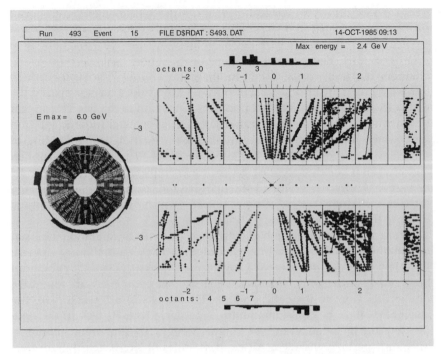

FIGURE 2.9. Display of a $\bar{p}p$ collision event from the 1985 engineering run. Courtesy of Fermi National Accelerator Laboratory.

the Central Muon Detector and the Beam-Beam Counters. The DAQ [data acquisition system] was sufficiently integrated to support the calibration and read out of these detector components and the level 1 trigger was operational." He concluded that the CDF physicists "are encouraged by the successes of the systems test/engineering run and are eager to bring the full CDF detector online" and he displayed a $\bar{p}p$ event (see Figure 2.9) (Eaton 1987, 447).

On the verge of real physics, CDF suddenly faced the threat of crippling financial shortfalls. In 1986, the DOE, responding to the Gramm-Rudman-Hollings deficit reduction act, demanded that Fermilab, including CDF, reduce its budget by 4.62%. Roy Schwitters noted that such a cut "would have a major impact on this year's run. It could jeopardize a major part of the detector, such as one side of the forward/backward detector, which would of course greatly compromise physics capability" (*CDF* 1/17/86). CDF's main concern was to "keep the run on schedule and its full equipment budget" (*CDF* 1/24/86).

Meanwhile, with the central detector in the assembly hall, and access to the endplug detectors in the collision hall regained, CDF took up the task of completing the assembly of the detector. Fermilab's Richard Kadel directed construction of the central tracking chamber, the delicate and closely spaced wires of which began to be strung late in the fall of 1985. In January 1986, CDF attached the endplug electromagnetic calorimeters (*CDF* 1/10/86). Transport of the heavy toroid magnets for the muon spectrometer was delayed due to damage of the rollers on the transport cart (*CDF* 1/24/86), but by mid-March half of the west toroid was in place (*CDF* 3/14/86). The other half was in as of April 4 (*CDF* 4/4/86).

At a DOE review held in March, CDF requested a $600,000 advance on their budget for the next year "to complete the detector by the run next fall." The response was "favorable." The fall run would be CDF's first opportunity to collect data with a basically complete detector and, consequently, their first chance to do real physics. Hence, the collaboration was determined to be ready for it (*CDF* 3/28/86). It was at this time as well that the "analysis working groups," also known as "algorithm groups," held their first meetings. These precursors to the later "physics groups" devoted themselves to developing analysis tools to identify specific event features: the jet group, the electron group, the missing E_T group, the secondary vertex group, and the muon group (*CDF* 3/28/86).

Since November 1984, trailers had been accumulating at the B-zero site. The growing population of CDF used these trailers for office space. In 1986, the management of Fermilab "officially declared the CDF trailers at B-zero an eyesore," and proposed a two-story office building to replace them (*CDF* 6/6/86). A week later lab administrators canceled this plan and suggested making room for CDF offices at Fermilab's main building, the fifteen-story Wilson Hall ("the high rise" in Fermilab jargon) (*CDF* 6/13/86). The administration withdrew that suggestion late in July. Instead, the five trailers at B-zero increased to eight and moved to the far side of the CDF assembly hall, hidden from the view of the administrative offices in Wilson Hall (*CDF* 8/1/86). Thus was the "eyesore" problem solved. The trailers have since increased in number and remain to this day.

By the middle of July 1986, CDF had installed the east (forward) toroid and calorimeter in the collision area, just prior to the closing of the shield door. Fermilab resumed operation of the main ring, although CDF gained badly needed access to the B-zero collision area intermittently until the end of August (*CDF* 7/18/86). CDF was scrambling to install their forward detector systems and test them before a collider run scheduled to begin in October (*CDF* 7/25/86). The forward electromagnetic and hadronic calorimeters were installed in the collision area in mid-August (*CDF* 8/15/86).

Plans for moving the central detector into the collision hall in late October ran into trouble, however. The accelerator was producing unexpectedly high radiation levels in the collision hall, which would have damaged the detector. For CDF, it was just as well, since the central detector was not completely assembled, anyway. The radiation-induced delay gave CDF time to install the central tracking chamber, vertex time projection chamber, and forward tracking chamber. Meanwhile the beam in the Tevatron reached 900 GeV (*CDF* 10/24/86). At the end of October, the accelerator operators had the radiation problem under control (*CDF* 10/31/86).

In small numbers, Fermilab stored antiprotons and collided them with protons in late November. Dennis Theriot, who had overseen the construction of the detector, announced a party to celebrate the detector's "completion" – a party that, fittingly, had to be postponed by about a week (*CDF* 12/5/86, 12/12/86). Finally, with the shield door open, transport of the central detector into the collision hall began. After the central detector was in place, the beryllium beam pipe of the detector was attached to the Tevatron beam pipe, and CDF began cooldown of the superconducting solenoid. The physicists had achieved a fairly complete assembly of the central and forward portions of the detector. The impending collider run would begin with a low-luminosity run to check out the detector as well as the Tevatron and antiproton source. A brief physics run would follow, CDF's first (*CDF* 12/19/86).

7. DISCUSSION

The physicists who joined CDF in its early years committed themselves to an endeavor that would not yield physics results for several years, if ever. Why would someone do this? The earliest members could only see their new project in vague outline: the construction of a detector with certain general features (large solid angle coverage, tracking ability over a large volume, maybe in a magnetic field, etc.) to look at events from colliding hadron beams (either protons on protons or protons on antiprotons, at yet-to-be-determined, but very high, energies) in order to pursue a few specific experimental goals, but open to possibilities still unknown, even unimagined.

From the conception of this general idea to the construction of an actual detector and the collection of data, CDF physicists had to make myriad decisions, each of which had implications for other decisions not yet made, or for decisions that had been made, but suddenly needed to be unmade. In the process, CDF continually checked and updated its goals. While these goals included some particulars such as the ability to detect Ws and Zs, these were subsumed within a general experimental program. Mel Shochet described the goal of finding the W and the Z as a "benchmark." "It was something that you could say, 'This we can do, and this is how well we can do it.'" But because the collisions to be produced at the Tevatron would be at much higher energies than had previously been achieved, "there were all sorts of new things that one might be able to see, and you wanted to have a detector which you felt was generally good in observing the elementary constituents" (Shochet 1995).

In some ways, the CDF detector constituted a bold new venture in experimental high energy physics. Its scale exceeded anything previously built. Furthermore, when CDF began planning their detector, they had no blueprint for detecting the products of head-on hadron collisions – Mark I was an electron-positron detector. Designed for lower energies, UA1, the first hadron collider detector, did not begin operation until the spring of 1982. The great bubble-chamber physicist Luis Alvarez had even declared that, because of the complex internal structure of hadrons, no more could be learned from banging hadrons together than could be learned by colliding garbage cans (Tollestrup 1998). However, the general pattern of the detector was not entirely without precedent: the Mark I at SLAC was an obvious predecessor. CDF's John Yoh claimed that any competent group of physicists with $50 million and the assignment to build a detector for high energy $\bar{p}p$ collisions would come up with something "pretty much like" the CDF detector. But Yoh added that some features of the detector, the value of which was not initially apparent to all or even most CDF members, proved very important: the calorimeter tower structure, the detector's hermeticity, and the silicon vertex detector (Yoh 1998a).

Geoffrey Fox had first pointed out the importance of having a fine enough segmentation of the calorimeter to identify jets and had pushed this idea through the collaboration. Increasing calorimeter resolution significantly increased the cost and complexity of the detector (Tollestrup 1995; Yoh 1998a). Having a fine-grained calorimeter turned out to be important, however, because the correct identification of jets would be crucial in gathering evidence of top decays. Also important was the *hermeticity* of the detector – the relative absence of cracks and gaps through which particles could escape undetected. Hermeticity mattered because identifying top events would require the detection of events in which neutrinos are produced. But neutrinos interact so rarely with matter that they are identified by the fact that they *do not* leave any trace in the detector: one infers that a neutrino has been produced when energy conservation requirements can be satisfied only by assuming that some particle has escaped detection. For this argument to be deployed reliably, though, other particles besides neutrinos must not routinely escape through cracks in the detector. This makes hermeticity – the absence of "leaks" in the detector – important. (On this score, D-zero would seek to outdo CDF. D-zero physicists consider CDF's detector to be rather leaky compared to theirs.) Finally, the need to identify events containing *b* quarks would make the silicon vertex detector crucial, but that is a subject for the next chapter.

The features of the CDF detector that allowed CDF members to pursue the experimental program that they gradually conceptualized all emerged from struggles within the collaboration. This struggle was aimed at building an *allgemeine* detector that would repay such a long-term and substantial commitment with a wide range of important experimental investigations. The story of the early years of CDF is a story of scientists doggedly pursuing the development of a means for answering questions. The detector itself was the focus: a machine that could be used to answer questions that had not yet been formulated, as well as some that had.

Allan Franklin has discussed the phenomenon of "instrumental loyalty." As Franklin puts it, experimental physicists often ask themselves not "What is the best physics experiment that can be done?" but "What is the best physics experiment that I can do with an already existing apparatus, or with a minor modification of that apparatus?" By using the same apparatus for a new experiment, physicists can take advantage of the "recycling of expertise," employing knowledge gained by previous use of that apparatus (Franklin 1997).

The point that I wish to make is closely related to Franklin's thesis of instrumental loyalty. When they began, CDF physicists were many years from being able to do any experiments. The question they faced was "What is the best apparatus that I can build that will enable me to do the best experiments, given all of the other constraints facing me?" Answering this question was especially difficult because of uncertainty surrounding both physical and administrative constraints in the early stages of the project.

What CDF set out to build was an apparatus with sufficient power and flexibility to *deserve* their loyalty over the course of numerous experiments, many of which no one had yet thought about, but all of which fit into a general line of investigation. Such a machine would also serve as a location for the accumulation of expertise that could be "recycled" in the service of these myriad experiments. One of the reasons everyone in the collaboration is given a say in the "blessing" process that each CDF paper must undergo is that expertise concerning the many complex detector subsystems is widely distributed and may be implicated in a variety of experimental analyses. A measurement of the mass of the W boson (Abe, Amidei, et al. 1995c) and a search for "second generation leptoquarks" (Abe, Amidei, et al. 1995b) involve different theoretical considerations, but both require the identification of muons, and hence both require the correct use of data from the central muon detectors. CDF has muon detector experts who typically built or worked on upgrading the muon detectors. Yet physics analyses that require the correct identification of muons may be pursued by CDF members who do *not* have such expertise. Similar comments apply to every piece of the detector. So the entire collaboration, that great conglomeration of varying types of physics expertise, gets to have its say on every result.

Closely related to instrumental loyalty is what might be thought of as *institutional loyalty*. An important factor in the performance of a particular experiment is establishing that the group one is working with has both the right kind of knowledge and the right kind of procedures for producing reliable results. This aspect of collaboration physics will be discussed in more detail in chapters to come.

Besides a concern with a general area of investigation, CDF certainly had an interest from the beginning in particular physical problems. The search for the *W* and the *Z* was frequently given as a rationale for the CDF detector in the early days, but largely because, according to Shochet, one could say of these particles, "that ought to be there" (Shochet 1995). CDF gradually realized that they were not going to discover the intermediate vector bosons. Instead, they sought to measure their properties better than had been done before, and, of course, they in time spoke of finding the top. So while CDF members in the early days regarded the ability to detect the *W* and *Z* as a constraint on detector design, the reasons for being able to do so shifted to the desire for the ability to do a variety of experiments calling for the very same capabilities as those needed for identifying intermediate vector bosons. One of these experiments was the top search itself.

Without having to rethink their detector in any fundamental way, CDF could adapt their experimental program to new questions and new searches. Their commitment was not so much to individual problems of the standard model as it was to a way of investigating a class of problems, an approach embodied in the CDF detector itself.

That detector evolved in response to many different factors: financial pressures, changes in the course pursued by Fermilab, increases in accelerator luminosity, problems posed by radiation, the availability of new technologies, the tenacity of collaboration members seeking to try out new technologies, and so on. Yet insofar as the CDF physicists' interests in "doing physics" shaped the CDF detector, those interests had more to do with the need to measure particular kinds of quantities than with the search for a particular particle. While changes in experimental plans might occasionally provoke changes to the detector, these were expensive, time-consuming, and impossible to carry out while the detector was in use. Hence, the detector needed to be able to serve diverse experimental purposes without undergoing radical changes in construction. Discoveries come and go, but the CDF physicists knew that whatever they set out to do with their detector, over the long run it would involve being able to identify electrons, muons, neutrinos, and hadron jets. They knew that they would need to identify the tracks of some of these particles, and that they would need to be able to measure certain of their properties, such as transverse momentum or total jet energy. Their detector had to be a good tool for all of these tasks.

Yet for the detector to be able to yield any useful data at all, the physicists in CDF had to be able to draw the information out of it, interpret it, and analyze it. In the next two chapters, I will describe the controversies and complexities that would plague the attempt to do this in the case of the top-quark analysis.

3

Doing Physics

CDF Closes in on the Top

The logic internal to science is not only to chase theory... there is also a logic internal to detector development.

Franco Bedeschi (1998)

1. WATCHING THE DETECTORS: PHYSICS BEGINS AT CDF

In January 1987, CDF began its first data-collecting run. They encountered problems immediately. Fermilab had kept the nonsuperconducting Main Ring in the same tunnel as the new Tevatron, and was using the older accelerator both as an injector for the Tevatron and to produce antiprotons. During antiproton production, a "spray" of radiation from the Main Ring, just 5 feet above, was interfering with the top four muon chambers of the CDF detector (*CDF* 1/9/87).

Because of the Main Ring radiation problem, CDF began data collection at a languid pace. For the week ending February 6, Fermilab produced about 50 hours of beam collisions. CDF accumulated about 0.1 inverse nanobarns (nb^{-1}) of data (*CDF* 2/6/87).[1] The collaboration promptly faced difficulties regarding the dissemination and interpretation of this data. At the February 13 CDF meeting, Roy Schwitters raised the issue of "public distribution of data," noting that "in the future discretion should be used before distributing information about perhaps less unambiguous events, particularly before most of the collaboration is aware of their existence" (*CDF* 2/13/87). Three weeks later a "speakers committee" was established "to provide speakers about CDF results in a fair and open manner." Furthermore, collaborators were required to receive approval for all talks given anywhere but at the collaborator's home institution. Anyone wishing to present physics results not yet released in preprint by the collaboration as a whole would need to present those results at a group meeting to have those results "blessed" by their collaborators (*CDF* 3/6/87). Although this established a system of

88

"quality control," it did not entirely prevent controversies over publicizing physics results, as we will see.

A couple of months into the run, CDF still did not have much data. The collaboration began a push to write as much data onto computer tape as possible before the scheduled end of the run on April 20, 1987. The accelerator division at Fermilab reduced the time devoted to accelerator studies, and CDF cut back on time spent exploring their detector. CDF requested that the lab extend the run into May so that enough data could be collected "to do significant measurements on *W*, *Z*, and of jet production up to p_t [transverse momentum] ~250 GeV" (*CDF* 3/27/87). The lab agreed to extend the run to May 11 (*CDF* 4/3/87). Most of the 33 nb^{-1} that CDF wrote to tape during this run were collected from March to the end of the run, amounting to a little under half of 72 nb^{-1} delivered by the accelerator.

A retrospective report stated that the object of the run "was to test and debug the data-acquisition system and to understand the characteristics of the tracking chambers and the calorimetry sufficiently well to do some preliminary physics measurements" (CDF 1987, 16). Significantly (for the smooth running of the collaboration), routines had been established for monitoring detector elements, establishing pedestal values, and calibration (Barnes 1987).

The detector was essentially complete for the 1987 run, and CDF described its components in a series of articles in *Nuclear Instruments and Methods in Physics Research* in 1987 and 1988. However, not every detector component had been fully functional during the 1987 run. The inoperative parts of the detector included half of the sampling planes in the forward calorimeter, part of the level two trigger, and some of the data acquisition system (ibid.).

The 1987 run was also the first real run for the collider. The luminosity achieved by the collider increased dramatically over the course of the run, exceeding 10^{29} cm^{-2}s^{-1} in the last two weeks of the run. The problem of radiation from the Main Ring had caused a 15% loss in "useful collision time" (ibid. 143). The collaboration planned to install two feet of iron shielding before the next run to solve this problem. CDF's goal for the next run was to collect 1 pb^{-1} (1000 nb^{-1}) (ibid., 148). (In fact, they would collect more than four times this much data during the next run.)

CDF had run into several problems during the 1987 run. Particles striking the accelerator magnets near the forward muon spectrometers were causing showers that falsely triggered those detectors. This problem would be remedied by means of more stringent track identification requirements (ibid., 148).

There were calorimeter problems as well. Some calorimeter components were excessively "noisy" (appearing to register low-level particle activity even when nothing was happening). Alvin Tollestrup headed a group trying to reduce calorimeter noise, but they needed first to identify which components were the source of the trouble. On March 17, Tollestrup asked G. P. Yeh of Fermilab to work on the problem. In this case, the key was to find the right way of looking at data that was already available. Yeh altered the "Lego" event display (see Figure 3.1)[2] used by CDF to look at calorimeter tower energies, so that the problem could be localized (Yeh, personal communication). Yeh and John Yoh used a similar approach in isolating the problem known as Texas towers. These appeared as anomalously large deposits of energy in single calorimeter towers (or occasionally two or more adjacent towers) and had been noticed first by collaborators from Texas A&M University. As early as April 1987, CDF physicists had discussed this problem and suggested that it might be caused by low-energy neutrons striking free protons (*CDF* 4/10/87). The dramatic increase in the energy of the proton that resulted from such a collision could mimic "up to 50 GeV of calorimeter energy in a single tower" (Barnes 1987, 149). CDF considered various solutions, including relying on the Texas towers' localized geometry to eliminate them using triggering requirements, or criteria applied offline to recorded data.

At the time of the 1987 run, the collaboration included 229 individuals from 17 institutions. About 12% of the individuals and 2 of the institutions (Pisa and Frascati) were Italian, and about 13% of the individuals and 2 of the institutions (KEK and Tsukuba) were Japanese. Over 20 of the graduate students involved expected to write Ph.D. theses based on data from this first run (Barnes 1987, 145).

2. IMPROVING THE DETECTOR – HOW THE SVX WAS WON

While the 1987 run progressed, the detector continued to evolve – at least on paper. One aspect of that evolution that would prove important to the top search was the effort to add a silicon vertex detector (SVX; see Figure 3.2), which was becoming unstoppable.

A charged particle passing through silicon will produce pairs of ions, dislodging electrons that will in turn produce more ion pairs. In a semiconductor detector such as the SVX, an electric field is used to separate positive and negative ions. The ions are then collected at electrodes, resulting in a signal proportional to the energy of the incident particle (Fernow 1986, 296–8). Drift chambers, for many years the workhorse of high energy physics for particle tracking and identification, operate on the same basic principle. Semiconductor detectors can achieve much higher resolutions than traditional drift chambers, in which resolution is limited by the space between wires. In silicon detectors, the wires are replaced with conducting strips etched onto the surface of silicon wafers. This etching technology is highly advanced, so silicon detectors can achieve resolutions on the order of 5 to 10 μm by placing strips very close together. Such high resolutions are especially advantageous close to the location of the proton-antiproton collision (the "primary vertex"), where particle tracks are too close to one

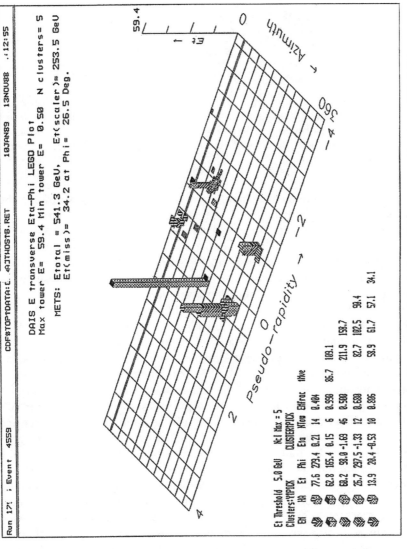

FIGURE 3.1. The Lego plot used in tackling the problem of Texas towers. Courtesy of G. P. Yeh.

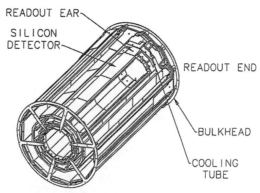

FIGURE 3.2. Schematic drawing of the SVX for run Ia. Reprinted from *Nuclear Instruments and Methods in Physics Research.* **A350**, Amidei, Azzi, et al., "The Silicon Vertex Detector of the Collider Detector at Fermilab," 75, Copyright (1994), with permission of Elsevier Science.

another for traditional drift chambers to be able to distinguish them. The proposal for the SVX was to place it in the very heart of CDF's detector.

The silicon vertex detector would come to play an important role in the analysis of the events yielding evidence for the top quark. In the wake of the top announcements of 1994 and 1995, *Scientific American* advertised on its cover an article on silicon detectors with the hyperbolic phrase "What found the top quark" (Litke and Schwarz 1995).

As with the detector as a whole, it would be misleading to say that the SVX was built for the purpose of tracking down the top quark. The SVX effort was initially supported by a small group from the University of Pisa, and in particular by Pisa physicist Aldo Menzione. Although these advocates of building the SVX certainly made their case in terms of its uses for doing physics, and although the detector would not have been built if others in CDF had thought it to be useless for physics purposes, the building of the SVX was not driven primarily by a desire for solving some specific, well-defined set of physics problems. Instead, the prime advocates of building the SVX were physicists who had used similar detectors in different settings in the past. They wanted to try their hand at making the new technology work in a more challenging setting. Hence, the SVX illustrates Peter Galison's point, that not only experiments, but instruments as well, have a "life of their own" (Galison 1997b). Franco Bedeschi, one of the builders of the SVX, expressed this point so neatly that I have used his words as an epigraph to this chapter: "The logic internal to science is not only to chase theory... there is also a logic internal to detector development" (Bedeschi 1998).

Most of the physicists from the University of Pisa who joined CDF had previously worked with a silicon microstrip detector in an experiment called NA1.[3] The NA1 (North Area 1) collaboration had done fixed-target experiments at CERN's Super Proton Synchrotron and had successfully used their Multi-Electrode Silicon Detector (MESD) in experiments such as the measurement of the lifetime of charmed particles (Alibini, Amendolia, et al. 1982). The MESD was a much simpler device than the SVX, and it was operating in a much less noisy environment. But even in 1980, while testing a silicon detector for use in fixed-target experiments at the SPS, the Pisa physicists were looking ahead to the colliding-beam experiments of the future: "owing to the fact that it can operate in a vacuum, MESD appears as a unique tool for small angle measurements in colliding beam experiments" (Amendolia, Batignani, et al. 1980, 460). The detector's use at CDF would not be restricted to small angle measurements.

The Pisa contingent at NA1 split primarily into two groups, both of which went on to do colliding-beam physics. One group joined ALEPH, a colliding-beam experiment at CERN's electron-positron collider LEP. The other group joined CDF, where they faced the challenge of making a silicon detector work in a colliding-hadron environment.

Along with fellow Pisa physicist C. Bradaschia and Tom Collins of Fermilab, Aldo Menzione wrote a CDF note (CDF-60) titled "Feasibility of Operating Silicon Detectors Inside the Collider Vacuum Pipe" in 1980, and the design report for the entire detector of 1981 did include a silicon vertex detector, largely at the urging of the Pisa group. But getting the SVX realized was difficult for many reasons.

The Pisa group originally planned to build everything themselves, since they had previous experience with silicon detectors. Besides, outside of the Pisa group and a few others "people really weren't very interested in [the SVX]" (Tollestrup 1998). The Pisa group's original plans relied on techniques that they could do themselves. Alvin Tollestrup recalled with a chuckle, "Aldo was going to do it in his kitchen. It was supposed to be very easy. I visited over there and they showed me their clean room [where the detector elements would be assembled], and I pointed out that there was a spider in it" (ibid., 1998).

A more serious problem was the reliability of the junction between the conducting material that constitutes the "wires" of the microstrip detector and the silicon wafer itself. At NA1, the Pisa group had used the "Schottky barrier" technique, whereby a "mask" is placed over the silicon surface, leaving exposed the areas on which one wishes to deposit conductor. The wafer is then placed in a vacuum chamber, and gold is evaporated into the chamber, where it is deposited onto the exposed surfaces of the silicon, forming a junction. Commercial chip manufacturers use a more sophisticated technique to implant the junction into the silicon, a process requiring very expensive specialized equipment. The Pisa group's research and development efforts revealed that the Schottky junction technique would not work in this ambitious use of a silicon microstrip detector. "It was too many detectors and too much work. And also, we did not feel comfortable with the quality of the junction" (Bedeschi 1998). The junction was not stable enough over time and deteriorated as oxygen penetrated the gold. In addition, the quality of the junction varied widely with the quality of the surface. Forming a Schottky junction required elaborate cleaning and preparation of the surface of the silicon, and Pisa struggled to maintain a uniform standard of purity in the chemicals that they used in preparing the silicon. Bedeschi observed, "Really a junction made with implantation is much more reliable, but for that we knew we had to go to a company" (ibid., 1998).

Finding a commercial manufacturer for these silicon strips posed its own problems. Silicon microstrip detectors are a kind of integrated circuit, but quite unlike any integrated circuits routinely manufactured at the time. Although much simpler than most integrated circuits in terms of circuit design, microstrip detectors are much bigger. SVX components were not something that a chip manufacturer could produce along its assembly lines. They required a special manufacturing process to produce a small number of silicon microstrip detectors. CDF managed to interest Micron Semiconductor, a very small English company. Similarly, ordinary silicon manufacturers could not produce the silicon for the detector, which required a resistivity at least ten times that of the silicon used in commercial integrated circuits. But the Pisa group already had a source for the silicon. German chemical manufacturer Wacker-Chemie had provided NA1 with "hyperpure," very high resistivity silicon for their MESD (Amendolia, Batignani, et al. 1980), and they also provided the silicon for the SVX.

In addition to these manufacturing problems, the Pisa veterans of the NA1 collaboration found that moving from fixed-target to colliding-beam uses of the new technology in itself posed a serious obstacle. The traditional approach to a particle detector is to have a separate cable for each channel of the detector. These cables then connect to the readout electronics, which are located outside of the area encompassed by the detector. This minimizes the exposure of the electronics to radiation and also keeps the electronics from interfering with the processes that the detector is being used to study. Likewise, as Franco Bedeschi explained, the early microstrip detectors at fixed-target experiments had large "fanouts" connecting each channel of the detector with a cable leading to external electronics. But in a colliding-beam experiment, these large chunks of hardware attached directly to the silicon detector would interfere with collision products headed for other components of the detector.

Thus the SVX faced a "data acquisition" problem, which generated skepticism in the collaboration about the viability of the SVX. The Pisa group tried to solve this problem by considering "strange cables," routed in clever ways through the heart of the detector, so that the problem, while bad, would not be as bad as people expected.[4] But, recalled Bedeschi, "that was before we started hearing about chips" (Bedeschi 1998).

Solving the data acquisition problem required a novel approach, and chips were the key. Instead of trying to find a clever way to get the information out of the detector to the electronics, the chip approach brought the electronics into the detector, by building the data acquisition electronics into a chip that would sit in the detector. The difficulty was to make the electronics small enough not to cause problems for the rest of the detector, but durable enough to withstand the intense radiation in the collision area. At SLAC, a physicist named Sherwood Parker had designed preamplifier chips for microstrip detectors, and Alvin Tollestrup gave him "some twenty thousand dollars" in research funds for a set of these "Parker chips" (Tollestrup 1998). But the Parker chip was not designed specifically for CDF's needs, and Menzione was looking for other approaches. The SVX group approached Lawrence Berkeley Laboratory, already a collaborating institution in CDF, which had both expertise and facilities for custom designing chips. As Bedeschi recalled, "We made an expedition to LBL. We had a meeting, and there were people from LBL, CDF... there was this guy Parker, and people from Pisa, and we had this day-long meeting, and there was a lot of excitement" (Bedeschi 1998). CDF member Bill Carrithers, from LBL, who would later become cospokesperson for CDF, knew of an electronics whiz at LBL named Stewart Kleinfelder. According to Tollestrup, "Bill thought that [Kleinfelder] could do a better job [than Parker], and I didn't think he could" (Tollestrup 1998).

Kleinfelder's LBL chip turned out to be just what the SVX effort needed. "It was the union of that custom chip development at Berkeley and the detector development at Pisa that made the thing possible" (Amidei 1995). With LBL's involvement, the SVX not only benefited from Kleinfelder's chip, but also from the work of Carl Haber, a "physicist type," who worked on testing the chip and integrating it with the rest of the detector, "and all of the infinite details which go with electronics" (Bedeschi 1998). For its part, Fermilab had Ray Yarema, who Tollestrup esteemed as a "first rate engineer." "Kleinfelder didn't have enough experience with systems to know how to get the grounds right and get all the bugs out, and Ray is superb at that kind of stuff" (Tollestrup 1998). But Tollestrup recalled that Kleinfelder and Yarema "wouldn't talk to each other," so Haber's coordinating efforts proved crucial.

Solving such technical problems was necessary, but not sufficient, to get the SVX incorporated into the CDF detector. Technical wizardry could not eliminate the risks associated with putting a completely new device, employing novel technology, in the middle of the detector. The Pisa-LBL group was asking the collaboration to let their ingenious new detector stand between the physics being studied and every other piece of the detector. To gain acceptance of the SVX, its proponents had to convince their colleagues that information from the SVX could supplement information from the rest of the detector, without destructively interfering with the functioning of the rest of the detector.

It was not an easy thing to sell. In the early stages of the SVX effort, the benefits of a silicon vertex detector were far from obvious, even to those who wanted to build one. Franco Bedeschi acknowledged that early on "the physics case for this device was not so clear" (Bedeschi 1998). One difficulty was that silicon detectors were considered useful for doing things believed to be difficult or impossible to do well at a hadron collider.

Silicon detectors were known to be very useful for two things: measuring particle lifetimes and "tagging" or identifying particles. The NA1 group had used their silicon detector to measure the lifetimes of charmed particles. They did this using an "active target" silicon detector bombarded with energetic photons. Charm decay candidates were selected by means of a spectrometer. They then used the silicon "telescope" to identify the exact place where the particle decayed (Alibini, Amendolia, et al. 1982). Once a particle's lifetime is known well enough, the same kind of measurement supports inferences in the opposite direction. Having measured how long it took a particle to decay, it can be identified (tagged) as a particle of a certain type. Silicon vertex detectors are especially well suited to make such identifications in the case of particles such as charm- or bottom-containing particles, which live just long enough to have decay vertices measurably far (given the high resolution achievable with silicon) from where they are produced.

Some in CDF doubted that one could make such measurements in a hadron collider, however, because of the high numbers of particles being produced in the collisions. The environment was considered too "noisy" for gathering useful information about individual b quarks. "There was a small number of folks who felt that there was no measurement advantage to be gained by doing precision tracking of the kind that the vertex detector would allow. . . . People did not see at first the utility of being able to measure or tag particle lifetimes" (Amidei 1995). Such measurements were thought to be useful for investigating the decays and properties of hadrons containing b quarks. But "at the time," Bedeschi recalled, "we did not think that CDF had any chance to do b physics. We thought that . . . the hadronic interaction environment was too confused to do b physics." That was the "common perception" (Bedeschi 1998).

Furthermore, the specifications of the 1981 design report led some to expect the mass resolution of the detector to be too coarse for effective discrimination between B mesons and background. Indeed, the experience of most experimentalists in high energy physics suggested that the actual performance of the detector would be worse than indicated by its design specifications. As it turned out, however, the mass resolution of the CDF detector proved better than that specified in the design report by a factor of two, as a result not just of the SVX but also of the quality of the construction of the central tracking chamber under Rich Kadel's direction.[5]

Nonetheless, Franco Bedeschi believed that the common perception that *b* physics was beyond the reach of CDF was not the result of any searching studies of the issue. "No one . . . had really looked into that. . . . Sometimes it happens, you know, people in a certain environment are convinced that something is impossible, and they don't even look at it. It was something that people thought you have to do it at e-plus/e-minus [colliders], where things are clean and nice" (Bedeschi 1998).

Furthermore, the usefulness of the SVX for the top search was far from obvious. As of the mid-1980s, many physicists believed that the top quark would have a smaller mass than the *W* boson. In the case of such a relatively light top quark, "you have lots of other handles to discover the top . . . the cross section is sufficiently high that you don't need a vertex detector to do that." In addition, the *b* quark produced in the decay of a light top quark would be "soft" (i.e., it would have a rather low momentum). It would consequently be more susceptible to scattering off of the beam pipe, thus degrading the resolution of the SVX and reducing the detector's efficiency (Bedeschi 1998).

Nonetheless, Menzione, Bedeschi, and other advocates of the SVX did come up with some arguments for the usefulness of the SVX. But these arguments were not clinchers, even to those making them, for whom the motivation for the detector was deeper than any particular physics application.

One interesting argument was to use the SVX to eliminate backgrounds in the top search. The technique suggested was in a way the opposite of how CDF actually used the SVX for the top search. The top quark would decay to a *b* quark and a *W* boson, regardless of its mass. If the mass of the top was greater than that of the *W*, the *W* would be a real particle, while if it was less, a "virtual" *W* would result. One prominent feature of some top quark decays then would be a high-momentum electron from the decay of the *W*. If, however, the *W* was a virtual particle, the momentum spectrum of those electrons would be less than if the *W* was real. Aldo Menzione recognized that one difficulty in identifying such electrons would be to separate them from electrons coming from the decay of *b* quarks, which would be produced directly from proton-antiproton collisions fairly often. In searching for the top at low masses, then, it was important to discriminate between electrons from virtual *W* decays and electrons from *b* decays. Menzione proposed using the SVX to "look at the impact parameter of the electron, and verify that it was coming from the origin, and not from the displaced vertex which would be typical of a *b*. So [Menzione] was going to anti-*b*-tag the event in this way" (Amidei 1995). In other words, instead of using the SVX to find the *b* quarks produced in top decays – the technique that proved crucial later on – Menzione was proposing to use the SVX to find *b* quarks in background events, and subtract those events from the sample.

Such arguments were not the prime mover in getting the SVX built, however. "For the top it is true that we were not sure it was going to help," acknowledged Franco Bedeschi. Dan Amidei called the anti-*b*-tagging proposal "a public relations argument that Aldo made up for building the SVX," which Amidei accepted "as one accepts the sort of, you know, catalogue of public relations arguments that you make when you give a proposal" (Amidei 1998). Pisa's Giorgio Belletini acknowledged that such arguments for the physics advantages of the SVX responded to the "friendly criticism" of skeptical colleagues with a rationale for the detector which, "even if it turned out to be totally false, at least it was not a priori false" (Bellettini 1998).

The Pisa group, and especially Menzione, were what Amidei called "detector guys," and Franco Bedeschi's comment about the "logic" of detector development could well be taken as a slogan for them. They desired to try out new detector technologies in new contexts to see if they could make them work. According to Amidei, "the reason Aldo wanted to do it was that it could be done. A silicon vertex detector at a hadron machine. It hadn't been done before and there was a space waiting to put it into" (Amidei 1998). And Bedeschi observed that "since these microstrip detectors had worked on fixed target experiments, for the people who were working on those to try to go the next step, which was harder from an experimental point of view, and to try to build such detectors for a colliding beam type of environment" was the most natural thing to do (Bedeschi 1998).

Such comments can mislead, however, by giving the impression that the motivation for building the SVX was little more than the kind of blind ambition that causes people to pursue "world's records" without regard to the importance of the task pursued. But the effort to built the first hadron collider silicon vertex detector was not like trying to be the first to consume 10 saltines in 30 seconds without drinking anything. Menzione viewed the effort in terms broadly historical and epistemological. He saw a historical basis for expecting the SVX to yield beneficial epistemological results, even if the exact nature of those benefits could not be pinned down in advance. Bedeschi credited Menzione with the "vision" of being able to look past the difficulties that related to specific applications of the new detector. "He carried on pushing, saying, 'look, it's true, now we have some difficulties to find a very strong physics case for this device. However, in science every time you introduce a qualitatively different new tool to look at nature, you always have gained substantially'" (Bedeschi 1998).

This does not yet explain why the SVX effort succeeded. No doubt the "public relations" arguments helped, but two other factors were crucial. One factor was sociological, the other factor concerned what we might call "detector philosophy."

The sociological factor was the effect on the group of the persistence of Aldo Menzione and his Pisa colleagues. "Aldo kept coming back," recalled Henry Frisch, "and eventually we slightly begrudgingly said okay" (Frisch 1995).

The second factor, often cited by CDF members, may have been more important. The SVX detector yielded the kind of information that CDF physicists cared about most: tracking information. As Amidei put it, "Being able to have the information down at the level of tracks, we all know at CDF, is an extremely powerful kind of way of getting at the world. . . . I think there was a deep appreciation in . . . the leaders at CDF that more tracking information is good" (Amidei 1998). Thus the Pisa group sought to install precisely the kind of new detector technology that could be appreciated by their colleagues. It fit well with CDF's emphasis on identification and tracking of individual particles.

A contrast can be drawn with the D-zero collaboration, which had a quite different detector philosophy. D-zero emphasized calorimetry. They sought to measure the energy of jets and the shape of particle events with great accuracy and efficiency. Thus, their calorimeter was much more finely segmented and better calibrated than CDF's calorimeter. However, the D-zero detector, which was built on a smaller budget than CDF's detector, had a smaller central tracking chamber, which functioned, according to D-zero member Mark Strovink, "more as a support" to the rest of the detector than as the fundamentally important instrument that the central tracker was in CDF (Grannis, Klima, et al. 1998). Unlike CDF, D-zero built their detector without a magnetic field, which allows momentum measurement and particle identification based on the curvature in a particle's track.

If the Pisa group had joined D-zero instead of CDF, it is unlikely that they would have succeeded in building their detector. They might not even have bothered proposing one, in the absence of a magnetic field. As D-zero member Rich Partridge observed, in the absence of a magnetic field, "the silicon vertex detector would do us very little good, because the low momentum [non-*b*-quark] tracks would scatter in the material and look like *b*s, and you would not get much worthwhile out of it. You need the magnetic field to identify the high momentum tracks." So a detector that yields useful information in one context might not function well in another. Yet underlying this difference of approach, the groups shared an interest in the same physics. Partridge added that "[i]n '84 [when D-zero was proposed], if you had put it to us, 'do you want to identify *b*s?' we would have said, 'of *course* we want to identify *b*s!' but the technology wasn't there to realistically propose something that [also] provided the good calorimetry that complements what CDF was doing" (Grannis, Klima, et al. 1998). They complement one another because, while D-zero lacked the tracking capacity of CDF, CDF's calorimetry was not nearly as good as D-zero's.

Lest one think of these two groups as working within incommensurable paradigms (Kuhn 1970) or speaking in different languages (Galison 1997b), let me add that these "detector philosophies" do not constitute binding worldviews in either group. As Partridge's comments indicate, members of each group are able readily to acknowledge and understand the relative advantages and disadvantages of the other collaboration's approach. It is precisely for this reason that, subsequent to the events covered here, each group "upgraded" their detector in ways that substantially incorporated strengths of the other detector. CDF upgraded their forward and backward calorimeters. D-zero put in a compact magnetic field and a silicon vertex detector. This merging of distinct detector philosophies is very much like the merger of the "image" and "logic" traditions chronicled by Peter Galison (1997b).[6] In both cases, physicists working in one tradition, recognizing what they could accomplish if they incorporated some of the tricks and technologies of their competitors, built something that went beyond what either group had built before.

By the time of the 1987 run, as a result of argument, persistence, and the right kind of idea for their context, the SVX advocates had managed to get their efforts taken seriously. An "algorithm group" called the Secondary Vertex Group formed. In February 1987, Franco Bedeschi reported on their progress at a CDF meeting, noting that the group "could use more people" (*CDF* 2/6/87). A month later, Aldo Menzione requested "a full time physicist residing in the U.S." for the SVX project (*CDF* 3/13/87). After the end of the run, the collaboration hired Dan Amidei, who was just finishing a postdoc with the University of Chicago and had been working on trigger electronics, to fill this position.

3. CDF COMES OF AGE: THE 1988–1989 RUN

As of February 1987, CDF expected to begin their next collider run in November of the same year (*CDF* 2/13/87), but in the middle of that summer the lab administration extended the fixed target run to December 15. This delayed the beginning of the collider run until the following February (*CDF* 7/24/87). Later, the lab again postponed the end of the fixed target run to January 27 (*CDF* 10/23/87) and then added another two weeks as that date approached. Lab director Leon Lederman himself came to the CDF meeting to justify that final postponement of CDF's data-taking efforts (*CDF* 1/22/88).

CDF used this period to fix problems and attend to the few elements left incomplete during the 1987 run. Software development and data processing were also important priorities during this period.

In anticipation of the coming flow of data, the collaboration instituted "physics analysis groups" (usually just called "physics groups") in addition to the existing algorithm groups (*CDF* 11/6/87). The algorithm groups lost importance over the next couple of years and finally were replaced entirely by the physics groups. The minimum bias, electroweak, QCD, and heavy flavors groups were the first physics groups. The top search was primarily the concern of the heavy flavors group, which also was responsible for *b* physics. The coconveners of the heavy flavors group were Mel Shochet from the University of Chicago and Brig Williams of the University of Pennsylvania.

Prospects for the upcoming collider run were haunted by financial problems, spawned (via politics) by the huge federal deficits of the late 1980s. CDF's goal for the run was to accumulate 1000 nb^{-1} (1 pb^{-1}) of data. However, reporting in October 1987 on the lab schedule – when the collider run was planned to begin the following April and continue until the end of the year – Roy Schwitters noted that "at present Lab funds are not sufficient for this; unless more money is allocated, there may be a shutdown period during the run" (*CDF* 10/23/87). The federal government had been running since October 1 on a series of emergency spending bills passed by Congress in the absence of a budget agreement with the Reagan administration, and a crisis in federal funding was looming. The Gramm-Rudman-Hollings deficit reduction act called for cuts of 8.5% in all nonmilitary federal spending (10.5% in military spending) on November 20 if a budget agreement was not reached. At the end of October, Alvin Tollestrup described the effects of such cuts as "unknown" but tried to avert collaboration-wide discouragement by noting that "The Lab does give a high priority to a large integrated luminosity CDF run" (*CDF* 10/30/87).

By mid-November, the failure of Congress and the Reagan administration to reach an agreement had forced an interruption of the fixed-target run, although the Lab remained open on the reduced funds made available by a "continuing resolution" passed in Congress (*CDF* 11/13/87). On November 20, Congress and the President reached an agreement on the budget, which was enacted over the next several weeks, and at the beginning of the new year, the Fermilab administration reassured CDF that "there will be sufficient funds to support a long collider run" (*CDF* 1/8/88).

However, the lab continued to delay the startup of the collider run. In order to allow the repair of some of the superconducting Tevatron magnets, the administration again revised the schedule. They would start up the Main Ring on May 25 and begin antiproton transfers to the Tevatron on June 6 (*CDF* 4/8/88).

On the eve of the new collider run, *CDF* suffered a near-disaster. On the afternoon of June 2, with the detector in the collision hall and the door closed, there was a fire on the detector. "Although fire damage was localized, dense smoke did fill the collision hall and escaped to the counting rooms via cable penetrations" (*CDF* 6/3/88). Subsequent investigation traced the source of the fire to an improperly tightened power lead connecting to a magnet coil on the front muon spectrometer (*CDF* 6/10/88).

The fire did not disrupt the schedule. A week later the accelerator was in "operations mode" (*CDF* 6/10/88), and by the end of the month, the lab had achieved storage of a stable colliding beam lasting 12 hours (*CDF* 7/1/88). Once stabilized, the accelerator began to perform well beyond expectations. Recall that the collaboration's goal for the run was to accumulate 1000 nb^{-1}, a 30-fold increase over the 1987 run. During a single week ending August 26, 1988, the accelerator delivered 75 nb^{-1} (*CDF* 8/26/88), and the total for the run as of one week later was 375 nb^{-1} (*CDF* 9/2/88). As before, however, the data recorded by the collaboration with their detector did not equal the luminosity delivered by the accelerator. For the success of the accelerator to translate into large accumulations of data, CDF's three-level trigger needed to be finalized. Only level one and part of level two had been complete for the previous run. A flurry of trigger work produced an operational level-three trigger by mid-September, and serious data taking began. CDF had 300 nb^{-1} on tape by the end of the month (*CDF* 9/30/88). They passed the 1000 nb^{-1} goal that had been set for the run on November 22, with the end theoretically six months away (*CDF* 12/2/88). CDF was finally using their detector as the data-producing machine it was intended to be.

Meanwhile collaboration members worked on changes they wanted to make to the detector after the run was over. In addition to work on the SVX, improvements to the muon detection system were in the works.

Physicists from Japan and the University of Illinois, where the original muon chambers had been built, were joined by a group from Harvard in proposing the muon upgrade. The muon chambers are the part of the detector farthest from the interaction region and provide the least angular coverage. The muon upgrade was designed to increase the area over which these important collision products could be detected. Presenting the upgrade proposal in September 1988, Tony Liss estimated the upgrade to cost $8 million (*CDF* 9/23/88).

This posed a problem. CDF already planned a number of upgrades in preparation for the next run, which they expected to produce luminosities far in excess of even the surprising success of the current run. Such high luminosities would exceed the capabilities of certain components of the CDF detector. CDF badly needed to upgrade their data acquisition system. The vertex time projection chamber would need to be replaced. The SVX remained to be built. Yet only $2 million was budgeted for CDF for the following year. Nonetheless, CDF presented the muon upgrade proposal to the Physics Advisory Committee, which questioned the feasibility of upgrading the muon detection system while carrying out the upgrades necessary to accommodate the increased luminosity of the next run. CDF replied that financial contributions from the Italians and Japanese would make all this possible (*CDF* 10/21/88).

Midway through the 1988–9 run, CDF requested an extension of the collider run. As of January 27, 1989, they had accumulated 2.1 pb^{-1} on tape. They wanted to run until they had a full 10 pb^{-1} (*CDF* 1/27/89). This gave the PAC something besides the muon upgrade to reject. CDF already had more than twice their target of 1 pb^{-1}, and they could expect a substantial chunk of additional data, even on the existing schedule. In February, Mel Shochet reported to the collaboration that the PAC had given initial approval to the muon upgrade but had decided to end the collider run on June 1, "due to power costs for running during the summer, among other reasons" (*CDF* 2/3/89).

Mel Shochet had by this time become a cospokesperson, replacing Roy Schwitters, who had left to become director of the ill-fated Superconducting Super Collider. Lederman had chosen Schwitters, and Shochet was chosen to replace him by the collaboration's advisory board, which afterward became the "executive board." The executive board then opened up the spokesperson positions to a kind of democratic election procedure every 2 years. According to Shochet, "We wanted the collaboration to decide who was going to lead the collaboration. . . . An awful lot of the hard work and bright ideas come from this cadre of young people, the postdocs, the senior graduate students, and they have a right to express their view of who they think ought to lead the collaboration" (Shochet 1995).

The procedure for choosing a new spokesperson usually took about three months. It began with open nominations. The executive board would appoint an election board, usually consisting of three people. After nominations were complete, a list of the nominees would go out to each institution. Each institution then held a separate meeting of its CDF physicists to establish an ordered ranking of these nominees. The election board chose from the list a ballot listing five to eight people, who were those most often supported by the institutions in their ranking of the nominees. This ballot would then be shown to the director of Fermilab, in order, according to a later spokesperson, "to make sure that there's nobody that he hates." The director's response was typically "I hate nobody" (Bellettini 1995). The ballot was distributed to each member of the collaboration. Each person would rank the candidates in order of preference, and the winner was the person most often preferred (Shochet 1995). Final tallies were kept confidential (Bellettini 1995). Spokespersons served 2-year terms, although Alvin Tollestrup's position only went up for election when Tollestrup himself chose to step down, at the end of 1992.

The institution of this electoral process completed the organizational structure of CDF. A brief description of that structure is in order. CDF physicists collaborated within the physics working groups to develop means of analyzing the data. Officially, any member could join any physics group, although typically a number of physicists from a single institution would join a working group together, to collaborate on a specific task. It was not unusual for a senior collaboration member to ask a young physicist to tackle a specific problem. It *was* unusual for the young physicist to refuse such a request.

Each of the working groups had two coconveners responsible for organizing the meetings of the group. Conveners of the physics groups served 2-year terms and were chosen by the spokespersons from a list of nominees solicited from the collaboration. The decisions of the spokespersons had to be presented for ratification to the executive board of the collaboration. The executive board consisted of two people from each of the collaborating institutions, plus the two spokespersons. Each institution on the executive board had one vote, except for Fermilab, which had two votes.

4. THE TOP SEARCH BEGINS: THE (VERY SHORT) RACE WITH UA2

Throughout the 1988–9 run, CDF was in a kind of race with the UA2 collaboration at CERN, which was also looking for the top. CERN's collider, the Super Proton-Antiproton Synchrotron, produced proton-antiproton collisions at a lower energy than the Tevatron collider, putting CERN at a disadvantage in looking for a potentially very massive top quark. Before the 1988–9 run ended, the race was over, not because CDF had crossed the finish line, but because UA2 was no longer in the race.

The way in which the top quark is produced depends on its mass. If the top had a mass less than that of the W boson, about 80 GeV/c^2, then it would be produced when the W decays to a t and a b, or it could be produced in proton-antiproton collisions by the fusion of gluons with gluons or the annihilation of very energetic quark-antiquark pairs. If it had a mass above that of the W boson, then it could not be produced by the decay of a W, but only by gluon-gluon fusion or quark-antiquark annihilation. Both UA1 and UA2 had been producing Ws for some time and, therefore, could be expected to produce top quarks from the decays of Ws, given a sufficiently low-mass top. However, CERN's nonsuperconducting Sp$\bar{p}$S, with its 630-GeV collisions, would produce massive particles such as the W at a lower rate than the 1.8-TeV Tevatron.

CDF won this "race" early in 1989, not by finding the top, but by showing that UA2 could not find it. The more massive the top quark is, the less often it is produced. This means that one can infer a lower limit on the mass of the top quark by failing to find it in a data set of a certain size. CDF quickly set a top mass limit that ruled out its discovery by UA2. To explain how they did this requires some preliminary understanding of how the search for the top quark relied on the top's decay products.

According to the standard model, if the top had a low enough mass to be produced by the decay of the W to a t and a $\bar{t}$, then the top would in turn decay to a b, a lepton, and a neutrino. Thus one would see in the final state a lepton (electron or muon), two jets (from the b and the $\bar{b}$), and an energy imbalance, or "missing transverse energy," from the undetected neutrino.

When top-antitop pairs are produced by gluon-gluon fusion or quark-antiquark annihilation, on the other hand, the top decays almost exclusively into a W^+ boson and a b quark, while the antitop decays to a W^- and a $\bar{b}$ (where the W is virtual if the top mass is less than that of the W). The b quarks then hadronize and the b-carrying hadrons quickly decay, producing jets. The two Ws can decay in several ways. For about 44% of $t\bar{t}$ decays, both will decay to quark-antiquark pairs. However, this final state looks so much like that of other processes that it is very difficult to find the top by looking in this decay mode.

Instead, CDF looked for the top by searching for final states produced by decay in the "lepton + jets" mode and the "dilepton" mode. In a lepton + jets decay (see Figure 3.3), one of the Ws produced from the $t\bar{t}$ pair decays into a quark-antiquark pair, resulting in jets, and the other decays

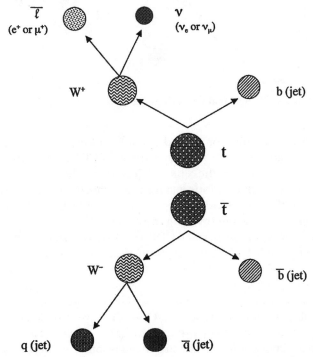

FIGURE 3.3. Schematic drawing of a lepton + jets decay of a $t\bar{t}$ pair. Alternatively, the W^+ may decay to a $q\bar{q}$ pair, and the W^- will decay to a ℓ (e^- or μ^-) and a $\bar{\nu}$ ($\bar{\nu}_e$ or $\bar{\nu}_\mu$).

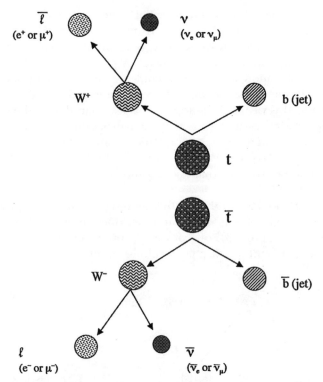

FIGURE 3.4. Schematic drawing of a dilepton decay of a $t\bar{t}$ pair.

leptonically to a lepton (e or μ) and a neutrino (decay to the elusive τ lepton, while possible, was not included in any of CDF's top search strategies until years later). Thus, in the lepton + jets mode, the final state of the top decay consists of a lepton, four jets from hadronized quarks (two from bs, and two from the quark-antiquark pair produced by the decay of one W), and missing transverse energy from the neutrino.

The other decay mode in which CDF searched for the top is the "dilepton" mode (see Figure 3.4). In this case, both Ws decay leptonically (to a lepton and its associated neutrino), and one finds in the final state two jets from the hadronized b quarks, two electrons or muons, and significant missing energy from two undetected neutrinos. The dilepton channel comprises three possible lepton combinations. One can have two electrons (e-e), two muons (μ-μ), or an electron and a muon (e-μ). Because electron-positron and muon-antimuon pairs can be produced by other processes, particularly decay of the neutral Z boson, this last (e-μ) possibility is the cleanest of all. There is little chance that this final state, or "signature," can be produced by other processes besides top decay. Likewise, the dilepton final states of top decays are more easily distinguished from other processes than lepton

+ jets final states. On the other hand, top decays into any of the dilepton final states should be rarer than into lepton + jets final states.

In February of 1989, while CDF was still collecting data, G. P. Yeh presented CDF's first results on the top search at the Second International Symposium on the Fourth Family of Quarks and Leptons. Yeh had joined CDF in 1985, after John Yoh arranged an interview for him with Alvin Tollestrup. Yeh recalled that Tollestrup had been reluctant to hire him at first "because he liked hardware persons," and Yeh had indicated that he was interested primarily in analysis and software (Yeh, personal communication). Yeh apparently won Tollestrup over. After he worked on the event display and calorimeter noise problems during the 1987 run, Tollestrup asked him in September of 1988 to join the top group. Yeh and a Chilean graduate student from Brandeis named Milciades Contreras developed the dilepton e-μ analysis between September and November of 1988. (This became the subject of Contreras's dissertation.) At the same time, Yeh, Contreras, and John Yoh developed an analysis for the lepton + jets channel that reconstructed the mass of events with an energetic lepton and two or more jets (Contreras 1995; Yeh 1995). This analysis could be used to search for a top quark with a mass less than that of the W by comparing the reconstructed mass of candidate events to what one expects from the W. If the top quark existed with a mass less than that of the W, then a peak in the reconstructed mass of events would appear at a mass less than that of the W.

Both Yeh and a representative of UA2 presented top results using such a "transverse mass" analysis in the lepton + jets channel. Yeh commented that after he gave his presentation "the top race was over" (Yeh 1995). He presented a comparison of the data to expectations for top quarks with masses in the range from 30 to 70 GeV/c^2 in the e-μ channel and in the e + jets channel using the reconstructed mass. Yeh concluded by noting, "On the basis of 2 pb^{-1} of data, we expect to see $e\mu$ events from $t\bar{t}$ decays, for top quark mass in the 30 to 60 GeV region. We observe zero candidates. Similarly, in the e + jets analysis, we also have not observed any signal from the top quark" (Yeh 1989). From their analysis of 2.5 pb^{-1} of data using a transverse mass reconstruction in the e + jets channel alone, UA2 likewise concluded that "there is no indication of top production for masses in the range $30 \leq m_{\text{top}} \leq 60$ GeV" (Parker 1989). These results ended the race with UA2. The mass of the top quark was so great that the lower-energy Sp$\bar{\text{p}}$S collider could not produce enough top quarks for UA2 to be able to distinguish them from background.

CDF stopped taking data at the end of May 1989, having accumulated an estimated 4.69 pb^{-1} of data on tape, out of 9.05 pb^{-1} delivered by the accelerator.[7] With a complete data set, CDF physicists could pursue more systematic top searches and publish specific mass limits. The first limit of 72 GeV/c^2 came from the dilepton search in the e-μ channel. Milciades Contreras read the paper presenting this analysis at the weekly CDF meeting on October 6, and the paper was published in *PRL* (Abe, Amidei, et al. 1990b). Meanwhile, Pekka Sinervo from University of Pennsylvania led a group that included Richard Hughes and Brig Williams in taking up the transverse mass approach. Combining the Pennsylvania group's results with the dilepton analysis, CDF published a top mass limit of 77 GeV/c^2 (Abe, Amidei, et al. 1990a; 1991).

Subsequent papers pushed the mass higher by including additional search modes. CDF's final limit based on data from the 1988–9 run was 91 GeV/c^2. They reached this limit by including all three channels in the dilepton final state: e-e, μ-μ, and e-μ. The dilepton search alone yielded a mass limit of 85 GeV/c^2. Also included for this limit, however, was CDF's first use of b-tagging in the search for top in the lepton + jets final state. Including this additional channel raised the limit to 91 GeV/c^2; this result appeared in both *PRL* and *PRD* (Abe, Amidei, et al. 1992a; 1992b).

The 91-GeV/c^2 mass limit was partly based on a method of b-tagging that looked for one of the expected products of the decay of the b quark, an isolated "soft" lepton, that is, a lepton with low *transverse momentum* (the component of the particle's momentum that is orthogonal to the beam line). For this search, events were selected with one energetic electron or muon, large missing energy, and at least two energetic jets. From among such events, those with an additional soft muon were tagged. This approach was developed initially by Claudio Campagnari, then with University of Chicago, and Paul Tipton, who was working for Berkeley at the time (Tipton 1995). This technique would be carried over into the analysis of the next run's data in the form of the "soft-lepton tagging," or SLT, analysis. Paul Tipton later received a Robert Wilson Fellowship to stay at Fermilab and take over coordination of the SVX construction effort when Dan Amidei was hired by University of Michigan. Claudio Campagnari, later hired by Fermilab, continued to work on the SLT analysis, and became a coconvener of the top group in 1992, just before run Ia started.

Just when the next run would begin remained uncertain late in 1989. A long fixed target run was scheduled, which the fixed-target experiments wanted to last as long as possible. CDF had requested that the collider run begin in March of 1991. But funding for the lab was also uncertain, and no schedule could be set without knowing how much money the lab would have. Another set of Gramm-Rudman cuts held uncertain consequences for both the schedule and the various CDF upgrade projects (*CDF* 10/27/89). At the end of 1989, the lab planned to have both CDF and D-zero in place and recording collisions by June of 1991 (*CDF* 1/5/90). The accelerator was shut down in late August of 1990 to allow CDF and D-zero time to work in the collision areas. The accelerator remained off until early December, when the lab restarted the collider for a series of performance studies. The lab planned a brief fixed-target run from March to August of 1991, with a brief pause to follow prior to the next collider run. The usual pattern of schedule changes would continue, however, and the actual delay in CDF's data collection would be considerably longer.

Budget uncertainties persisted because of budget conflicts in Washington and the looming Gramm-Rudman Act. The minutes to a CDF meeting in late September 1990 noted that "if Fermilab does not receive spending authorization, i.e., a budget or continuing resolution is not passed by Congress, Fermilab will close on October 3." In the case of a budget or continuing resolution being passed with Gramm-Rudman cuts applied, the lab would close for four days later in October. "The schedule will slip until the budget has been passed" (*CDF* 9/28/90).

The schedule slipped further when inadequate radiation shielding at B-zero and other Fermilab locations forced a delay of the beginning of the fixed-target run. That the berm surrounding the beam at B-zero turned out to be 2 feet shallower than anyone had thought contributed to the problems (*CDF* 3/29/91). The lab repaired this deficit by adding more dirt and a concrete wall to shield the B-zero assembly building. While a goal of 20 pb^{-1} of data on tape for the next run was set in May of 1991 (*CDF* 5/24/91), a sense of despondency brought on by the long collision-less period set in, seeping even into the CDF meeting minutes: "The completion of a second year totally bereft of luminosity was marked with the consumption of several cakes" (*CDF* 5/31/91).

Meanwhile the upgrades progressed. Here is a brief summary of these changes to the CDF detector: (1) The VTPC, a drift chamber with resolution of about 1 mm, was removed and replaced with the combination of the vertex drift chamber, a modified version of the VTPC adapted for higher luminosity, and the silicon vertex detector. (2) The central muon upgrade was installed, consisting of a 0.6-m layer of steel positioned outside of the existing muon chambers, with another four layers of muon chambers just outside of the steel. The steel served to prevent "punchthrough" of energetic hadrons into the muon chambers, so that tracks in the muon chambers would really be muons. (3) The central muon extension was added, a set of four free-standing conical arches containing drift chambers to provide muon detection in the region $0.6 < |\eta| < 1.0$. (4) A set of proportional chambers known as the "central preradiator" was inserted between the solenoid and the electromagnetic calorimeter to provide information on the early development of electromagnetic showers. (5) The data acquisition system was upgraded to accommodate the higher luminosity of the upcoming run.

The period from May 1989 to June 1992 represented a long dry spell without data for CDF. It may have been just what their neighbors at D-zero needed. The D-zero collaboration had gotten a much later start, and although by

1989 they had built much of their hardware, D-zero needed to work hard and fast to develop physics software and to study their detector's operation. D-zero also wanted to find the top quark, and just like CDF, they preferred to find it first. They would use a quite different detector, however. The D-zero detector began as a calorimeter-based detector without a central magnetic field (the design choice rejected by CDF), and without a silicon vertex detector. D-zero would also be taking data for the first time in the next run. Both collaborations would be kept very busy.

5. RUN IA AND THE DPF EVENT

In June of 1992, CDF's long wait ended, and the run designated "Ia" began. As of June 5, the detector was producing data (*CDF* 6/5/92), and on June 12 a status report stated that "97% of channels are working, detectors are meeting performance specs, and approaching final calibration" (*CDF* 6/12/92). Not long after that report, however, the collider had to be shut down so that CDF could repair the VTX, which, due to broken wires and cabling problems, had 8 bad modules out of a total of 28 (*CDF* 6/26/92). While this interrupted data taking for a couple of weeks, it was apparent that on the whole the detector and the accelerator were functioning well. Still, problems arose occasionally. For example, the lab terminated collisions when it appeared that the SVX was being exposed to too much radiation (*CDF* 10/9/92). Active data taking occupied 80% of one late August weekend, with 90% efficiency during some individual shifts (*CDF* 8/28/92). In early December, CDF had collected nearly 4 pb^{-1} of data on tape (*CDF* 12/4/92), but another brief shutdown came in January, and there was talk of using the opportunity to fix the two remaining problem modules in the VTX, a proposal that met with "little enthusiasm" from anyone except those responsible for the VTX itself, as it would require pulling out the plug calorimeters and disturbing the SVX just when its performance was beginning to meet expectations (*CDF* 1/15/93).

The collider reached a record luminosity of 7.48×10^{30} cm^{-2}s^{-1} in March (*CDF* 3/12/93), and when the run was terminated at the end of May, CDF had about 21 pb^{-1} on tape out of 28.9 pb^{-1} delivered by the collider (*CDF* 5/29/93). Was the top somewhere in those 21 pb^{-1}?

Answering that question required knowing how to look for the top, and many different ideas circulated in the collaboration about how to conduct that search. Advocates of different approaches engaged in a lengthy competitive process of developing, refining, and advertising their favored analyses. But a sense of urgency had set in late in 1992, largely as a result of a single event.

On October 29, 1992, just prior to the meeting of the Division of Particles and Fields (DPF) of the American Physical Society, CDF recorded an event

Run 41540, Event 127085

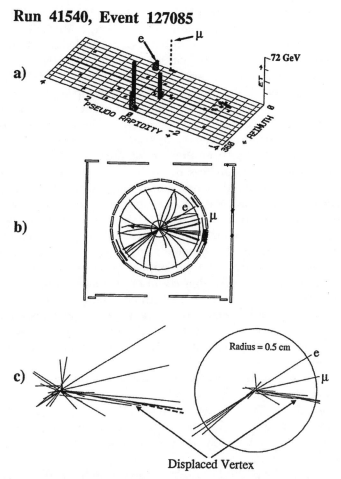

FIGURE 3.5. Three views of the "DPF" event: (a) shows a Lego plot, representing E_T as measured by the calorimeter, in the η-ϕ plane; (b) displays hits in the muon chambers and the central muon upgrade, as well as tracks in the central tracking chamber, in the r-ϕ plane; (c) gives an r-ϕ view of the SVX tracks, with a detail of the jet containing a displaced vertex; track length is proportional to p_T (Abe, Amidei, et al. 1994b, 2978).

that looked strikingly like an e-μ top decay (see Figure 3.5). This event made an impression on people, according to some, simply on the grounds that it "really looked like top" (Jensen 1995). Because the e-μ decay channel had very little background, it was improbable for this event to be something other than a top-quark event. In fact, although it was not known at the time, one of the jets in this event would be tagged by both the SVX and SLT b-tagging algorithms, which are described in Section 6.2 of this chapter. "It looked in many ways like the ideal top candidate" (Liss 1995). Milciades Contreras recalled that some collaboration members even regarded the event by itself as worthy of publication as a top-quark event (Contreras 1995).

The collaboration, however, was not prepared to go so far. The event was not quite perfect. If the event were a genuine top-antitop decay, then one should be able to reconstruct the two sides of the decay process as the decay of two equally massive particles, the top and its antiparticle. To carry out such a reconstruction of this event, "you have to push the jet energies a little bit" (Jensen 1995). Given the uncertainties on the measurement of jet energies using the calorimeters, this is possible, but, in the lingo of high energy physics, this made the event somewhat less than "golden." In any case, many CDF physicists were skeptical about the possibility of a "golden event" (i.e., one event that all by itself establishes the existence of some new phenomenon). The collaboration was simply not prepared to base a discovery as important as the top quark on a single event, no matter how unambiguous that single event might appear. Working in previously unexplored experimental territory, they could not achieve enough confidence that they had ruled out all other possible interpretations of a single event to make a definitive pronouncement that it was indeed a top-quark event. (As I will describe in Chapter 4, Section 11, this debate over discoveries based on single events would be recapitulated at D-zero, where the same conclusion would be reached.)

This nearly golden event was discussed at a collaboration meeting to bless presentations for the meeting of the Division of Particles and Fields, and the collaboration decided that the event, which subsequently came to be known as "the DPF event," would be shown at the meeting, but only "qualitatively discussed in a dispassionate way" (*CDF* 11/6/92). No claim would be made as to what the event signified.

Although the DPF event did not by itself constitute the discovery of the top quark in the minds of CDF physicists, it stimulated the ambitions of CDF members eager to play a role in the discovery of the top. It "set off a whole flurry of activity" (Liss 1995). The event came early in the run, leaving open the possibility that a significant number of additional candidate events might turn up during the remainder of the run. Also, the dilepton decays were expected to be rarer than decays in other modes, leading some to think, in the words of Fermilab's Drasko Jovanovic, "if we have one e-μ event, we should have a dozen or so other events" in different top decay channels (Jovanovic 1995). For those working on methods of searching for the top quark in various channels, the top events possibly already lying in the data and those that might join them over the next months were a powerful incentive to push their analyses forward. To describe the efforts of these different groups, we need to back up a little, to the efforts to find the top in the 1988–9 (run 0) data.

6. TOP SEARCH STRATEGIES

Efforts to find top decays at CDF fell roughly into two categories: counting experiments and "kinematic" searches. The counting experiments looked for the top quark by defining a candidate event in terms of a set of requirements, or "cuts," that an event must satisfy. These cuts were meant to select for certain features of events that are indicative of top decays. For any given set of cuts, a certain number of non–top quark events could be expected to meet those requirements, and so the search for the top would look for an excess number of candidate events beyond this expected background. The evaluation of any such excess would turn on the assessment of how probable such an excess would be on the assumption that only background processes were being detected. That is to say, counting experiments constitute a kind of *significance test*. Discussion of the evidential strength of a significance test often turns on the quantitative assessment of the *significance level* of the results. This quantity represents, roughly, the probability of getting a result such as one has found in one's data, assuming that the phenomenon one is testing for does not occur.[8] A smaller significance level indicates a lower probability and (other things being equal) stronger evidence in support of the phenomenon in question. (The notion of statistical significance plays a central role in this episode and in the philosophical discussion to come.) The dilepton and soft lepton searches that served as the basis of the mass limits published on the 1988–9 data are examples of counting experiments, the assessment of which involved just such significance calculations.

The kinematic searches took a different approach. In a kinematic search, one took events selected using cuts on the data as one might for a counting experiment, but then tried to squeeze additional information out of those events. The basic idea was to select individual candidate events and, by attempting to reconstruct them as top-quark events based on the momenta and energies of their components, to compare the *likelihood* of the hypothesis that those events were top quarks to the likelihood of their being background events. The likelihood of a hypothesis on given data is defined to be the probability of those data on the assumption of that hypothesis.[9] Likelihood methods are a means of choosing amongst competing hypotheses on given data, by comparing the relative probability of the data on the various hypotheses. Typically, the technique selects the hypothesis that maximizes the relative likelihoods (i.e., the hypothesis under which the probability of the data is maximized relative to other hypotheses under consideration). In the kinematic searches at CDF, such comparisons of likelihoods required the use of Monte Carlo computer models of both the background and top decays at various hypothetical top masses.

Although a philosophical debate has arisen over the soundness of significance tests as opposed to likelihood analyses for the purposes of scientific inference, I will postpone discussion of these issues until the last two chapters of this book. Here I will simply describe the different approaches to the top search as they developed within CDF. Disputes within CDF over which top search strategy to adopt seem not to have had any direct connection with the philosophical disagreements in question.[10]

6.1. Likelihood ("Kinematical") Analyses

Three main "kinematic" top searches can be identified within CDF. One group working on a kinematic analysis was led by physicists from the University of Pisa, particularly Hans Grassmann and Marina Cobal, another was the work of Kuni Kondo, from Tsukuba University, and the third was a joint effort of Krzysztof Sliwa, of Tufts University, with two theorists from outside the collaboration: Gary Goldstein, a colleague of Sliwa's at Tufts, and Richard Dalitz, a well-known Oxford theorist. All three kinematic analyses began as attempts to understand the 1988–9 data. As with all of the top analysis efforts, physicists pursuing these avenues periodically presented their results at the meetings of the top group, where the strengths and weaknesses of particular methods were discussed, and where physicists tried to persuade others of the advantages of their own way of doing things.

In any attempt to use likelihoods to discriminate between hypotheses, a crucial problem is to choose a sensible feature of the data for differentiating those hypotheses. CDF was blessed with extraordinarily rich data for each event. Likelihood methods required choosing some feature of those data that would effectively and reliably discriminate between top signal events and background. A physicist attempting to develop such an approach had to choose a feature for evaluating likelihoods that was known to vary between background events and signal events in some predictable way. The various kinematic methods proposed in CDF can be distinguished in part by the different kinds of information that they used.

The Pisa Event Structure Analysis. In a top group meeting in 1990, Pisa's Hans Grassmann described a likelihood approach to finding the top that he had been working on with others, including Marina Cobal. The approach advocated by Grassmann used features of the "event structure" of lepton + jets events that would tend to be quite different for top decays compared to background W + jets events (hence this came to be known as the "event structure analysis"). Specifically, the authors of this approach were looking at the quantity $\cos \theta^*$, where θ^* is defined as the angle between the path of an outgoing particle product and the incoming proton beam in the rest frame of the center of mass for the event. $\cos \theta^*$ is lower (θ^* is greater) for top events than for background W + jets events, when one looks at angles for the "leading" (most energetic) jets.

Using this information, one can look at the likelihood, for a given event's combination of event structure measurements, of either the assumption that it is a W + jets background event or that it is a lepton + jets top decay event. That is, one can use the probability distribution for each individual feature of the event (the energy and angle of the most energetic jet, the energy and angle of the second most energetic jet, etc.) to determine how probable each particular feature is under either hypothesis, and then multiply those probabilities to arrive at the joint probabilities of the event as a whole. In other words, one is interested in comparing the probability of getting an event with the particular event structure obtained, on the assumption that it was selected from background, with the probability of getting the same result on the assumption that it was selected from a population consisting of a combination of top decay events and background. Grassmann and his colleagues proposed using the log of the ratio of these two likelihoods to describe individual events as "more top-like" or "less top-like."

Cobal, Grassmann, and S. Leone, another Pisa physicist, published the idea of using $\cos \theta^*$ to take advantage of the differing event structure for top and background events in a paper submitted to the Italian Journal *Il Nuovo Cimento* in June 1993 (Cobal, Grassmann, and Leone 1994). In that paper, however, they proposed to use $\cos \theta^*$ as a quantity on which to set cuts, for the purpose of a counting experiment. With or without the likelihood function, however, this approach relied heavily on the use of Monte Carlo computer models for the distribution of the event structure parameters for both the top signal and the W + jets background. Such Monte Carlo programs tend to be very long and complicated, and at the time many CDF physicists were skeptical of the reliability of Monte Carlo results for purposes of calculating top backgrounds. That skepticism carried over to the event structure analysis, as well as the other kinematic approaches.

Kuni Kondo's Dynamical Likelihood Method. Kuni Kondo preferred to describe his analysis as "dynamical" rather than "kinematical." The CDF detector measured certain kinematic features of events: the energy of the jets and the momenta of leptons, for example. The missing energy carried off by neutrinos could be inferred by the transverse energy needed to satisfy energy conservation constraints. A kinematic reconstruction of an event as a top-quark decay would attempt to fit those kinematic features to the masses of the particles produced in a top decay – by, for example, trying to determine whether a pair of jets looked like it came from the decay of a particle with a mass of around 80 GeV/c^2 (the mass of the W), and a lepton-neutrino pair looked like it came from a particle with the same mass, and another two jets looked like they came from the decay of particles with the same mass as the b. An event that could be analyzed and interpreted in this way would constitute a (kinematically) reconstructed lepton + jets top candidate event.

Kuni Kondo's method would give a kinematical reconstruction of events and then calculate the likelihood of that reconstruction using the dynamics of the hypothesized decay process (Kondo 1988, 4126). This would involve using the measurements made by the detector to "calculate from a given event the probability function of the dynamical variables, e.g. masses, decay widths and ratios of coupling constants, involved in the process" (Kondo 1991, 836). Kondo wanted to develop a general method for using "cross section formulae to reconstruct the event and find out the parton-quark–level process" (Kondo 1995). In doing so, he was not just interested in the top quark. Looking back on the development of his technique, he said that he wanted to develop a formalism for searching for supersymmetry (SUSY) particles at hadron colliders. Describing SUSY as "a theorist's dream, an experimentalist's nightmare," Kondo noted that SUSY particles, if they exist, will decay mostly into quark, rather than lepton, final states, meaning SUSY searches require effective methods for measuring and interpreting the jets produced by quark hadronization processes. Although theory describes particle processes at the parton level for both quarks and leptons, quarks pose a special problem for experimentalists: leptons (except neutrinos) are detected directly, but quarks are typically detected indirectly, by measurements of jets. Measuring jet energies is difficult, and figuring out which quark a jet came from is even more difficult. Consequently, searching for both SUSY particles and the top requires measuring jets, and Kondo sought a general method for relating all event features, including jets, back to parton-level dynamics. Quark fragmentation is a statistical process, and Kondo sought to use the statistics of the process to relate given jet energies to particular quark processes via the assessment of the likelihoods of those processes. So he named his technique the Dynamical Likelihood Method (DLM), because it attempted to relate probabilistically all event features to parton-level dynamics (Kondo 1995).

Kondo began to work on this method prior to CDF's first real physics run and published a paper on his method in the *Journal of the Physical Society of Japan* in 1988. "People said it's a very nice method, and encouraged *me* to work on that," recalled Kondo, clearly indicating that this encouragement did not go so far as to constitute actual assistance. While Krys Sliwa and others developed similar kinds of analyses, Kondo's own technique remained a mostly solitary effort.

Kondo had the opportunity to try out his method on the lone dilepton candidate found during the 1988–9 run. Kondo found that the event could be reconstructed with his method as the decay of a top-antitop pair, with a top mass of around 130 GeV/c^2, but "it has a very broad error," he commented later, and allowed that he did not know "if that particular event was a top or not" (Kondo 1995).

In any case, Kondo wrote a paper presenting this analysis. He wanted to publish the paper in *PRL*, to secure credit for the method that he had developed. He was concerned about similarities between his own method and the method developed by Krys Sliwa with Gary Goldstein and Richard Dalitz. "I proposed [DLM] myself, but almost exactly the [same] methodology was proposed by Dalitz and Goldstein, and I'm not sure how that happened. . . . They published a paper, so I wanted to have credit for the method" (Kondo 1995). This paper met resistance from the collaboration, who felt it too closely resembled a top discovery claim, although Kondo later insisted that he was careful to avoid any such implication. The spokespersons, Alvin Tollestrup and Mel Shochet, were, Kondo said, "rather embarrassed, why I wanted to publish." Knowing that anything published in *PRL* would be considered CDF work, Kondo "didn't make such a trouble," and instead published the paper in the *Journal of the Physical Society of Japan* (Kondo 1995). In that paper, which Kondo coauthored with Takeshi Chikamatsu and Shin-Hong Kim, both Tsukuba physicists in the CDF collaboration, the authors commented on the dilepton event: "The event can be interpreted as from the $t\bar{t}$ production, both t and $\bar{t}$ decaying in semileptonic modes. There is no experimental confirmation, however, of b and/or $\bar{b}$ quarks in this event, and an identity of the event is unknown" (Kondo, Chikamatsu, and Kim 1993, 1177).

As with the event structure analysis, Kondo's method elicited skepticism because of its heavy reliance on Monte Carlo calculations. The perceived complexity of his technique posed another obstacle. Other collaboration members found it difficult to understand. For the technique to be accepted, other physicists had to be able to think through it and believe that it worked. Kondo did not succeed in persuading them to make that kind of commitment.[11] Dan Amidei, noting the strong preference at CDF for counting experiments over likelihood techniques, observed, "High energy physicists in a way are a very prosaic bunch. You *count* something. You measure how many times you count it, and then you compare that to how many times you expected to count it. People are willing to devote great parts of their life to building hardware to count things, but they're not willing to devote excess neural capacity to understanding a new mathematical formalism" (Amidei 1995).

The Dalitz-Goldstein-Sliwa Method. Kondo had published his paper in part to establish authorship of the method, with one eye on the similar method proposed by Krys Sliwa, Richard Dalitz, and Gary Goldstein. The Sliwa-Dalitz-Goldstein method has a peculiar history of its own. Some obstacles that stood in its way were common to all of the likelihood approaches discussed here. Sliwa, Dalitz, and Goldstein encountered additional difficulties, however, that were unique to their situation as a joint venture of physicists from inside and outside of the collaboration. Their story reveals some of the dilemmas of modern collaboration physics, in which limits are imposed on the ability of physicists to communicate with colleagues outside the collaboration. This feature of collaboration in high energy physics deserves attention, because it runs counter to the norm of openly sharing the results of research that some have regarded as a crucial feature of science (see Merton 1973).

Dalitz and Goldstein had developed a method for the kinematical analysis of dilepton $t\bar{t}$ decays. Sliwa believed their method could be made more reliable by adapting it for the more common lepton + jets events. Whereas dilepton decays would produce two neutrinos, lepton + jets decays would result in just one neutrino, so that one could use kinematical constraints to solve for the energy of that lone neutrino, rather than trying to infer neutrino energies from what Sliwa referred to as "the worst experimentally measured variable in CDF, the missing transverse energy" (Sliwa, personal communication). Sliwa had employed a similar strategy in solving the problem of reconstructing events with undetected neutrinos in his earlier work, in Fermilab's Tagged Photon Spectrometer Collaboration, on the semileptonic decays of the charmed D^0 meson (Anjos, Appel, et al. 1989).

Although Dalitz and Goldstein had already published their method for dilepton events, and Sliwa believed that he could make the necessary modifications on his own, Sliwa decided to work on the project with Goldstein and Dalitz. He had already conversed with the two theorists when they were working on their original method. In 1991, they had asked for his help in interpreting the errors on jet energies in CDF data because they were trying to apply their method to the first CDF dilepton event, the same published e-μ event from the 1988–9 data that Kondo analyzed around the same time (Dalitz and Goldstein 1992a; 1992b). Since Goldstein was a colleague of Sliwa's at Tufts, he concluded that collaborating with the two theorists would be the "best thing to do" (Sliwa, personal communication).

Like Kondo's technique, the method attempted to reconstruct selected events as top decays using constraints imposed by the hypothesized top decay mechanism. In both dilepton and lepton + jets top events, complete reconstruction is impossible. In dilepton events, there are two undetected neutrinos; in lepton + jets events, one neutrino escapes undetected. Although the top-quark decay will produce jets both via the production of b and $\bar{b}$ quarks and (in lepton + jets decays) in the $q\bar{q}$ decay of a W boson, one does not know which jet to assign to which final state parton (b, $\bar{b}$, W^+, or W^-) because of the unknown flavors of the jets. As a consequence, one must calculate a likelihood of the top hypothesis for each possible assignment (of energies to neutrinos and of jet energies to individual quark products) and for a hypothetical top-quark mass.

One distinctive aspect of the Dalitz-Goldstein-Sliwa method (Goldstein, Sliwa, and Dalitz 1993) was how it handled the problem of associating measured features of events with underlying parton-level processes. Other methods used the kinematics of events based on all possible assignments of jets and leptons to top decay products at the parton level, and then added the probability distributions (for top decay hypotheses for that event with various top masses) of all those possible reconstructions into a single distribution. Only the likelihood evaluated for all possible reconstructions was taken as a measure of how "top-like" the event was. The Dalitz-Goldstein-Sliwa method, however, did not treat all possible solutions equally in this way.

Instead, it regarded each possible assignment of measured jets and leptons to parton-level objects separately at a first stage of analysis, assigning each reconstruction a "weight" based on how well it satisfied certain constraints – a measure of how "top-like" that reconstruction was. These constraints included, for example, information about the angular distribution of the leptons in the rest frame of the $t\bar{t}$ pair and the requirement that it allow a reconstruction of the initial $t\bar{t}$ pair as having a very small net transverse momentum. Then a likelihood distribution (over a range of possible top masses) was produced for that reconstruction. The likelihood distributions for different reconstructions were thus weighted for "top-likeness" before being added together to yield a total likelihood distribution for the event as a whole. Reconstructions that fared poorly according to the kinematic constraints would make relatively small contributions to the overall distribution. Top-quark decays should show up as a peak in that distribution at the mass of the top.[12]

In February 1992, Goldstein and Sliwa were invited to present their method for dileptons at a meeting of the heavy flavors group (the precursor to the top group), and they presented their analysis of the published e-μ event (Top Working Group, 2/6/92).[13] Sliwa later recalled that "I was surprised at the reaction to this talk on the subject. I thought it was hostile, as if [Dalitz and Goldstein] have done something wrong and improper, although the event had been published already" (Sliwa, personal communication). Goldstein and Sliwa again presented results at a heavy flavors group meeting in April, in which Goldstein explained how the technique could be extended to apply to lepton + jets events, and Sliwa showed the results of applying the technique to data, showing a bump in the probability distribution at a top-quark mass of about 120 GeV/c^2 (*Top* 4/16/92). After that, according to Sliwa, "Goldstein was told not to come to Top Group meetings," a decision that Sliwa believed was "highly improper" (Sliwa, personal communication).

In May, Goldstein, Sliwa, and Dalitz released two CDF notes, numbered 1750 and 1751, to the collaboration (Goldstein, Sliwa, and Dalitz 1992; Sliwa, Goldstein, and Dalitz 1992). They submitted the first of these papers, after some revisions, to *Physical Review D* in June 1992, at the start of run Ia. It appeared in print as "Observing Top-quark Production at the Fermilab Tevatron" the following February. In this paper, Goldstein, Sliwa, and Dalitz explained the method, with studies of Monte Carlo–produced data but no analysis of CDF's actual experimental data. The authors concluded that "[i]f real events contained a top-quark contribution in the mass range above 120 GeV (as [CDF's 1988–9] dilepton event might suggest), this procedure could provide a dramatic demonstration of the existence of such a top quark" (Goldstein, Sliwa, and Dalitz 1993, 971).

The other CDF note that Goldstein, Sliwa, and Dalitz wrote in May 1992 goes further, presenting results of an analysis of data from the 1988–9 data. Titled "Search for $t\bar{t}$ Events in the 'Semileptonic' Mode, or, On Observation

of Top Quark Production at the Tevatron," the paper presented "a preliminary result" of applying the method to the 4.1 pb^{-1} of data from the 1988–9
run (Sliwa, Goldstein, and Dalitz 1992, 1). They applied the analysis to events
identified as $W \rightarrow e\nu$ or $W \rightarrow e\mu$ decays by the "standard" CDF selection criteria. To these events, they then applied additional cuts requiring four jets
with specified momentum values, cuts on the transverse energy of the lepton,
and missing energy (from the undetected neutrino). Three sets of cuts were
presented, with varying degrees of strictness. From the data thus selected, a
probability distribution was produced, as well as a corresponding distribution based on a Monte Carlo simulation. "The plots show very clearly a well
separated enhancement around $M_t = 135$ GeV in the accumulated probability distributions, as expected from the Monte Carlo studies" (ibid., 7).
The authors also presented their results "in a more conventional framework" by showing how many events pass the three sets of cuts, compared to
expected backgrounds calculated by Monte Carlo. For each set, there was
a slight excess, "consistent with expected numbers." However, the authors
regarded this as secondary: "more importantly it is remarkable that those
events satisfy all our kinematic criteria for real top production" (ibid., 7).

The top mass estimates from the Dalitz-Goldstein-Sliwa analysis, which
consistently fell into the 130–140 GeV/c^2 range, were considerably lower
than the later estimate of 174 GeV/c^2 that appeared in CDF's paper claiming
evidence for the top quark (the Evidence paper), as well as all subsequent
estimates based on additional data. Some CDF physicists later indicated
that when Sliwa's analysis was run on events identified by the subsequent
standard CDF top analysis as background, it yielded a peak in the same 130–
140 GeV/c^2 region, from which they concluded that the method identified
background events as top candidates all along (Gerdes 1995).[14]

Because Goldstein and Dalitz had developed the original version of the
method for dilepton events, and because he believed that early results suggested the potential for some kind of discovery based on the method in
the future, Sliwa asked that the two theorists be listed as coauthors on any
paper based on the Dalitz-Goldstein-Sliwa method. This, according to Sliwa,
"was not enough for Goldstein, who became overexcited thinking that the
discovery has been made." Sliwa alleged that Goldstein wrote to CDF's executive board, unbeknownst to Sliwa, and "sounded as if he did not realize
that the statistical significance of my very interesting results was yet to be
determined" (Sliwa, personal communication). The executive board denied Sliwa's request for coauthor status for Dalitz and Goldstein. In a CDF
meeting, the policy behind this refusal was affirmed: "policy is that only
collaborators have access to raw data and the author list. Other people can
be referenced, acknowledged, and given special thanks" (*CDF* 6/12/92).

Then, a very strange thing happened; it took an already tense and somewhat antagonistic situation and transformed it into an awkwardly public dispute regarding appropriate conduct in the context of collaboration physics.

"A claim that the top quark has been found is being suppressed by scientists at the Fermilab particle physics centre near Chicago," announced an article in the British magazine *New Scientist,* dated June 27, 1992. Dalitz, Goldstein, and Sliwa appeared in the article as a "rival group," the publication of whose paper CDF was "blocking," and the author reported Goldstein saying that he was "'quite confident' that they have discovered the existence and the mass of the (Top) quark." The article quoted Alvin Tollestrup insisting that "[t]he two people who are not members of CDF should not have had access to the data," and that "[p]eople not sufficiently expert in interpreting the data have jumped to conclusions." The article concluded, somewhat archly, "If Dalitz turns out to be correct, the $68 million spent on upgrading the Tevatron will have been wasted; and the main credit for finding the particle will go to Dalitz, a scientist outside Fermilab" (Bown 1992, 10).[15] A similar article appeared in the British trade publication *The Engineer.*

CDF physicists discussed these articles at a CDF meeting on July 10. Mel Shochet, reporting on the results of an executive board meeting, noted that letters responding to the articles had been sent to the editors of both publications (*CDF* 7/10/92). An article recounting the episode and describing the response of CDF physicists to the initial articles appeared in the July 24 issue of *Science* (Flam 1992). That article described the position of Dalitz and Goldstein in more reserved terms: "They think the patterns they found look suspiciously, though not unambiguously, like the top quark and deserve further analysis." CDF, the article noted, "isn't impressed" by Dalitz and Goldstein's claims. The *Science* article described Dalitz and Goldstein as having "gained access to unpublished, partially analyzed CDF data only when Krys Sliwa, a member of the group, shared it without telling his colleagues." The article recounted how the results of the Sliwa-Goldstein-Dalitz analysis were presented to CDF and found wanting, and how Goldstein and Dalitz were subsequently excluded from CDF top group meetings. The author quoted Tollestrup saying, "They haven't put in the effort and commitment that others have," and included comments from Tollestrup and Shochet on how difficult it would be for outsiders unfamiliar with the subtleties and complexities of the detector and the data to successfully analyze those data.

One premise of the *Science* article was that Sliwa "shared [data] without telling his colleagues." This point was emphasized several times: "Goldstein says Sliwa shared unpublished data from experimental runs of 3 years before to help them try out the method." "Shochet says CDF member Sliwa violated an unwritten code of ethics by sharing data with outsiders." Even Sliwa's unwillingness to be interviewed for the article was described in a way that presumed the truth of this premise: "Sliwa chose not to comment to *Science* about his decision to share the data" (Flam 1992).

Sliwa rejected this claim in his communications with me. As he described it, "I have done the data analysis myself, although I did not keep the results secret from Goldstein." He said that, in response to the article, he wrote a letter of protest to *Science* "to defend my integrity" against the charge of sharing unpublished CDF data. Sliwa alleged that in the *Science* article Goldstein "made false statements about his gaining access to the CDF data and analysing it" (Sliwa, personal communication). In his letter to *Science*, Sliwa stated that, although the three physicists had worked together on developing the method, "I proceeded to apply this modified method to the analysis of CDF experimental data." Sliwa denied that he had made substantive information about CDF's unpublished data available to Dalitz and Goldstein:

I made available to Goldstein summary files of lepton and jet momenta for a small number of interesting events identified by my analysis and informed him of some preliminary results of my CDF data analysis. I did not believe that sharing my events with a colleague who had co-developed the method "violated an unwritten code of ethics." ... I have not given access to either the actual raw CDF data or to any subset of raw or "processed data" which could provide the basis of any conclusive physics analysis.

Sliwa also affirmed in this letter his respect for the established CDF procedures for making physics results public through the "blessing" procedure (Sliwa 1992).

Goldstein then wrote a letter of his own to *Science*, in which he emphasized that "the work on modifying the method of Dalitz and Goldstein and applying that method to data was a three-way collaborative effort." Goldstein denied both the sense of furtiveness and secrecy conveyed by the *Science* article ("Sliwa ... shared [the data] without telling his colleagues"), and Sliwa's suggestion in his letter that, although the method was developed jointly with Dalitz and Goldstein, it was Sliwa alone who, using unpublished CDF data, applied the method ("I proceeded to apply this modified method to the analysis of CDF experimental data"). On the first point, Goldstein wrote, "It was known from our first presentation that our work involved some of their corrected data. However there was never any intention to publish that data without CDF approval." Regarding the extent and nature of his and Dalitz's involvement in producing results using the method, Goldstein recounted how the method itself had been developed "[t]hrough studies of some real data, and many studies of simulated data," and how the method had then been applied "to data that had been processed and corrected." He cited the two CDF notes that the three physicists had coauthored and described work on the analysis as a "synergistic and collaborative process, by no means the exclusive effort of any one of us" (Goldstein 1992).

I do not aim to settle the dispute over what happened or whether anything improper occurred regarding this method and its application. I do wish to consider the effects these events may have had on CDF physicists' responses to the Dalitz-Goldstein-Sliwa analysis.

Those responses constitute a complex phenomenon, and the reasons for individual reactions are to a degree inscrutable. Different parties give different accounts, and the best I can hope to do is to air the different perspectives.

It seems likely that Sliwa's attempts to develop a kinematical likelihood method for finding the top would have received a cool response even without the complications brought on by his involvement with Dalitz and Goldstein. After all, like the other "kinematical" methods, it relied on a complex mathematical apparatus and required a comparison with a model of the background that relied heavily on Monte Carlo calculations. In the opinions of some CDF physicists, these facts stood in the way of the Dalitz-Goldstein-Sliwa technique no less than the other kinematical methods.

But some CDF physicists have also acknowledged that the unpleasant atmosphere generated by the controversy surrounding Sliwa's work hampered progress on the Dalitz-Goldstein-Sliwa method. According to Dave Gerdes, "Krys really never got the time of day after [the appearance of the articles in *New Scientist* and *Science*]. . . . [He] took it very personally, and responded very personally" (Gerdes 1995). Dan Amidei expressed the opinion that Sliwa was "way too defensive in his presentation all along, and . . . that just turned people off" (Amidei 1995). Drasko Jovanovic, who had hired Sliwa when the latter was still a postdoc, believed that he was "spurned by the rest of the collaboration" "because he was acting singly, and not in a larger collaboration. . . . Probably to this day he is a very unhappy individual, because he never managed to get enough people in the group to listen to him." Jovanovic noted that, although Sliwa's "mass was wrong, . . . he had peaks which were believable," and he considered Sliwa "the first to start to use . . . kinematic quantities" in the top search at CDF (Jovanovic 1995).

Sliwa remained skeptical about the identification of top candidates via the standard CDF top analysis in his 1996 communication with me. He wrote that he was "not really convinced that CDF has found top," and expressed disappointment at his colleagues' response to his ideas: "Although what is required to understand the technique is nothing more than some knowledge of relativistic kinematics, parton model geometry and basic particle physics, at times it seemed to be a barrier too high for many to even try. . . . I think it should suffice to say that, so far, only one person, outside my small Tufts CDF Group, has asked whether he could obtain a copy of my program and use it" (Sliwa, personal communication).

6.2. Counting Experiments

The Dilepton Search. Such obstacles did not stand in the way of the efforts to develop counting experiments to look for the top quark, which were also evolving throughout the period following the 1988–9 run. In Section 4,

I described the origins of the dilepton search, which drove the mass limit papers published on the 1988–9 data. G. P. Yeh and Milciades Contreras led the further development of the search for dilepton events. Once the mass limits obtained from the 1988–9 data eliminated the possibility of a low-mass top quark, they updated the analysis to look for a heavier, rarer top, with stricter requirements to eliminate more background. They also modified the analysis to include data from the additional muon coverage gained in run Ia. Otherwise, the dilepton analysis retained most of its features from the 1988–9 run.

Leptons + Jets Searches: SLT and SVX. In addition to the dilepton search, which was not expected to turn up large numbers of events, members of the top group were developing techniques to take more advantage of the lepton + jets decay channel, in which one of the Ws produced in the decay of the $t\bar{t}$ pair decays to a quark-antiquark pair, while the other W decays to a lepton and neutrino. Two means of tapping into this channel were evolving: soft-lepton tagging and secondary vertex tagging (SVX).[16] One physical principle provided the basis for both of these efforts. In the lepton + jets channel, one looks for an energetic lepton and missing energy (an undetected neutrino) from the decay of one of the two Ws, as well as energetic jets from the decay of the two b quarks and from the quark-antiquark decay of the other W. But not all Ws are produced by top decays, and there is a significant lepton + jets background caused by processes in which Ws are produced in conjunction with jets. In a top decay, however, two of those jets will be due to the decay of the b and $\bar{b}$ quarks produced by the $t\bar{t}$ pair.[17] Since b quarks will otherwise be quite rare, this provides a way of distinguishing top events from most of the W+ jets background. CDF sought to identify top events by tagging lepton + jets events as events with b quarks.

The SLT strategy sought to identify instances of a common mode of b decay that results in a lepton and a neutrino, along with a jet. In such decays, the lepton from the b decay will tend to be less energetic (softer) than the lepton produced by a W decay. The SLT search looked for lepton + jets events with a soft muon or electron as an indication of a b quark. This involved using the same kind of information used to identify the higher energy leptons sought in both dilepton and lepton + jets events, but requiring lower values for the transverse momentum of the leptons thus identified.

CDF had already used a form of soft lepton tagging on the 1988–9 data. As mentioned previously, the 91-GeV/c^2 mass limit that emerged from the earlier data was based in part on an analysis, developed by Paul Tipton and Claudio Campagnari, that looked for events with an energetic lepton, two or more energetic jets, and a soft muon from the decay of the b quark. Paul Tipton had gone on to become SVX project coordinator, but Claudio Campagnari continued to work on the SLT analysis and was joined by Fermilab physicist Avi Yagil. They set out to refine the analysis for a heavier top and to include soft electrons as well as soft muons in their search. Campagnari and Yagil worked on the SLT analysis rather independently of the rest of the collaboration. Indeed, they acquired a reputation for invisibility while their analysis was under development. One colleague described Campagnari and Yagil's approach to working on physics problems as: "you lock yourself in an office with a lot of cigarettes . . . [and] don't pay any attention to any request for communication or meeting . . . and you get a lot of work done" (Amidei 1995). One result of this approach, as will be seen, was that controversies regarding the details of the analysis did not erupt until Campagnari and Yagil emerged from their smoky offices with their analysis substantially complete.

To tag b events with the SVX, CDF made use of the fact that bs have a characteristic lifetime. Although it does not take very long for such particles to decay (their mean lifetime is on the order of 10^{-12} seconds), they tend to be produced at high momenta in top-quark decays. Hence, the distance they travel before decaying is measurable using the silicon vertex detector. The point at which the b decays is called a secondary vertex: the products of the b decay radiate out from a point a short distance from the point of origin for the other products of the collision. The SVX search looked for lepton + jets events in which a secondary vertex could be found at a distance from the primary vertex that was consistent with the lifetime of a b quark with high momentum (see Figure 3.6).

The atmosphere in the SVX b-tag working group was different from that in other groups working on the top search. Not only were more people involved, three *different* groups were pursuing distinct approaches. The competition among the three groups was so fierce that Dan Amidei, who

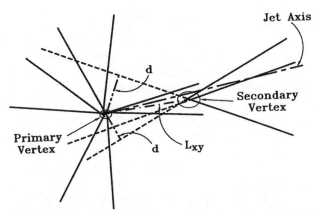

FIGURE 3.6. A schematic representation in the r-ϕ plane of an event with a secondary vertex. Solid lines represent charged-particle tracks reconstructed from SVX data. The proton-antiproton interaction takes place at the primary vertex. L_{xy} is the r-ϕ projection of the distance traveled by a b quark from the primary vertex before decaying at the secondary vertex (Abe, Amidei, et al. 1994b, 2986).

was convener of the *b*-tag group at the time, described them as "separate sects" (Amidei 1995). A group of physicists from the University of Rochester, including Paul Tipton, Brian Winer, and Richard Hughes, along with Rochester graduate student Gordon Watts, developed an algorithm known as jet vertexing (JETVTX). University of Michigan physicists, including Dave Gerdes and Dan Amidei, developed the jet-probability algorithm. The first display of a displaced vertex using SVX data (using a display program developed by G. P. Yeh; see Figure 3.7) was based on an algorithm called *d*-*φ*, developed largely by Simone Dell'Agnello, then a graduate student, along with Franco Bedeschi and other University of Pisa physicists (Yeh 1998a).

All three groups had designed their algorithms to do the same thing: identify events with a secondary vertex characteristic of a *b* decay. Knowing that such an approach promised to yield a powerful means of digging a top signal out of the data, the authors of each algorithm wanted their own creation to be the algorithm that found the top quark. As a result, the atmosphere in the *b*-tagging group became something of a "feeding frenzy" during run Ia (Gerdes 1995). Members of the various factions frequently ran their algorithms on the data to see how many events they tagged and compared their results to see how many, and which, events each algorithm tagged. Such peeking at the data was especially common during the early stages of development of the algorithms.

This situation came about in part because of the open nature of the collaboration. In principle, any member of CDF could work on any piece of physics she (but more commonly he) wished to. In practice, most high energy physics problems require the coordinated efforts of more than one person. The top quark was on everyone's mind at the beginning of run Ia, and many considered SVX *b*-tagging the most promising way to make a significant, high-profile contribution to finding the top. Yeh, Contreras, and Yoh had pioneered the dilepton search several years before, and any new work on it would amount to the unglamorous task of updating it for the new run. Likewise, the soft lepton approach to *b*-tagging had its roots in the previous run and was known to be the province of Campagnari and Yagil. But the silicon vertex detector was a high-profile upgrade that would generate data for a qualitatively novel approach to *b*-tagging in run Ia. Hence, the field was wide open for anyone who could convince others that he had the best idea for using this information. Secondary vertex tagging was the new frontier in the top search.

One might think that the people who built the SVX detector would automatically be those best situated to serve this function. Many prominent members of the *b*-tag working group had in fact served time on the SVX construction project: Franco Bedeschi, Dan Amidei, and Paul Tipton for example. Veteran CDF member Henry Frisch even held back from working

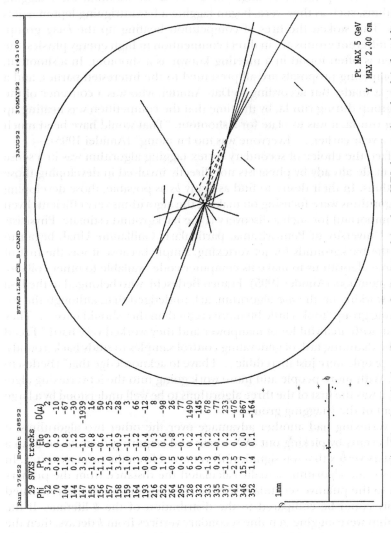

FIGURE 3.7. The first display of a secondary vertex based on SVX data, using the d-ϕ algorithm. Courtesy of G. P. Yeh.

on this piece of physics because he thought "the guys who built the SVX ought to be the guys to exploit it" (Frisch 1995).

Such restraint was not widespread. As run Ia progressed, a discovery based on the incoming data began to seem quite plausible. First, the DPF event, while not utterly convincing to most collaboration members, was a hopeful sign that the top discovery might be at hand. In addition, the b-tagging algorithms, such as they were, began tagging a few intriguing lepton + jets events. This stoked the fires of competition heating up the b-tag group. When different groups are in direct competition in high energy physics, the solution is often found in a meeting known as a shootout. In a shootout, the competing proposals are all presented to the interested parties, and a choice is made. But according to Dan Amidei, who was a convener of the b-tag group during run Ia, by the time that the competition was heating up during run Ia, it was too late for a shootout. "That would have been much better a year earlier. . . . Everyone was too far along" (Amidei 1995).

In fact, the choice of secondary vertex tagging algorithm was in a sense being made already by physicists not directly involved in developing those algorithms. In their desire to find as many bs as possible, those developing the algorithms were focusing on making the algorithms very efficient. Even more important for such a discovery is the background estimate. Physicists at the University of Pennsylvania, particularly Guillaume Unal, began to study the backgrounds for jet vertexing, simply because it was the first of the three algorithms to make its computer code available to other collaboration members (Amidei 1995). Franco Bedeschi, who belonged to the Pisa group working on the d-ϕ algorithm, acknowledged that, although the jet-vertexing group "took a little bit more credit than they should have. . . . They came in with an awful lot of manpower and they worked very hard." Faced with the daunting task of generating control samples to study backgrounds, "these people were just incredible . . . I have to acknowledge that" (Bedeschi 1998). With more people and more work going into the jet-vertexing algorithm, it was the first of the three algorithms to be well understood by a large segment of the b-tagging group.

Jet vertexing had another advantage over the other two algorithms: it tagged events by picking out SVX tracks and requiring them to be fit to a secondary vertex that was significantly removed from the primary vertex. In doing so, the algorithm actually calculated the distance from the primary vertex to the putative secondary vertex. The distribution of this calculated quantity could be compared to the distribution of the b lifetime. If the algorithm were tagging genuine secondary vertices from b decays, then the two distributions should agree (see Figure 3.8). Hence the algorithm directly generated information that was useful for a straightforward and powerful test of its own reliability.

The jet-probability algorithm, on the other hand, calculated for each track in a jet the impact parameter d (distance to the primary vertex at the

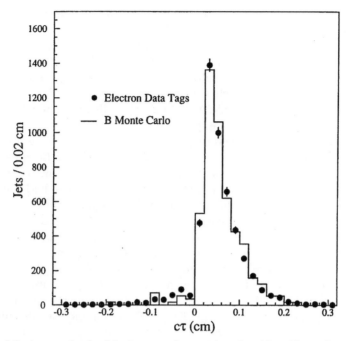

FIGURE 3.8. A cross-check of the jet-vertexing tagging algorithm. Shown is a comparison of the $c\tau$ distribution (where τ is the effective lifetime of the particle producing the secondary vertex) for secondary vertices tagged in control sample data using jet vertexing (dots) with the expected distribution of $c\tau$ for b quarks based on Monte Carlo simulations (solid line) (Abe, Amidei, et al. 1994b, 2989).

nearest point extrapolated from the track, see Figure 3.6). The algorithm then used that information to calculate a probability for each track in the jet, on the assumption that the track originated at the primary vertex. The probabilities for individual tracks were combined to form a joint probability for the jet as a whole. Jets with a very low joint probability on the assumption of having originated from the primary vertex were tagged. Note that this algorithm proceeded without identifying any specific secondary vertex. The d-ϕ algorithm tagged events by means of the correlation between the impact parameter d and azimuthal angle ϕ in b decays, but it also did not uniquely determine a secondary vertex. While the latter two algorithms used information related to the distribution of the decay length for b quarks, only jet vertexing automatically yielded a quantity directly comparable to such a distribution.

Thus, while in principle one could perform a check on the jet-probability or d-ϕ algorithms similar to that described earlier for jet vertexing, the comparison with an easily understood physical parameter of b quarks was not as direct or straightforward. The jet-vertexing algorithm was judged, even by some people who had worked on competing algorithms, to be "the easiest to explain, and . . . [the one that] yielded the most concrete physics information in the form of a decay lifetime plot" (Gerdes 1995). Members of the *b*-tag group recognized that in presenting a top discovery based on SVX *b*-tags "you would need some really concise, pithy way to show the world that this algorithm was finding what it was supposed to" (Amidei 1995). Such a test of the algorithm's reliability fell out naturally from the jet-vertexing algorithm and not from the competitors.

Consequently, the group chose jet vertexing as the official algorithm for giving top results. According to Dan Amidei, "Everyone saw the unassailable logic" of the choice. For those who had worked on the other algorithms, "the sociological downside" was that the Rochester physicists who had developed jet vertexing gave all the presentations on SVX *b*-tagging to the collaboration when the top results were divulged (Amidei 1995). On the other hand, to some observers of the *b*-tag group, the issues were not so clear-cut. G. P. Yeh maintained that the *d*-ϕ algorithm was both the first algorithm to work, and the "simplest to understand mathematically" (Yeh 1998a).[18]

For all the excitement in the *b*-tagging group, it was not obvious, as run Ia progressed, that CDF would be able to claim a top-quark discovery based on the data collected in that run. The DPF event alone was not sufficient for a discovery announcement, and as long as the other analyses were in flux, had poorly understood backgrounds, and had not been run on the full set of data from run Ia, it was not clear whether much else would turn up. Recollections of the mood during this period vary. According to Tony Liss, the DPF event was followed by a long period in which "we saw nothing new. . . . People started to get a little depressed over the prospects of finding a signal" (Liss 1995). Hans Jensen recalled that during the development of the tagging algorithms, "It was . . . clear as one proceeded that actually there were events that were tagged that looked a lot like what you expect" from the top quark (Jensen 1995), and veteran Fermilab physicist Drasko Jovanovic recalled that "events [were] coming in a statistical fashion, one per month, but there will be two months of nothing, and then sometimes two in a week. . . . It was obeying in the periodicity [a] statistical distribution" (Jovanovic 1995).

7. CONCLUSION

That the work of scientists is riddled with contingency has become a truism within science studies. In this chapter, I have described several developments within CDF that instantiate that truism, but also cast it in a different light than is usually shone upon it. Many factors influencing the development of CDF's detector and the analysis they used to search for evidence of the top quark seem to be at best distantly related to the reliability of CDF's scientific conclusions: the stubborn persistence of Aldo Menzione in pushing the SVX detector; the good fortune of having within the collaboration a group from LBL with access to the expertise needed to integrate the SVX electronics into the detector itself; the barriers of complexity, distance, language, and personal and political tensions that stood in the way of the various kinematic analyses; the bandwagon effect by which the Rochester group's jet-vertexing algorithm benefited from the availability of its computer code to those wishing to work on backgrounds. Can anyone seriously claim that each of these forces was inherently epistemically virtuous? Certainly not.

I do claim, however, that through all the conflicts and confluences of egos, allegiances, and interests that ran through the CDF collaboration, these experimenters, *as* experimenters, shared a certain epistemic goal: they sought not only to gather information about particle processes generated by the accelerator's collisions but also to gather information in a manner that would enable them to develop models of how their experiments worked, test the adequacy of those models, and draw inferences from those models that would help to answer the kinds of questions that the detector had been built to address. Of course, among those questions was the question of whether the top quark did exist, and what its properties were if it did. Even if that shared epistemic goal did not take priority over all other considerations at all times for every investigator, it was a goal that dominated and shaped the ways in which other considerations could be invoked, and that could not be relinquished entirely without abandoning the experimental enterprise altogether.

The fact that Aldo Menzione's persistence helped get the SVX incorporated into the CDF detector, for example, illustrates nicely how scientific developments can turn on highly individual personal commitments. But note also how important it was that CDF had the kind of "detector philosophy" that made a detector like the SVX a suitable choice. Such a detector philosophy is not merely a matter of subjective aesthetic preference but rests upon prior choices regarding investigative strategy, as described in Chapter 2. Typically, it reflects the previous experiences and expertise of those building the detector. Given the kind of detector CDF had built thus far, a detector like the SVX, if it worked (and it might well not have) would yield more powerful tests for CDF's questions.

One can hardly deny that intracollaboration politics played an enormous role in shaping the collaboration's analytic strategies to be used in pursuing the top quark. But those politics must also be viewed within the context of a shared epistemic goal. Indeed, they were one aspect of a multiparty conflict over how that goal was to be met. The three major attempts to use kinematic information to find the top quark, in addition to their individual difficulties, shared two obstacles – they all were somewhat more complicated than the counting experiments (from the perspective of experimental particle physicists), and they all relied on Monte Carlo calculations of backgrounds. Thus it was more difficult for them to satisfy two necessary conditions: the requirement that members of the collaboration be able to understand the analysis and the requirement that members of the collaboration trust the analysis. (The latter demanded the former, but the former was not sufficient for the latter.) There is no reason to think the kinematic analyses could not in principle have met these conditions, but the barriers that the kinematic analyses would have had to overcome to satisfy these epistemic requirements compounded with their political difficulties. To make a complex analysis work, and be convincing, is often a matter of having enough people putting in enough time to work through the various problems that come up in developing an analysis. (Recall Kuni Kondo's comment – "People said it's a very nice method, and encouraged *me* to work on that" – and how Krys Sliwa "never really got the time of day" in the wake of the controversies over the Dalitz-Goldstein-Sliwa analysis.) The collaboration's energies, following the path of least resistance, went naturally to the counting experiments, which could be more easily understood, and whose reliability could be more easily verified, at least at first glance. Accepting an analysis involved not so much a will to believe as a will to solve and a will to understand. To the extent that loss of political prestige in the group hampered the efforts of particular group members, it did so largely by making others in the collaboration less inclined to invest the effort necessary to gain an adequate understanding of the methods proposed (an investment considered to be unduly burdensome by some in the collaboration, although Krys Sliwa, for example, insisted that his method was not, after all, so difficult to understand). Some collaboration members even maintain that the political difficulties surrounding the Dalitz-Goldstein-Sliwa method in the end made no difference because the method itself was flawed.

Being able to model the experiment, being able to test that model, being able to make evidential assessments based on the results of the experiment – these, I claim, were shared epistemic goals within the collaboration, sometimes glimpsed only dimly on the far horizon, through a haze of collaboration politics. That such aims nevertheless helped constitute CDF's experimental enterprise is a claim that I have only begun justifying, and much more must be said to clarify what it means. Our best clues will emerge from attending to disagreements within the collaboration over the attainment of these goals.

4

Writing Up the Evidence

The Evolution of a Result

But in fundamental experiments, such as we have here, it is not enough to indicate the general meaning of an experiment, to describe the instruments used in performing it, and to tell in a general way that it has yielded the result expected; it is indispensable to go into the details of the experiment itself, to say how often it has been repeated, how the conditions were modified, and what the effect of these modifications has been; in a word, to compose a sort of brief of all the circumstances permitting the reader to sit in judgment on the degree of reliability and certainty of the result.

Wilhelm Weber, *Electrodynamische Maassbestimmungen* (Leipzig, 1846).
Quoted in Duhem [1914] 1954, 197.

On April 26, 1994, science reporters from the international press thronged the lobby of Wilson Hall at Fermilab to hear presentations from CDF and D-zero representatives on the results of their top-quark searches (see Figure 4.1). The occasion was CDF's release of a paper describing "evidence for top quark production." D-zero did not claim to have established such evidence.

In this chapter, I will describe the development of CDF's "Evidence for Top Quark Production" paper (the Evidence paper). Writing up and publishing the results of experiments serves several purposes for scientists. Obviously, scientists use publications to communicate their results to a broader audience. Less obvious is the determinative role of the research report: scientists engaged in preparing their results for publication are sometimes simultaneously determining the exact nature of the result of their experiment, and testing the assumptions underlying that determination. This aspect of scientific procedure stands out prominently in the development of CDF's Evidence paper.

FIGURE 4.1. G. P. Yeh talking to reporters at the joint CDF/D-zero press conference, April 26, 1994. Courtesy of Fermi National Accelerator Laboratory.

1. THE LEPTON-PHOTON MEETING AND THE ORIGINS OF THE EVIDENCE PAPER

CDF members developed the analyses for finding the top decays described in the previous chapter while simultaneously collecting data during run Ia. For most of the run, any thought of making a top discovery claim rested entirely on speculation. That situation changed, but subtly, as the end of the run approached. The XVI International Symposium on Lepton and Photon Interactions was to be held on August 10–15 in Ithaca, New York. Paul Tipton was slated to present CDF's top search results. Consequently, CDF physicists working on the top search needed to finalize those results for presentation at a July collaboration meeting so that the collaboration as a whole could approve Tipton's presentation. In May, using a partial data set, the SVX search had found three tagged events with jet vertexing against a background of approximately one (Tipton 1995). Just prior to the July meeting, applying the jet-vertexing algorithm to the entire Ia data set still tagged just three events. Jet probability tagged three as well, and the d-ϕ algorithm tagged two. At this time, the dilepton analysis identified two candidate events – both e-μ decays – with an expected background of about half an event. The SLT analysis came out with seven tagged events with a background of about three.

The three counting experiment groups – dilepton, SVX, and SLT – had deliberately not been discussing their findings with one another. Hence, when they all presented their results at the July meeting, it was news to everyone – whether they noticed or not. Mel Shochet later recalled of that meeting, "It was late in the day on Friday, the second day of a collaboration meeting, and . . . nobody paid much attention. My eyes opened up, and so did Alvin's, and we looked at each other because we were counting" as representatives of the counting experiment groups presented their results. None of the three counting experiments by itself was very exciting, but adding them together "it looked like there very well might be something there." Alvin Tollestrup closed the meeting with a reference to the "exciting results" that the collaboration was seeing, and Mel Shochet heard a "buzz around the auditorium: 'What's he talking about? What exciting? Where? In what analysis?'" (Shochet 1995).

Hans Jensen recalled walking back to the CDF site from the meeting at Wilson Hall with Harvard physicist John Huth and discussing the results. They both felt that, based on the combined results, "maybe we really have something here" (Jensen 1995). Tony Liss recalled Tollestrup, Shochet, and Jensen stopping him in the B-zero parking lot after the meeting and asking, "What does Paul think he's going to show at Lepton-Photon?" Liss replied that he thought Tipton wanted to show the full top results (Liss, personal communication). Paul Tipton later confirmed that he had *wanted* to present the complete results but did not expect the collaboration to allow him to do so (Tipton, personal communication).

Deciding what Tipton ought to say and show at the conference required considering potential repercussions of his presentation. Jensen recalled that he and Huth met with Paul Tipton, with the top group coconvener Tony Liss, and with the collaboration spokespersons Bill Carrithers and Mel Shochet to discuss this problem (Jensen 1995). It was clear that, whatever the views of individuals close to the top analysis efforts, the collaboration as a whole was unprepared to present these results as a top discovery. More work was necessary before everyone in the collaboration could be convinced that they were seeing real top decays. On the other hand, if Tipton presented the numbers of candidate events compared to background without an accompanying declaration that this was the top quark, it was likely that "the world would probably interpret this as the first evidence for top" (Tipton 1995). CDF was loathe to allow physicists outside the collaboration to "discover" the top in CDF's data by attending a conference and seeing Tipton's presentation. Also, those who had worked on the top analysis realized that presenting the full results, with or without an accompanying discovery claim, when the collaboration had not had time to study and understand the multifaceted analysis used for the top search, would not be, in Paul Tipton's words, "fair to our collaborators" outside of the top group (Tipton 1995). Once the results were made public, each member of the collaboration would face questions about them from other physicists. The collaboration needed time to become prepared for such questions.

Tipton's Ithaca presentation was therefore based on a partial data set. He avoided bold or provocative remarks (Tipton, personal communication). By not presenting the entire analysis, the collaboration avoided both making a discovery announcement that they were not ready to defend and presenting without such an announcement information that others might interpret as a discovery, thus "letting the world discover top for us" (Jensen, personal communication).

The introduction to the published version of Tipton's presentation states, "Due to the preliminary nature of this work, no combined limit (or discovery) will be quoted from the sum of these analyses" (Tipton 1994, 465). Tipton presented two dilepton candidate events, two SLT candidates (from the first half of the Ia data set only), and three SVX candidates, with expected backgrounds of 1.3 ± 0.2, 1.8, and 1.2 ± 0.2, respectively. These events did not include any of the muon events identified with the newly installed muon extension. Tipton projected that ongoing improvements in the SVX track-recognition routines and the discrimination between pions and low-energy electrons would result in an "approximately 40% increase in $t\bar{t}$ acceptance time integrated luminosity over what has been used for this presentation" (ibid., 477). He presented a mass limit of $113 \text{ GeV}/c^2$ for the top, based on the dilepton channel alone. Tipton concluded, "We have not exploited the full luminosity, and there are improvements underway to the analyses which will yield rather large increases in acceptance for $t\bar{t}$ events. Stay tuned" (ibid., 477).

Of course, Tipton knew a little about what his audience should stay tuned for. He knew that the entire data yielded seven rather than two SLT tags, and that CDF had more than three SVX tags. The SVX b-tagging group had improved the SVX tracking code, raising the number of tags to five for the jet-vertexing algorithm. Jet probability tagged six. Shortly thereafter, minor improvements to the tracking code and the jet-vertexing algorithm reversed those numbers. Jet vertexing yielded six tags, and the jet-probability algorithm tagged five. The top group had already chosen jet vertexing as the "official" CDF b-tagging algorithm, so CDF now had three more tagged events in the SVX search than Tipton reported at the Lepton-Photon conference. One of the additional tags was the result of the extended muon coverage made possible by the central muon upgrade and central muon extension that had been installed prior to run Ia, and the other two resulted from improvements in the SVX tracking code (Tipton, personal communication).

At the time of the Lepton-Photon conference, those working on the top analysis "went into high gear" (Liss, personal communication). They wanted to get the top results published. To understand how they pursued this goal, it will help to understand a little about how CDF collectively made decisions on publication.

On the one hand, CDF physicists wanted to put their findings into print. On the other hand, they did not want the embarrassment of publishing errors. Given the complexity of the analyses typically offered in CDF papers, the complexity of the data that served as the basis of those analyses, and the widely distributed expertise regarding the instrument that produced those data, an obvious and important means of reducing the chance of publishing an erroneous result was to give all CDF members an opportunity to criticize collaboration papers. Hence CDF sought to make the paper-writing process extremely open within the collaboration. As the case of the Evidence paper will show, controversy could arise when CDF members perceived that this aim was not being met.

The paper-writing process typically began within the physics working groups. The membership of these groups was open to anyone interested, as was the choice of work to be done. CDF had no formal mechanism for assigning tasks or for controlling the membership of the working groups. Each working group had at least one convener, who was responsible for scheduling meetings and for keeping minutes. Conveners also had responsibility for formally initiating the paper-writing process.

That process began when working group conveners approached the spokespersons and asked them to appoint "godparents" for the paper. These godparents would serve as internal referees, offering criticisms, and working with the group until they were ready to present the paper to the collaboration. CDF modeled this organization of the paper-writing process explicitly on the use of outside referees by journals and grant providers. Collaboration physicists perceived the approval of the godparents as a very stiff requirement. After the godparents had been satisfied – and sometimes they never were – the working group could present the paper to the entire collaboration. The paper was then subject to criticisms and questions from every member of the collaboration. All collaboration members could find the paper posted in a computer file, and then post their individual questions or criticisms in a separate file. Here again the collaboration could decline to approve the paper, meaning that its authors could neither submit it to a journal nor present it at a conference (Frisch, personal communication; Barnett, personal communication). CDF physicists regarded this process of internal criticism as a rigorous screening for weak analyses or hasty conclusions. As one senior collaboration member put it, "If you get past the collaboration anything you run into on the outside is pretty mild" (Frisch, personal communication).

By the time of the Lepton-Photon meeting, top group conveners Tony Liss and Claudio Campagnari had already asked for and been assigned godparents for the write-up of the top results. Alvin Tollestrup led the godparent committee. In one respect the procedure was unorthodox. The appointment of godparents came earlier in the process than was typical. Ordinarily, the spokespersons appointed godparents only after the originating authors had completed a draft of the paper, or at least after they had substantially completed the analysis to be used in producing a result. The top godparents were appointed during the run, while the analysis was still taking shape. John Huth, one of the top godparents, noted that they got more involved than most godparents in guiding the analysis, although "not in a tremendously strong way" (Huth 1995). Tony Liss, however, recalled the godparents getting "much more closely involved in this analysis than normal godparents do." He observed that godparents "are supposed to be a somewhat removed group who can look objectively at the analysis. [The top godparents] were really part of the writing of the paper" (Liss 1995). John Huth maintained that this close involvement helped to avoid some "potentially serious issues" and thus was "really positive" (Huth 1995). Krys Sliwa, on the other hand, objected that "[w]ith the Godparents for the top analysis becoming a part of the closed analysis group the principle of independent internal review has been abandoned" (Sliwa, personal communication).

The top authors and godparents initially followed a strategy to produce a series of short papers on different aspects of the top search. They proposed to write one *PRL* paper each on the dilepton search, the SLT-tagging lepton + jets search, and the SVX-tagging lepton + jets search. The godparents insisted that, in addition to these three papers, a fourth *PRL* paper should present a mass analysis, since if these events really were top-quark events, then estimates of the top mass based on individual events should cluster around a single value. They intended the fourth paper also to summarize the combined results of the three searches, give an estimate of the cross section, and perhaps briefly discuss the kinematic features of the candidate events. Milciades Contreras was primarily responsible for the dilepton paper; Avi Yagil and Claudio Campagnari wrote the bulk of the SLT paper; the SVX paper was the work of Simone Dell'Agnello, Dave Gerdes, and Brian Winer; and the mass analysis and summary fell to Tony Liss.[1]

At least, this is the story told by most of the physicists involved in the process. A dissenting view is that there were five *PRLs* that constituted the evidence for the top quark; the fifth one was a paper based on kinematical evidence that had already been written but not approved for publication. Such was the recollection of Pisa physicist and later CDF spokesperson Giorgio Bellettini. As recounted in the previous chapter, Pisa physicists had been involved in developing a kinematic search for the top. "There was a ... paper on the kinematical evidence which was there ... before the other papers and was not approved for publication" (Bellettini 1995).

Part of the difficulty here is that, as John Huth pointed out, there was no consensus on how to proceed, and so CDF had no well-defined "official" plan as such. Top godparents John Huth and Alvin Tollestrup were both ambivalent about how to present the top results, and what to include in those results (Huth 1995; Tollestrup 1995). In any case, the conveners of the top group were pushing ahead with four *PRL* articles on the top, without the kinematic evidence. They intended to present these four articles to the collaboration at its October 1993 meeting to get them blessed and then to send them in for publication.

According to Liss, the papers were written in a hectic atmosphere: "We were going pretty fast, and we weren't taking the collaboration with us." He conceded that they did not share much information with other collaboration members during the writing process and that his own contribution, the summary and mass analysis, was in the worst shape. It would have been unreasonable to expect that work to be more polished on the schedule they were following, however, since Liss had been forced to leave much of this paper unwritten until the last minute. He could not complete his paper until the other parts of the analysis, on which his conclusions depended, were finalized (Liss, personal communication).

2. THE "OCTOBER MASSACRE" AND WRITING THE EVIDENCE PAPER

These papers were heading into an ambush, however, because of growing unhappiness with the way that the top group and their godparents planned to present the results of CDF's top search. Henry Frisch felt that publishing "four incomprehensible papers back to back in *Phys. Rev. Letters* that raised more questions than they answered" was "the wrong approach" and would potentially "generate a tremendous backlash from our solid state colleagues." Frisch attempted to raise this issue during one of the regular weekly CDF meetings, but he was told that there was no time on the agenda. "I finally went underground. I just decided I would do it via e-mail. . . . It was guerrilla warfare finally" (Frisch 1995).

Such measures set the stage for what Tony Liss later called the "October massacre." Collaboration members received copies of the four papers, which the top godparents then presented at the October collaboration meeting. "Basically, the collaboration hated these papers," recalled Paul Tipton. "They all had the same introduction . . . appeared to have not enough detail, left a lot of open questions." The fourth article by Liss, which was supposed to tie everything together and make sense of it all, "wasn't even really finished" (Tipton 1995). Collaboration members believed that *PRL* would never allow the collaboration to publish four papers strung together, and expressed their opinion that, in Tipton's words, "these papers suck" (Tipton, personal communication). The collaboration decided, over the objections of the authors of the four *PRL* papers, to submit instead one long paper to *PRD* that would give a complete account of the analysis, integrate the three counting experiments, and present an improved mass analysis. CDF could then clarify the relationships between the counting experiments. A brief summary submitted to *PRL* would accompany the *PRD* article. Spokespersons Mel Shochet and Bill Carrithers asked Tony Liss to put the *PRD* paper together. He responded that he would not assume responsibility for it alone, as he had for the aborted fourth *PRL*; Paul Tipton joined him in taking on this task (Liss, personal communication).

Liss and Tipton led a core group of about ten people in the arduous task of writing the text of the paper. This core group met a couple of times a week between the October collaboration meeting and Christmastime, when they released the first draft to the collaboration. This draft bore the same title as the paper that CDF eventually published: "Evidence of Top Quark Production in $\bar{p}p$ Collisions at $\sqrt{s}$ = 1.8 TeV" (Liss, personal communication).[2] Each member institution jointly posted their criticisms and questions regarding the paper in a computer file open to the collaboration. Liss and Tipton were responsible for responding to all of these, although many of the answers were written by other members of the core group. Rather than sending these replies directly to the questioners, Liss and Tipton took the unusual step of sending them to the godparents, who reviewed the responses before sending them on to the collaboration, sometimes intervening to impose a more courteous tone on the replies. Those working on the paper, beleaguered and feeling the pressure to produce quickly, sometimes felt annoyed at questions they received. "It was difficult in many cases not to send back snide remarks after you've been working on something so hard, [and] someone sends you some comment you think is really foolish.... It was good that we could write our snide comments and it was good that they didn't actually get back to [the collaboration]" (Liss 1995).

The Evidence paper generated an unusually large amount of interest among CDF members, some of whom scrutinized each sentence. Some of the comments called for clarification of a calculation. Others asked the authors to include a particular cross-check on the analysis. Numerous disputes centered on the inclusion of plots and figures.

At a January 13, 1994, collaboration meeting, Alvin Tollestrup outlined work done in response to questions from collaborators. This included cross-checks examining the effects of 12 different variables on the SVX tagging rate as well as investigation of an excess of tagged Z events, a problem that is discussed in more detail in Section 4.

A second draft responding to the first round of criticisms circulated about a month after the first (Liss, personal communication). A debate arose over the title. A set of informal and approximate conventions governs titles of experimental papers in particle physics. A paper with a title beginning "Search for..." typically reports a negative result from the search for some new phenomenon. When an article presents a claim to have definitively established the existence of the phenomenon in question, the title often begins with "Observation of..." or "Discovery of..." Some members of the collaboration thought that the top results were not strong enough to justify any title more suggestive than simply "Search for Top Quark" – the kind of title CDF used in reporting negative results and mass limits. Paul Tipton believed that "our biggest risk would be that we were viewed as trying to make too much of this . . . if you soft sell this thing it's going to fly a lot better," so the weaker claim of a "search" was more appropriate (Tipton 1995). Giorgio Bellettini, on the other hand, believed that a title indicating a stronger claim than "Evidence of Top Quark Production" was justified. Based in part on his belief that CDF should include the kinematical evidence in the statistical significance of the result, he was "not convinced that we should just talk of . . . evidence" (Bellettini 1995).

Moreover, the paper's authors needed to win over some collaboration members who did not believe that the collaboration should publish the results at all. The excess of candidate events over background was not very large, and the total number of events was small. Some CDF physicists expressed concerns over publishing on the basis of such small statistics. "Some people thought there was definitely a top signal, and others were getting in the way. Others thought it was marginal and being railroaded through" (Frisch, personal communication). Krys Sliwa belonged to the second group. Noting that the size of the effect under discussion was roughly equivalent to 2.8 standard deviations, Sliwa observed that during the earlier period of bubble chamber physics, "anything below 5σ significance was ignored as a possible fluctuation. It is clear that the standards have significantly deteriorated since. None of the already published CDF papers on top quark should have been submitted yet, judged by the 'old' standards."[3] Sliwa also complained that discrepancies appeared in some distributions when the top candidates in the data were compared with Monte Carlo–generated top-antitop events (Sliwa, personal communication).

These facts did not bother most collaboration members sufficiently to cause them to object to publishing the results as part of an evidential claim in support of the top-quark hypothesis. However several other issues became points of contention, some of which were resolved to everyone's satisfaction, others of which remained subjects of controversy. In the next few sections, I will examine a number of these issues, some of which sparked heated controversies within the collaboration. Specifically, I will discuss: whether the SVX *b*-tagging algorithm was chosen in a biased manner (Section 3); why the data showed an excess of tagged *Z* events (Section 4); the appropriate p_T cut for the SLT analysis (Section 5); a candidate event disputed by authors of different analyses known as the "pseudo *e-μ* event" (Section 6); what kind of discussion of the kinematics of the candidate events to present (Section 7); and whether to count events or "tags" (Section 8).

3. THE CHOICE OF THE SVX TAGGING ALGORITHM

The fact that the jet-vertexing algorithm used in reporting the results of the SVX search yielded more tags than its competitors raised some eyebrows within the collaboration. This certainly made the results appear stronger, but it prompted some to question whether the algorithm was chosen because it was superior to the others, or because it had the good luck to happen upon the largest number of candidate events. The top group replied that they chose jet vertexing because it directly yielded a distribution that could be compared to the b lifetime and had better-studied backgrounds; furthermore, at the time they chose it, jet vertexing did *not* tag the largest number of events. The algorithm picked up an additional event as a result of subsequent improvements in the tracking code used for the SVX detector. This reply satisfied most collaboration members. They concluded that the choice was made in an honest and unbiased manner.

However, some considered such a claim dubious and believed that jet vertexing was chosen precisely because it maximized the apparent signal, putting the best "spin" on the data. Some questioned whether jet vertexing would have been retained as the tagger of choice if one of the other algorithms turned out to tag more candidate events.

Reinforcing such doubts were hints in the data that the jet-vertexing algorithm itself might be biased. For example, more events were tagged by both the SVX and SLT searches than one would expect based on Monte Carlo simulations of top decays. That could indicate a bias resulting either from tuning jet vertexing directly on events in the data (choosing cuts to include particular events as candidates), or from choosing the SVX tagger on the basis of maximizing the number of candidate events. But the excess could also be bad (or good?) luck, or reflect an inaccurate Monte Carlo model. Hence, no one could make a conclusive argument that the SVX tagging algorithm had been directly or indirectly tuned on the signal. Yet during "run Ib," with more data, the question of bias in the jet-vertexing algorithm resurfaced, as I will explain in the next chapter.

4. THE EXCESS OF TAGGED Z EVENTS

Physicists both in and out of the top group demanded checks on the tagging results using various background samples in order to test the top group's assumptions regarding the efficiency of the counting experiment algorithms. Similarities between the production mechanisms for W and Z bosons suggested that one way to study the efficiency of the b-tagging algorithms was to run them on a sample of events with multiple jets and Zs. This sample would resemble that in which CDF looked for top quarks but should not actually contain any top quarks (a standard model top quark would not decay into the Z boson). Hence it should not exhibit any excess in Z events tagged for secondary vertices beyond what one expects from background. Finding such an excess would suggest that something besides the top quark might be contributing to the excess of top candidate events.

Unfortunately for CDF, they did find a small excess of tagged events in the Z + jets sample. The sample itself was quite small. Using tight cuts to select Z candidates, the Ia data contained only five Z events with three or more jets. Among five such events, top group physicists predicted that they should find an average of 0.31 ± 0.05 SVX tags and 0.33 ± 0.03 SLT tags (0.64 ± 0.06 total), based on the assumptions that they had used in calculating the backgrounds for the lepton + jets searches. Although no SLT tags were found in the five Z + jets events, they found two SVX tags. They faced two possible explanations for this excess: either they had failed to take into account some source of b quarks in calculating backgrounds, or it was just bad luck, an upward fluctuation in one of the backgrounds that *had* been taken into account. Taking the former possibility seriously would cast doubt on the validity of interpreting the excess lepton + jets candidate events in the SLT and SVX searches as top decays.

Those meager two events generated considerable worry amongst top group members. They explored the problem at length while writing the first draft of the Evidence paper. At a top group meeting held on the day after the October massacre, Pisa's Franco Bedeschi gave a presentation on this issue and considered the possibility that at least one of the tagged Z events was a "bad tag." But they could not make the events go away quite so easily, and the problem came up again a month later. In a presentation to the top group, Rochester physicist Brian Winer showed that, on the group's assumptions about background sources, such a discrepancy had a probability of 14%, which was judged a high enough probability to regard the tagged Zs as an ordinary statistical fluctuation (*Top* 11/18/93).

However, the tagged Z events continued to bother some collaboration members, who considered the problem unresolved. The events were ultimately included in the Evidence paper as a failed cross-check, with the comment: "Although statistically limited, the excess of tagged Z + 3 or more jet events could potentially signal a (non-$t\bar{t}$) source of heavy-flavor production in association with a vector boson, which exceeds our background predictions'" (Abe, Amidei, et al. 1994b, 3003). (This anomaly would disappear in the higher statistics of run Ib, suggesting that it was indeed simply bad luck, a statistical fluctuation that went the wrong way.)

5. THE SLT p_T CUT

At the January 13, 1994 meeting in which the collaboration discussed the first draft of the Evidence paper, Claudio Campagnari responded to a criticism of the SLT analysis that deserves some attention, both because of the methodological issues it raises and because of the controversy it stirred up within the collaboration.

The SLT analysis tagged bs by looking for soft leptons from the decay of b quarks. The controversy concerned one of the parameters used in this analysis, the transverse momentum p_T of the lepton. Soft lepton tagging had been used, for soft muons, in the earlier top mass limit paper based on the 1988–9 data, in which the muons were required to have p_T between 2 and 15 GeV/c. (By contrast, for the dilepton search, a lepton had to have $p_T \geq 20$ GeV/c.)

The SLT search had originated when CDF was still searching for a top quark that could possibly have a mass just slightly larger than the W. That situation changed as run Ia progressed. It became increasingly obvious that the top quark probably had considerably greater mass than they had earlier expected. This had implications for the SLT search. If the mass of the top exceeds that of the W (around 80 GeV/c^2) by just a little, then when the top decays, almost all of the energy made available by that decay goes into making a W, and only a small amount of energy remains for the b – not enough to give it a really hard kick. Accordingly, the lepton produced when that b decays would in turn have only a small amount of momentum. The greater the difference between the top quark mass and the W boson mass, the greater the momentum of the b quark that is produced in top decays, and the greater the momentum of the lepton produced when the b decays. In other words, the more massive the top, the harder the soft leptons would be.

The dispute that emerged concerned whether to increase the lower threshold of the soft lepton p_T requirement to 4 GeV/c. Much of the background for the SLT search would have soft leptons with p_T between 2 and 4 GeV/c. Raising the cut would consequently eliminate a substantial amount of background. As it happened, however, it would also eliminate much of the excess over background that CDF could report for the SLT search. This issue became a point of contention for G. P. Yeh.

As one of the principal authors of the dilepton analysis, Yeh was busy answering questions and working with his students prior to the release to the collaboration of the first draft of the Evidence paper, just before Christmas 1993. Over Christmas, Yeh read the other analyses. "Within an hour or so [of reading the SLT results], I was curious [about] what the seven candidates looked like, what's their p_T. . . . There was no plot in the paper to show that." In addition, the paper had no plot showing the expected p_T distribution of signal and background. Yeh requested these plots, and, he reported, Avi Yagil replied that Yeh could produce them himself based on information in documents already available to the collaboration. Since he "didn't think that was the right reply," Yeh went to the godparents and asked that the plots for the expected and observed p_T distributions be provided for the SLT analysis (Yeh 1995).

Yeh's request was favorably received by at least one godparent, Cathy Newman-Holmes of Fermilab, and when the collaboration had its next meeting in January, Yeh prevailed in getting the plots included in the paper. These plots, which Yeh described as "standard" for any analysis of this sort, came to be known as "G. P. plots" because of his insistence on their importance (Yeh 1995; personal communication).

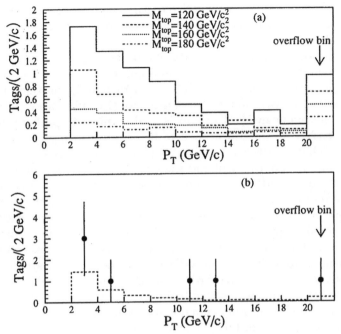

FIGURE 4.2. The G. P. plots: (a) shows the expected p_T spectra of SLT-tagged soft leptons based on Monte Carlo predictions for various top masses; (b) displays the p_T spectrum of SLT-tagged leptons in the data (dots) compared to the spectrum of the expected background (dotted line). Compare the expected background in the 2–4 GeV/c^2 bin in (b) to the expected signal in that bin from a high mass top (bottom two curves) in (a) (Abe, Amidei, et al. 1994b, 3001).

These plots (see Figure 4.2) showed a fairly high expectation value for background in the soft lepton p_T spectrum between 2 and 4 GeV/c, while indicating that only a small amount of signal could be expected in that same portion of the spectrum, given a massive top quark such as the one reported on in the Evidence paper (about 174 GeV/c^2). Furthermore, three of the seven SLT candidate events fell into precisely that background-dominated portion of the spectrum. Yeh succeeded in getting the plots included in the paper, but what these plots showed did not satisfy him. "In the end … the 2 GeV cut was not good. … 170 GeV top does not give you such low p_T events" (Yeh 1995).

The plots raised a troubling question: Given that the top group had considered moving the soft lepton p_T cut to 4 GeV/c, what was the reason for keeping it at 2 GeV/c? In light of the several candidate events that would have been missed if the cut had been raised, some worried that the SLT authors kept the cut at 2 GeV/c to avoid losing those events. Selecting cuts on the basis of whether or not they deliver a desirable number of candidate events is a worrisome potential source of bias in particle physics experiments. High energy particle experimentalists have a name for the practice: "tuning on the signal."

The authors of the SLT analysis, Claudio Campagnari and Avi Yagil, defended their choice of cuts and the results of their analysis. The 1992 mass limit paper had used a cut of 2 GeV/c, and Yagil and Campagnari argued that the new analysis should cover a mass range continuous with that covered by the old analysis. "The Run Ia analysis picked up where the 1992 analysis ended, so we also needed to be sensitive to top quarks around 100 GeV. That's partly why we used 2 GeV once again" (Campagnari, personal communication). Yagil and Campagnari maintained that they had chosen 2 GeV/c, and there was no justification for changing that decision after the results were in, which would amount to tuning *out* the signal. Furthermore, although one measure of optimization, the ratio of expected signal over the square root of expected background, favored moving the cut from 2 to 4 GeV/c, another measure, the signal over background ratio, remained roughly constant for both choices. Campagnari considered it a "judgment call" (Campagnari, personal communication). In addition, while the background was indeed higher between 2 and 4 GeV/c, they also understood this background better, so it had smaller systematic errors.

Thus, the defense of the SLT analysis rested on three claims: (1) the soft lepton p_T cut had been chosen to be 2 GeV/c prior to analyzing the data; (2) according to one plausible measure of optimization, they found no significant difference between the 2- and 4-GeV/c choices; and (3) by placing the cut at 2 GeV/c, they reduced the systematic error in the SLT search.

Responses to the controversy varied. Paul Tipton, who had worked with Campagnari on the soft muon analysis for the 1992 mass limit paper, was satisfied that the SLT cuts had not been tuned on the signal, but had been properly predesignated: "We set our cuts in advance, and you toss the dice, and you get what you get.... 2 GeV was the right answer because we hadn't tuned our cuts on the data; there was no reason to, after looking at the data, tune our cuts to get rid of top.... I think Claudio thought some of the proponents of changing the cuts didn't really want to find it." The dilepton analysis had been the primary engine behind mass limits, and those involved in it had developed, some thought, a mentality for setting strong limits on the top mass, which favored *not* finding candidate events (Tipton 1995).

Others in the collaboration had reservations about the p_T cut but did not place great importance on the issue. Milciades Contreras, who had worked closely with G. P. Yeh on the dilepton analysis, stated that "[m]y feeling was that the cut was perhaps a little bit too low, and that perhaps we were... letting luck play in our favor." Yet, he insisted, the difference in question was only a matter of "shades." "I'm not concerned really about that, and I could live with that cut perfectly well" (Contreras 1995).

Some lodged harsher criticisms. As G. P. Yeh saw it, the fact that the
signal-to-background ratio for the SLT search was less than 1.0 violated a
commitment to keep that ratio higher than 1.0 for each part of the analysis
(Yeh 1998b). The collaboration affirmed this commitment, he believed, in
the introduction to the Evidence paper: "After imposing selection requirements, a signal-to-background ratio greater than 1.0 can be achieved with
reasonable efficiency for $t\bar{t}$" (Abe, Amidei, et al. 1994b, 2968).

In addition, not everyone accepted the claim that the SLT group set
their p_T cut in advance. In January 1993, midway through run Ia, the SLT
group issued a CDF note, number 1961. In that paper, the authors examined
$9\,\mathrm{pb}^{-1}$ worth of data, nearly half of the total data reported on in the Evidence
paper. They evaluated the efficiency of the SLT algorithm using both a 2-
and a 4-GeV/c p_T cut and concluded that "[i]n raising the p_T threshold
from 2 to 4 GeV, efficiency of the tag is reduced by a factor between 35%
for a 100 GeV top, and 20% for a 140 GeV top.... the background will
however be reduced by a factor of two" (Campagnari, Farhat, et al. 1993a,
15). Further into the note, the authors even suggest they may be in favor
of having the cut at 4 GeV/c: "raising the p_T cut to 4 GeV/c reduces the
background by a factor of two with a modest decrease in acceptance. A
decent signal to noise [ratio] can be achieved by requiring at least three jets"
(ibid., 16). Later studies of SLT backgrounds and efficiencies, released to
the collaboration in July and September of 1993, continued to show results
using both 2- and 4-GeV/c cuts (Campagnari, Farhat, et al. 1993b; 1993c;
1993d).

This increased the suspicions of some critics. Having looked at half the
data, they argued, the SLT authors were still trying to decide between putting
the cut at two and putting it at four, although they were looking at Monte
Carlo simulations of top quarks with masses of 100, 120, and 140 GeV/c^2.
At the end of the run, some ambiguity still remained as to the finality of
their decision. If in fact the cut had been chosen in advance of analyzing
the full Ia data set, it was not clear exactly when the choice had been made.
Fermilab physicist Morris Binkley regarded the choice with some skepticism.
Describing the choice of SLT algorithms as a choice from a "continuum" of
algorithms, one for each possible p_T cut, Binkley noted, "I believe that choice
was made after we'd seen the data.... Certainly that choice was discussed
extensively after we'd seen the data. Some people may have in their minds
decided that they were going to make the cut at 2 GeV before they saw the
data; I'm not sure" (Binkley 1995).

However, most members of the collaboration found Campagnari and
Yagil's arguments persuasive enough that they, like Milciades Contreras,
could "live with" the analysis as it stood. The SLT analysis became part of
the Evidence paper with the 2-GeV/c cut in place, accompanied by the G. P.
plots. In Chapter 5, I will again address the question of bias in both the SVX
and SLT analyses, in the light of information that became known during

run Ib, and in Chapter 7 I will discuss the problem of bias and "tuning on the signal" in the context of a philosophical account of statistical testing.

6. THE "PSEUDO e-μ EVENT"

Amid the heightened tensions of a possibly impending discovery, the debate over the SLT analysis raised sensitive questions about the motivations of various collaborators. Some suspected the authors of the dilepton analysis of motives rooted in bad feelings from a previous dispute regarding a single event. Dan Amidei, who recalled having dubbed it the "pseudo e-μ event" (the name stuck), described the event: "it had a muon, it had two jets, it had an object which for all intents and purposes was an electron – a perfect dilepton event. The electron-like object did not quite pass the electron cuts.... You couldn't imagine that thing was anything other than an electron. But it didn't pass the cuts that we agreed upon" (Amidei 1998).

The problem again concerns when and how decisions were made. In the section of the Evidence paper describing the dilepton analysis, CDF notes that in an earlier section of the paper "we discussed electron and muon selection, and both strict and loose criteria to select central leptons were defined. We require each event to have at least one central lepton passing the strict cuts" (Abe, Amidei, et al. 1994b, 2976). The strict and loose cuts for electrons in the dilepton analysis are given in Table 4.1. CDF had three banks of muon detectors in the central detector region: the central muon chambers (CMU), the central muon upgrade (CMP), and the central muon extension (CMX) (see Figure 2.5). The strict muon cuts required that a candidate be recorded in the CMU/CMP detectors. A muon candidate recorded in the CMX detectors (at a greater distance from the collision point along the beam line) would not qualify. The loose muon cuts allowed particles leaving tracks in the CMX detectors to count as muons. Since the

TABLE 4.1. *Cuts Used to Select Central Electrons for the Lepton + Jets Search, Along with "Strict" and "Loose" Cuts Used for Central Electrons in the Dilepton Search*

Variable	$e +$ Jets Cut	Dilep Strict Cuts	Dilep Loose Cuts
E_{had}/E_{EM}	<0.05	<0.05	$<0.055 + 0.045E_T/100$
E/P	<1.5	<2.0	<4.0
L_{shr}	<0.2	<0.2	<0.2
$|\Delta x|$	<1.5 cm	<1.5 cm	<1.5 cm
$|\Delta z|$	<3.0 cm	<3.0 cm	<3.0 cm
χ^2_{strip}	<10	<15	
z-vertex match	<5.0 cm		
Isolation	$I_{cal}/E_T(e)$ <0.1	I_{trk} <3 GeV/c	I_{trk} <3 GeV/c
Efficiency	$(84 \pm 2)\%$	$(87 \pm 1)\%$	$(94 \pm 1)\%$

Source: Abe, Amidei, et al. (1994b, 2973).

dilepton analysis required at least one central lepton (either electron or muon) to pass strict cuts, an event that had a muon candidate passing the strict cuts and an electron candidate passing the loose cuts should have qualified. However, according to G. P. Yeh, at the time that the pseudo e-μ event appeared, "we had done the other part, of including tight electrons and loose muons. But we forgot to include tight muons, loose electrons" (Yeh 1998b). When they applied the tight muon/loose electron cuts to the data, the pseudo e-μ event passed, but otherwise it did not qualify as a dilepton candidate.

According to Dan Amidei, "There ensued from that a very bitter argument in the collaboration about whether those guys were tuning their cuts" by choosing the loose cuts in such a way as to guarantee inclusion of the pseudo e-μ event in their sample (Amidei 1998). Complicating the dispute was the fact that the event otherwise qualified as an SLT candidate. Since the SLT and dilepton analyses both looked for events with two leptons, but were supposed to be looking for completely different patterns of top decay, no event could qualify for both. Hence, one of the requirements for SLT candidates was that they *fail* the cuts for the dilepton search. Since the pseudo e-μ event passed all the other requirements for being an SLT candidate, it would automatically be in the SLT sample provided only that it failed to make it into the dilepton sample. Melissa Franklin, a CDF member from Harvard, recalled the debate as "insanity. . . . People would get up and say 'I think that event should go in this analysis!' 'No I want it in my analysis!' [It was] nuts, completely nuts" (Franklin 1998).

G. P. Yeh denied having strong feelings about whether to include the pseudo e-μ event in the dilepton sample. "I thought we could go either way. I didn't feel strongly about it." Others in the collaboration insisted that, since everyone had seen the Ia data by the time the overlooked class of dilepton events was addressed, the cuts should be left as they were. According to Yeh, his response to this verdict was "okay, fine" (Yeh 1998b).

Although Dan Amidei regarded this decision as a victory for "conservatives" within the collaboration, the decision ultimately enhanced the size of the sample of candidate events for the search as a whole. For reasons discussed later, events tagged by both the SLT and SVX b-tagging algorithms were counted twice. The pseudo e-μ event not only passed the SLT cuts but was also tagged by the SVX algorithm; therefore, by being excluded from the dilepton sample, it counted double in the final statistical assessment of the top-quark evidence. As another consequence of the episode, according to Amidei, those who had been in favor of relaxing the dilepton cuts to include the pseudo e-μ event "lost some of their influence [on] the process" (Amidei 1998).

Nevertheless, the pseudo e-μ event would become a dilepton event again in the run Ib analysis, when changes in the SVX tagger removed it from the SVX sample.

7. THE ROLE OF KINEMATICAL EVIDENCE

Perhaps the most contentious issue of all within the collaboration concerning the content of the Evidence paper was the relevance of the kinematical features of the candidate events. Among the three main strategies to identify top decays by reconstructing the kinematics of events described in the previous chapter, the "event structure" analysis developed by Pisa physicists had the most influential support. CDF included many Italian collaborators, whose vital contributions to the collaboration included indispensable work on the SVX. They strongly supported the event structure analysis. However, the counting experiments had set off the drive to publish a top claim, and many people who had worked on the counting experiments did not regard the event structure analysis as highly credible. In determining what to include in the Evidence paper, the collaboration faced a range of options with respect to the kinematical information. At one extreme the collaboration could decline to discuss the kinematics entirely. At the other extreme, one could treat the kinematic analysis on a par with the counting experiments and derive a statistical significance from the kinematics to combine with that of the counting experiments in an overall significance estimate. They eventually chose a middle course, but in the process, according to Alvin Tollestrup, the disagreements "almost split the organization open" (Tollestrup 1995).

Giorgio Bellettini believed strongly that the kinematical information yielded evidence for the top and deserved a prominent place in the presentation of the results. Prior to the "October massacre," some CDF members had written a *PRL*-length article presenting the kinematic analysis, and Bellettini believed that it deserved publication alongside the other four articles under preparation. "My position was that since it was work well-analyzed, quoted within the errors and uncertainties, it ought to be published together with the other four. But unfortunately I was in the minority" (Bellettini 1995).

When the collaboration chose instead to write a large *PRD* article, Bellettini wrote a chapter for it presenting "all the kinematical evidence that we have." However, the collaboration did not adopt Bellettini's version of the chapter on kinematics, choosing instead a version that "did not elaborate much on the kinematics. It just showed the energy distribution of the various jets without working out any consequence" such as a statistical significance, although one could derive such a number from the event structure analysis. Looking back on the rejection of his version of the kinematical section of the Evidence paper in favor of a truncated version, Bellettini stated that "[t]his was a very hard period, in which I think we made a mistake" (ibid.).

John Yoh also wanted to use kinematic features of the top candidate events as part of the evidence for the top quark, but in a different way, employing a quantity called H_T, defined as the scalar sum of the E_T of the jets, the p_T of leptons, and the missing E_T carried off by neutrinos in a given event. (D-zero later used a cut on H_T to separate top signal from background in a counting experiment analysis.) Yoh described H_T as "a very good discriminator for top." However, this technique was not viewed favorably by the leaders of the top group. As a result, in Yoh's view, the collaboration chose not to pursue a "powerful approach" to finding the top, which "almost certainly would have provided an additional 1–2 sigma of significance to the signal," although he acknowledged that the approach was dependent on a Monte Carlo model and "should perhaps be presented only as additional evidence, and perhaps not the main-stream" (Yoh, personal communication).

John Huth had already served as a godparent for the previously developed kinematic analyses and became a godparent for the Evidence paper partly because of that prior experience. He later recalled that people who had been working on the counting experiments tended to wish for the kinematic information to be excluded from the paper. By contrast, people in the collaboration who had not been involved in developing top search algorithms tended to favor including that information. "What happened was there was a bit of a groundswell of interest in the kinematics that in the end put it in, in a somewhat muted form. . . . There was a lot of divisiveness there."

As Huth recalled, attitudes toward the kinematics became polarized. "Some people even resisted the idea of even looking at kinematic information . . . in fact actively discouraged it." On the other hand there were others working on the kinematic analysis who "thought they'd already discovered [the top], based on the kinematic information." In Huth's opinion, one of the important tasks for the Evidence paper godparents was to "get both sides to sort of back away from their rather strong positions to take a much more rational view of the kinematics" (Huth 1995).

Alvin Tollestrup, as leader of the godparent committee, asked Steve Geer, a Fermilab physicist, to lead the effort on the Evidence paper's presentation of kinematics, and Geer produced a pared-down version of the event structure analysis. According to Tollestrup, the task was politically difficult, because of the widely differing opinions within the collaboration, but "it came through really well-balanced" (Tollestrup 1998). Collaboration members who had been working on the counting experiments accepted the resulting analysis, but the reasons for acceptance varied. Some regarded the analysis as genuinely relevant, while others accepted it as a political expedient to keep the paper-writing process moving forward.

Many people not directly involved in the kinematic analyses tended to regard them as a kind of cross-check or support for the counting experiments. According to this view, the kinematic analysis did not by itself provide evidence for the top quark, but gave one more confidence in the results of the counting experiments. The counting experiments yielded the actual evidence for the top.

Bruce Barnett of Johns Hopkins believed that the most important step to convince people that CDF had evidence for the top quark was to show an excess of events with high energy leptons, with the proper number of jets, and with *b* quarks in those jets. "To do that you needed the SVX and the SLT" counting experiments. The kinematic studies "indicated that the events looked unusual. They looked more like top than they looked just like background." This information simply allowed one to feel more comfortable presenting statistical significance calculations for the counting experiments. The kinematics "made you feel much more confident," so that "you don't have to drink as much Maalox" before presenting the top results to skeptical physicists at a conference. In that sense, according to Barnett, the kinematics were comparable to other reassuring features of the data not included in the statistical significance, such as the *b*-tags in the dilepton sample (Barnett 1996).

Mel Shochet expressed a similar attitude. He believed that the "overwhelming majority" of the collaboration agreed with him that the kinematic analyses were "not the way you discover the top quark." Those analyses "found events with *W*s in which the jets were more energetic than one expected from a Monte Carlo simulation of the background. That presents three alternatives." One is that the Monte Carlo calculation is mistaken. Another is that the data contains *something* with a large mass that is causing more energetic jets than expected, but that something is not the top quark. Finally, it is possible that the top quark really is there. "And in that analysis there was no clear way of addressing" which of these three alternatives was correct. The kinematic analysis was "additional, supporting evidence that what we saw was consistent with the top picture, but it was not in my view prima facie evidence for the top quark." Nor could one combine a statistical significance from the kinematic analysis with that from the counting experiments, in Shochet's opinion, since "we did not feel that the study of systematic uncertainties was sufficiently far along so that the event structure paper could be put on the same quantitative level as the counting experiment. Moreover, it was clearly correlated with the counting experiment." Consequently, one could not simply multiply the two significances, and the collaboration did not understand the correlations well enough to know how properly to combine their significances (Shochet 1995).

Hans Jensen also regarded the kinematics as "checks" on the results of the counting experiments. The kinematical analysis itself, however, relied on "things you don't completely understand. You're talking about response to jets, and Monte Carlos for backgrounds that you don't have completely under control, so I think many people were reluctant to say that this is really part of the evidence." Although one might quote a smaller significance level by combining somehow the kinematics with the counting experiments, "most people felt that this is not the straight and narrow path that convinces you that this is top" (Jensen 1995).

Some regarded the kinematics as dispensable to the Evidence paper. According to Milciades Contreras, "the Evidence paper . . . would not have suffered much with the exclusion of the kinematics section. However, this was work that was done by some of our collaborators, it added a little bit more information, and the consensus was to include it" (Contreras 1995). Paul Tipton believed that the kinematic approaches to finding the top "somewhat got in the way" by providing a point of contention between those who believed the data constituted evidence and those who were skeptical. Many of those who believed that the Ia data yielded real evidence for the top quark, according to Tipton, wanted to include the kinematic results in the paper. Skeptics used the difficulties of the kinematic analyses to argue that any kind of claim about the top would be premature. "So in the end of the day I think they held things up" (Tipton 1995). Henry Frisch described the kinematic analyses as "peripheral." "Comparisons with Monte Carlos do not mean much to me. It has to be data-to-data comparison in general. . . . [The kinematics section of the Evidence paper] was put there partly for political reasons, partly to mollify people. The essence of it was in the tagging. . . . It was nice that [the kinematics] was there, sort of, but mostly it was put there just to make people go away. . . . The tagging was the key" (Frisch 1995).

As late as March 31, 1994, on the very brink of releasing the Evidence paper, some members of the top group still questioned whether kinematics, or even a mass estimate, should be included in the evidence paper at all. At a top group meeting on that date, some members advocated including just those sections that described the counting experiments, along with an assessment of the statistical significance of the counting experiment results and a measurement of the $t\bar{t}$ production cross section. They worried that the sections on the kinematics and mass estimate had not been finalized to everyone's satisfaction and could not be made ready in time for inclusion in the paper. However, at the end of the discussion, the group remained committed to including a brief discussion of the kinematics. They decided to declare the physics content of the paper closed to further substantial modification. The scope of the Evidence paper would neither expand beyond nor diminish below the work already agreed to and documented (*Top* 3/31/94).

If physicists working on the counting experiments took a skeptical attitude toward the kinematics, and those working on the kinematic efforts tended to think that they showed convincing evidence for the top, still others had reservations about placing too much importance on any one aspect of the analysis, including the counting experiments. John Huth noted that "the requirements for a discovery are really . . . multiple signatures. . . . It's a complex object you're trying to demonstrate you've seen, and so you can't just do it with one single variable." He felt that the counting experiments gave one "just a number," without confirming "other aspects of the properties of the thing, which was also important" (Huth 1995).

Morris Binkley had studied the kinematics of top events carefully. He considered the relative likelihood variable used by the event structure analysis "a very nice variable," but just as he suspected that the counting experiments might be partially tuned on the signal and hence biased, he also believed that the kinematic analysis that was shown in the Evidence paper was "biased with statistical fluctuations and gave people a biased view of what the kinematics looked like." So while he agreed with those who had worked on the counting experiments that it would have been a mistake to give a significance based on the kinematics in the Evidence paper, he also considered it a mistake to give a single significance estimate based on the counting experiments. A more accurate picture would emerge, according to Binkley, from showing a variety of test statistics for finding the top, using a variety of sets of cuts to select events (Binkley 1995). He proposed a plot showing such a variety of cuts for the kinematic analysis in the Evidence paper, which was rejected, on the eve of the paper's release, as too complex (*Top* 3/31/94).

8. DOUBLE-TAGGED EVENTS AND SIGNIFICANCE: WHAT TO COUNT?

One aspect of the top analysis that had to be justified to the rest of the collaboration was the method of counting used to calculate the significance of the counting experiments. Twelve events were selected as candidates by at least one of the three counting experiments, with a background expectation of about 5.7 events, yielding a significance estimate of 1.6×10^{-2}. However, three of those events were picked by both the SVX and SLT analyses. Many top group members felt that these "double-tagged" events meant that CDF had stronger evidence of having found the top than the significance estimate of 1.6×10^{-2} indicated.

One could, however, count tags instead of events, which yielded 15 counts (13 lepton + jets tags plus 2 dilepton events) with a background expectation of about 5.96. These numbers yield a considerably lower probability of getting such a result in the absence of any top quark production (2.6×10^{-3} rather than 1.6×10^{-2}). As justification for counting tags in this way, top group members noted that the double-tagged events were simply much less likely to occur in the absence of real b quarks than were single-tagged events. They estimated the probability of double-tagging an event with b-quarks to be 70 times greater than the probability of double-tagging an event lacking b quarks, which was calculated to be about 0.05% (*Top* 1/20/94).

However, the authors of the top analysis had to fight the question of double-tagged events on two opposite fronts. The double-tagged events made the evidence for top decays in the data stronger at the same time that they raised troubling questions. The top group estimated that they should expect, on average, 1.3 double-tagged events based on the assumption that the data contained a mixture of top events and background, whereas 3 events were actually double-tagged (an outcome with a probability of 0.14) (*Top* 1/20/94). This excess over expectations raised the possibility that they did not understand the tagging algorithm efficiencies as well as they should. If the algorithms were pulling in more candidate events than expected, then perhaps this meant that their background estimates were too low. Possibly some physics process other than top decays or the anticipated sources of background was contributing to the *b*-tagging rate. The excess of tagged *Z* events only strengthened this concern. The excess double-tagged events could also indicate biases in the counting experiments themselves, a by-product of having chosen cuts specifically to capture particular events.

To address the problem required a reexamination of the reasoning behind the 1.3 expectation value for double-tags assuming a combination of top decays and background. This number had been based on a Monte Carlo simulation of both top decays and background. But the background estimate for the SVX counting experiment relied on a method that calculated the probability of an SVX tag based on a number of measured features (parameters) of the events to which it was applied. Applied to the events in CDF's data sample, this method estimated how much background could be expected in events with *those particular features* found in the sample. Guillaume Unal used a similar technique to reevaluate the expectation value for double-tags in CDF's data. He applied a parametrization similar to that used in the SVX background estimate to the 7 events tagged by the SLT algorithm, assuming a mixture of top signal and background, and arrived at an expectation value of 1.8 ± 1.3 double-tagged events. Based on this number, the probability of finding 3 or more double-tags came out to be 0.30. Having narrowed the gap between top quark–based expectations and reality, Unal concluded that the rate of double-tags posed no problem for interpreting the excess as coming from top decays (*Top* 2/24/94).

Taken by itself, the decision to counts tags rather than events seems not to have generated great controversy within the collaboration (although one of the referees for the *Physical Review* questioned the justification for this choice; *Top* 6/10/94). A side effect of the decision, however, was to enhance the influence of the SLT results on the significance calculation. The SLT search had the lowest signal-to-noise ratio of the three counting experiments. Yet in the statistical significance estimate, each SLT tag counted the same as an SVX tag or a dilepton event. It followed that if, as some suspected, the large number of SLT candidates was partly the result of an upward fluctuation in the background for that search, that fluctuation would enhance the estimated statistical significance of the combined results even more than that of the results of the SLT analysis alone. CDF would abandon this method of combining results in run Ib, as I will explain in the next chapter.

9. ENDLESS MEETINGS AND NONCONVERGENT PROCESSES

While CDF attempted to resolve such disagreements, the physics community eagerly wished to know what was happening with the search for top at CDF. As long as the group had failed to find the top quark, they would present their data at conferences and discuss its implications (for the top mass, for instance). At the Lepton–Photon conference, Paul Tipton had told his audience to "stay tuned," and they had. But until the collaboration had reached consensus on their paper, CDF could not present anything at all. As a result, they stopped discussing their data at conferences, and even canceled some scheduled talks (Barnett, personal communication; Liss, personal communication). This tactic eventually drew attention even in *The Economist*, which noted the meaningful silence in a short piece titled "Quarks Don't Bark," inferring that a major development must be imminent (*Economist* 1994).

In the face of the controversies outlined here, the collaboration struggled to reach a consensus about the paper's contents. The collaboration received the first draft just before Christmas 1993. Replies and complaints came back, and another draft soon appeared, but some people were skeptical that this would yield a final resolution. Henry Frisch described the process as "ping pong . . . somebody takes a paper, works on it, puts it over the net, people don't like it, they send it back over the net. That's a nonconvergent process. You can't write a paper that way" (Frisch 1995). Frisch felt that the process had been too closed, that people in the collaboration not involved in the writing process but intensely interested in this major discovery needed more opportunity to participate. Others disagreed with him. "Mel [Shochet] felt [the process] was already open. I felt it was very far from open" (Frisch 1995). The process was under the control of only a small portion of the many people whose work had contributed to the top results, as Frisch saw it. Melissa Franklin recalled, "You would ask questions in meetings and people were really kind of testy. There were a number of people who felt they were in control of the top, and they were in charge, and you were just basically bugging them" (Franklin 1998). Alvin Tollestrup denied that decisions were made behind closed doors, and insisted on the necessity of smaller, orderly meetings to avoid big, open meetings in which "you'd just have chaos, with screaming and shouting." According to Tollestrup, "We pulled together as an advisory committee people who were writing the various sections. We had to have some organization where you could sort of get the goddamn paper written. . . . I just felt that that was better done in a small group of people. . . . [The meetings] weren't closed but they weren't announced" (Tollestrup 1998).

Frisch proposed "open daily meetings." He argued that allowing people to feel that they have an opportunity to participate would remove one source of resistance to the paper. Frisch's idea was to say to the collaboration, "'Everybody's invited, but let me warn you, we're going fast, and it's just going to end at some point. But don't say you weren't told. . . . If you care, participate.' . . . Then people don't have the moral outrage of saying, 'It happened behind closed doors,' or 'I wasn't asked,' or 'I'm more conservative than you are'" (Frisch 1995). The collaboration never held daily meetings but eventually did hold open meetings two or three times a week. At first the attendance was quite heavy, but it declined as time went on. "Most people get bored and wander away. . . . The point is the opportunity is there" (ibid.).

By April an approximate consensus had formed within the group to go ahead with an announcement that they were submitting a paper to *PRD* titled "Evidence for Top Quark Production in $p\bar{p}$ Collisions at $\sqrt{s} = 1.8$ TeV." This is not to say that everyone felt satisfied with every part of the paper. However, unhappy dissenters faced the realization that their colleagues were ready to go forward with the paper, and their only real recourse would be to remove their names from the paper. Such a form of protest is officially allowed within CDF, but it is a step that is perceived to weaken one's influence within the collaboration, making it a drastic step to be taken only as a last recourse. Furthermore, some physicists who were critical of parts of the analysis believed that other parts were sufficiently strong to compensate. John Yoh, for example, did not feel that he needed to take his name off of the Evidence paper in spite of his concerns about the SLT analysis because he regarded the kinematics of the events as a source of evidence. Consequently, he agreed with the central claim of the paper that CDF had found evidence for the top. "I was convinced that the conclusion we reached in the paper is correct and supported by the evidence, despite the concern I have with one of the analyses" (Yoh, personal communication). In the end, only one physicist, Eric Wicklund of Fermilab, took his name off of the paper.

CDF announced the release of their paper at Fermilab on April 26, 1994. The *New York Times* reported the event on its front page (Broad 1994). The Evidence paper actually appeared in *PRD* in September (see Figure 4.3).

It is an enormous article. In the *PRD*, it takes up 60 densely packed pages. This in itself posed a problem. While the *PRD* article served the purpose of satisfying those in the high energy physics community interested in knowing in great detail how CDF produced their result, "there is another group in . . . the broader physics community that really just wants the bigger picture" (Shochet 1995). To convey the relevant information to this less specialized group, some members proposed "a four page executive summary for the *Physical Review Letters*, which serves both audiences" (ibid.). Getting that *PRL* written was, according to Henry Frisch, "a second battle." After six months spent producing the *PRD* article, Tony Liss and Paul Tipton "were really tired, and they didn't want to hear about it." So the *PRL* was written "with rubber cement and scissors" in the conference room in the high energy physics building at University of Chicago by Mike Albrow, Henry Frisch, Dave Gerdes, and Weiming Yao (Frisch 1995). They gave the paper the same title

158

PHYSICAL REVIEW D   VOLUME 50, NUMBER 5   1 SEPTEMBER 1994

ARTICLES

Evidence for top quark production in $\bar{p}p$ collisions at $\sqrt{s}$ = 1.8 TeV

F. Abe,[13] M. G. Albrow,[7] S. R. Amendolia,[23] D. Amidei,[16] J. Antos,[28] C. Anway-Wiese,[4]
G. Apollinari,[26] H. Areti,[7] P. Auchincloss,[25] M. Austern,[14] F. Azfar,[21] P. Azzi,[20] N. Bacchetta,[18]
W. Badgett,[16] M. W. Bailey,[24] J. Bao,[34] P. de Barbaro,[25] A. Barbaro-Galtieri,[14] V. E. Barnes,[24] B. A. Barnett,[12]
P. Bartalini,[23] G. Bauer,[15] T. Baumann,[9] F. Bedeschi,[23] S. Behrends,[2] S. Belforte,[23] G. Bellettini,[23]
J. Bellinger,[33] D. Benjamin,[32] J. Benlloch,[15] J. Bensinger,[2] D. Benton,[21] A. Beretvas,[7] J. P. Berge,[7]
S. Bertolucci,[8] A. Bhatti,[26] K. Biery,[11] M. Binkley,[7] F. Bird,[29] D. Bisello,[20] R. E. Blair,[1]
C. Blocker,[29] A. Bodek,[25] V. Bolognesi,[23] D. Bortoletto,[24] C. Boswell,[12] T. Boulos,[14] G. Brandenburg,[9]
E. Buckley-Geer,[7] H. S. Budd,[25] K. Burkett,[16] G. Busetto,[20] A. Byon-Wagner,[7] K. L. Byrum,[1] C. Campagnari,[7]
M. Campbell,[16] A. Caner,[7] W. Carithers,[14] D. Carlsmith,[33] A. Castro,[20] Y. Cen,[21] F. Cervelli,[23]
J. Chapman,[16] M.-T. Cheng,[28] G. Chiarelli,[8] T. Chikamatsu,[31] S. Cihangir,[7] A. G. Clark,[23] M. Cobal,[23]
M. Contreras,[5] J. Conway,[27] J. Cooper,[7] M. Cordelli,[8] D. P. Coupal,[29] D. Crane,[7] J. D. Cunningham,[2]
T. Daniels,[15] F. DeJongh,[7] S. Delchamps,[7] S. Dell'Agnello,[23] M. Dell'Orso,[23] L. Demortier,[26] B. Denby,[23]
M. Deninno,[3] P. F. Derwent,[16] T. Devlin,[27] M. Dickson,[25] S. Donati,[23] R. B. Drucker,[14] A. Dunn,[16]
K. Einsweiler,[14] J. E. Elias,[7] R. Ely,[14] E. Engels, Jr.,[22] S. Eno,[5] D. Errede,[10] S. Errede,[10]
Q. Fan,[25] B. Farhat,[15] I. Fiori,[3] B. Flaugher,[7] G. W. Foster,[7] M. Franklin,[9] M. Frautschi,[18]
J. Freeman,[7] J. Friedman,[15] H. Frisch,[5] A. Fry,[29] T. A. Fuess,[1] Y. Fukui,[13] S. Funaki,[31]
G. Gagliardi,[23] S. Galeotti,[23] M. Gallinaro,[20] A. F. Garfinkel,[24] S. Geer,[7] D. W. Gerdes,[16] P. Giannetti,[23]
N. Giokaris,[26] P. Giromini,[8] L. Gladney,[21] D. Glenzinski,[12] M. Gold,[18] J. Gonzalez,[21] A. Gordon,[9]
A. T. Goshaw,[6] K. Goulianos,[26] H. Grassmann,[6] A. Grewal,[21] G. Grieco,[23] L. Groer,[27] C. Grosso-Pilcher,[5]
C. Haber,[14] S. R. Hahn,[7] R. Hamilton,[9] R. Handler,[33] R. M. Hans,[34] K. Hara,[31] B. Harral,[21]
R. M. Harris,[7] S. A. Hauger,[6] J. Hauser,[4] C. Hawk,[27] J. Heinrich,[21] D. Hennessy,[6] R. Hollebeek,[21]
L. Holloway,[10] A. Hölscher,[11] S. Hong,[16] G. Houk,[21] P. Hu,[22] B. T. Huffman,[22] R. Hughes,[25]
P. Hurst,[9] J. Huston,[17] J. Huth,[9] J. Hylen,[7] M. Incagli,[23] J. Incandela,[7] H. Iso,[31]
H. Jensen,[7] C. P. Jessop,[9] U. Joshi,[7] R. W. Kadel,[14] E. Kajfasz,[7,*] T. Kamon,[30] T. Kaneko,[31]
D. A. Kardelis,[10] H. Kasha,[34] Y. Kato,[19] L. Keeble,[30] R. D. Kennedy,[27] R. Kephart,[7] P. Kesten,[14]
D. Kestenbaum,[9] R. M. Keup,[10] H. Keutelian,[7] F. Keyvan,[4] D. H. Kim,[7] H. S. Kim,[11] S. B. Kim,[16]
S. H. Kim,[31] Y. K. Kim,[14] L. Kirsch,[2] P. Koehn,[25] K. Kondo,[31] J. Konigsberg,[9] S. Kopp,[5]
K. Kordas,[11] W. Koska,[7] E. Kovacs,[7,*] W. Kowald,[6] R. Krasberg,[16] J. Kroll,[7] M. Kruse,[24]
S. E. Kuhlmann,[1] E. Kuns,[27] A. T. Laasanen,[24] S. Lammel,[4] J. I. Lamoureux,[33] T. LeCompte,[10] S. Leone,[23]
J. D. Lewis,[7] P. Limon,[7] M. Lindgren,[4] T. M. Liss,[10] N. Lockyer,[21] O. Long,[21] M. Loreti,[20]
E. H. Low,[21] J. Lu,[30] D. Lucchesi,[23] C. B. Luchini,[10] P. Lukens,[7] J. Lys,[14] P. Maas,[33]
K. Maeshima,[7] A. Maghakian,[26] P. Maksimovic,[15] M. Mangano,[23] J. Mansour,[17] M. Mariotti,[23] J. P. Marriner,[7]
A. Martin,[10] J. A. J. Matthews,[18] R. Mattingly,[2] P. McIntyre,[30] P. Melese,[26] A. Menzione,[23] E. Meschi,[23]
G. Michail,[9] S. Mikamo,[13] M. Miller,[5] R. Miller,[17] T. Mimashi,[31] S. Miscetti,[8] M. Mishina,[13]
H. Mitsushio,[31] S. Miyashita,[31] Y. Morita,[13] S. Moulding,[26] J. Mueller,[27] A. Mukherjee,[7] T. Muller,[4]
P. Musgrave,[11] L. F. Nakae,[29] I. Nakano,[31] C. Nelson,[7] D. Neuberger,[4] C. Newman-Holmes,[7] L. Nodulman,[1]
S. Ogawa,[31] S. H. Oh,[6] K. E. Ohl,[34] R. Oishi,[31] T. Okusawa,[19] C. Pagliarone,[23] R. Paoletti,[23]
V. Papadimitriou,[7] S. Park,[7] J. Patrick,[7] G. Pauletta,[23] M. Paulini,[14] L. Pescara,[20] M. D. Peters,[14]
T. J. Phillips,[6] G. Piacentino,[3] M. Pillai,[25] R. Plunkett,[7] L. Pondrom,[33] N. Produit,[14] J. Proudfoot,[1]
F. Ptohos,[9] G. Punzi,[23] K. Ragan,[11] F. Rimondi,[3] L. Ristori,[23] M. Roach-Bellino,[32] W. J. Robertson,[6]
T. Rodrigo,[7] J. Romano,[5] L. Rosenson,[15] W. K. Sakumoto,[25] D. Saltzberg,[5] A. Sansoni,[8] V. Scarpine,[30]
A. Schindler,[14] P. Schlabach,[9] E. E. Schmidt,[7] M. P. Schmidt,[34] O. Schneider,[14] G. F. Sciacca,[23] A. Scribano,[23]
S. Segler,[7] S. Seidel,[18] Y. Seiya,[31] G. Sganos,[11] A. Sgolacchia,[3] M. Shapiro,[14] N. M. Shaw,[24]
Q. Shen,[24] P. F. Shepard,[22] M. Shimojima,[31] M. Shochet,[5] J. Siegrist,[29] A. Sill,[7,*] P. Sinervo,[11]
P. Singh,[22] J. Skarha,[12] K. Sliwa,[32] D. A. Smith,[23] F. D. Snider,[12] L. Song,[7] T. Song,[16]
J. Spalding,[7] L. Spiegel,[7] P. Sphicas,[15] A. Spies,[12] L. Stanco,[20] J. Steele,[33] A. Stefanini,[23]
K. Strahl,[11] J. Strait,[7] D. Stuart,[7] G. Sullivan,[5] K. Sumorok,[15] R. L. Swartz,[33,*] T. Takahashi,[19]
K. Takikawa,[31] F. Tartarelli,[23] W. Taylor,[11] Y. Teramoto,[19] S. Tether,[15] D. Theriot,[7] J. Thomas,[29]
T. L. Thomas,[18] R. Thun,[16] M. Timko,[32] P. Tipton,[25] A. Titov,[26] S. Tkaczyk,[7] A. Tollefson,[7] A. Tollestrup,[7]
J. Tonnison,[24] J. F. de Troconiz,[9] J. Tseng,[12] M. Turcotte,[29] N. Turini,[3] N. Uemura,[31] F. Ukegawa,[21]
G. Unal,[21] S. van den Brink,[22] S. Vejcik III,[16] R. Vidal,[7] M. Vondracek,[10] R. G. Wagner,[1] R. L. Wagner,[7]

---

*Visitors.

0556-2821/94/50(5)/2966(61)/$06.00   50   2966   ©1994 The American Physical Society

(94)

FIGURE 4.3. The first two pages of the Evidence paper (Abe, Amidei, et al. 1994b, 2966–7).

50   EVIDENCE FOR TOP QUARK PRODUCTION IN $\bar{p}p$ ...   2967

N. Wainer,[7] R. C. Walker,[25] G. Wang,[23] J. Wang,[5] M. J. Wang,[28] Q. F. Wang,[26] A. Warburton,[11]
G. Watts,[25] T. Watts,[27] R. Webb,[30] C. Wendt,[33] H. Wenzel,[14] W. C. Wester III,[14] T. Westhusing,[10]
A. B. Wicklund,[1] R. Wilkinson,[21] H. H. Williams,[21] P. Wilson,[5] B. L. Winer,[25] J. Wolinski,[30] D. Y. Wu,[16]
X. Wu,[23] J. Wyss,[20] A. Yagil,[7] W. Yao,[14] K. Yasuoka,[31] Y. Ye,[11] G. P. Yeh,[7]
P. Yeh,[28] M. Yin,[6] J. Yoh,[7] T. Yoshida,[19] D. Yovanovitch,[7] I. Yu,[34] J. C. Yun,[7]
A. Zanetti,[23] F. Zetti,[23] L. Zhang,[33] S. Zhang,[15] W. Zhang,[21] and S. Zucchelli[3]

(CDF Collaboration)

[1] *Argonne National Laboratory, Argonne, Illinois 60439*
[2] *Brandeis University, Waltham, Massachusetts 02254*
[3] *Istituto Nazionale di Fisica Nucleare, University of Bologna, I-40126 Bologna, Italy*
[4] *University of California at Los Angeles, Los Angeles, California 90024*
[5] *University of Chicago, Chicago, Illinois 60637*
[6] *Duke University, Durham, North Carolina 27708*
[7] *Fermi National Accelerator Laboratory, Batavia, Illinois 60510*
[8] *Laboratori Nazionali di Frascati, Istituto Nazionale di Fisica Nucleare, I-00044 Frascati, Italy*
[9] *Harvard University, Cambridge, Massachusetts 02138*
[10] *University of Illinois, Urbana, Illinois 61801*
[11] *Institute of Particle Physics, McGill University, Montreal, Canada H3A 2T8
and University of Toronto, Toronto, Canada M5S 1A7*
[12] *The Johns Hopkins University, Baltimore, Maryland 21218*
[13] *National Laboratory for High Energy Physics (KEK), Tsukuba, Ibaraki 305, Japan*
[14] *Lawrence Berkeley Laboratory, Berkeley, California 94720*
[15] *Massachusetts Institute of Technology, Cambridge, Massachusetts 02139*
[16] *University of Michigan, Ann Arbor, Michigan 48109*
[17] *Michigan State University, East Lansing, Michigan 48824*
[18] *University of New Mexico, Albuquerque, New Mexico 87131*
[19] *Osaka City University, Osaka 588, Japan*
[20] *Universita di Padova, Instituto Nazionale di Fisica Nucleare, Sezione di Padova, I-35131 Padova, Italy*
[21] *University of Pennsylvania, Philadelphia, Pennsylvania 19104*
[22] *University of Pittsburgh, Pittsburgh, Pennsylvania 15260*
[23] *Istituto Nazionale di Fisica Nucleare, University and Scuola Normale Superiore of Pisa, I-56100 Pisa, Italy*
[24] *Purdue University, West Lafayette, Indiana 47907*
[25] *University of Rochester, Rochester, New York 14627*
[26] *Rockefeller University, New York, New York 10021*
[27] *Rutgers University, Piscataway, New Jersey 08854*
[28] *Academia Sinica, Taiwan 11529, Republic of China*
[29] *Superconducting Super Collider Laboratory, Dallas, Texas 75237*
[30] *Texas A&M University, College Station, Texas 77843*
[31] *University of Tsukuba, Tsukuba, Ibaraki 305, Japan*
[32] *Tufts University, Medford, Massachusetts 02155*
[33] *University of Wisconsin, Madison, Wisconsin 53706*
[34] *Yale University, New Haven, Connecticut 06511*
(Received 25 April 1994)

We present the results of a search for the top quark in 19.3 pb^{-1} of $\bar{p}p$ collisions at $\sqrt{s}=1.8$ TeV. The data were collected at the Fermilab Tevatron collider using the Collider Detector at Fermilab (CDF). The search includes standard model $t\bar{t}$ decays to final states $ee\nu\bar{\nu}$, $e\mu\nu\bar{\nu}$, and $\mu\mu\nu\bar{\nu}$ as well as $e+\nu+$jets or $\mu+\nu+$jets. In the $(e,\mu)+\nu+$jets channel we search for b quarks from t decays via secondary vertex identification and via semileptonic decays of the b and cascade c quarks. In the dilepton final states we find two events with a background of $0.56^{+0.25}_{-0.13}$ events. In the $e,\mu+\nu+$jets channel with a b identified via a secondary vertex, we find six events with a background of 2.3±0.3. With a b identified via a semileptonic decay, we find seven events with a background of 3.1±0.3. The secondary vertex and semileptonic-decay samples have three events in common. The probability that the observed yield is consistent with the background is estimated to be 0.26%. The statistics are too limited to firmly establish the existence of the top quark; however, a natural interpretation of the excess is that it is due to $t\bar{t}$ production. We present several cross-checks. Some support this hypothesis; others do not. Under the assumption that the excess yield over background is due to $t\bar{t}$, constrained fitting on a subset of the events yields a mass of $174\pm10^{+13}_{-12}$ GeV/c^2 for the top quark. The $t\bar{t}$ cross section, using this top quark mass to compute the acceptance, is measured to be $13.9^{+6.1}_{-4.8}$ pb.

PACS number(s): 14.65.Ha, 13.85.Ni, 13.85.Qk

FIGURE 4.3 *(continued)*   (95)

as the *PRD* and submitted it in May to *PRL*, where it appeared in July (Abe, Amidei, et al. 1994a).

10. EVIDENCE FOR THE TOP QUARK: THE PAPER

The Evidence paper has ten major sections. The paper begins with an introduction explaining the current status of the search for the top quark and describing the top's major decay modes, the three search modes employed, and the major sources of background. A description of the CDF detector follows in section two. Section three describes the methods used to identify and model jets and high- p_T leptons. The dilepton search is presented in section four, and the lepton + jets search, in section five, which contains separate subsections for the SLT and SVX b-tagging methods. Section six discusses the results of the three counting experiments and estimates their statistical significance. Section seven contains an analysis of the $t\bar{t}$ production cross section. The lepton + jets search yielded 52 events with W candidates and 3 or more jets prior to b-tagging. The kinematic properties of these 52 events are discussed in section eight to compare to predictions with and without the top quark hypothesis. In section nine, CDF estimates the mass of the top to be $M_{top} = 174 \pm 10^{+13}_{-12}$ GeV/c^2. Combined with the analysis of production cross section presented in section seven, this yields a top-antitop production cross section estimate of $\sigma_{t\bar{t}} = 13.9^{+6.1}_{-4.8}$ pb. Section ten presents conclusions.

Section six constitutes the nexus of the paper's evidential argument. Here the authors assess, first for each search individually and then for the combined result, the statistical significance of the results. That is, CDF attempts to calculate the probability **P** of getting as many or more than the observed number of candidate events on the assumption that there are no top decays present in the data. CDF compares the number of candidate events with the expected background for each search and for the combined results.

In the dilepton search, CDF finds two events. The background calculation gives an expected $0.56^{+0.25}_{-0.13}$ events. The SVX search yields six candidate events with expected background of 2.30 ± 0.29 events. Seven events are found by the SLT search. The expected background is 3.1 ± 0.3 events. This gives values for **P** of $\mathbf{P}_{dil} = 0.12$, $\mathbf{P}_{SVX} = 0.032$, and $\mathbf{P}_{SLT} = 0.041$. These numbers are not much different from what they had been at the time of the October massacre. The numbers of observed candidate events are identical. Careful study of backgrounds had increased the estimates for the SVX and dilepton searches by small amounts: the dilepton background by 0.01 events, the SVX background by 0.30. On the whole, the group understood the backgrounds for the counting experiments much better than they had the previous fall.

Combining the three searches yields a total of 12 events. The background expectation is calculated to be $5.7^{+0.49}_{-0.44}$. (This number includes the effect of subtracting an expected mean of 0.26 SVX-SLT double-tags, and, because of the changes in the counting experiment backgrounds, is a little higher than the 5.5 background expectation reported in the October *PRL* version of the total results (CDF 1993c).) This would lead to a significance level of $P_{combined} = 0.016$.

CDF, however, argues that 0.016 is in fact an overestimate of $P_{combined}$. Double-tagged events are much less likely to be mistags than single-tagged events. Because mistags are a major source of background for this search, "effective background for the three double-tagged events is therefore considerably smaller than for the other seven [lepton + jets] events" (Abe, Amidei, et al. 1994b, 3004). To justify counting tags rather than events, CDF attempts to demonstrate that they understand the potential correlations between the two tagging algorithms well and have accounted for them in the parameters employed in those algorithms.

But calculating $P_{combined}$ for the 15 events with $5.96^{+0.49}_{-0.44}$ background now requires a more sophisticated technique. This is because of a shared source of background for the SVX and SLT searches: events in which Ws and "heavy flavors" (*b* and *c* quarks) are directly produced by proton-antiproton collisions. A fluctuation in this background for one search necessarily means a fluctuation for the other. The top group worked out the technique for taking this effect into account in the course of developing the Evidence paper from its beginnings as four *PRL*s in October 1993. At that time, Tony Liss calculated the significance level based on the observed number of tags using a simpler version of the technique used later. The calculation started with a Gaussian (normal) distribution centered on 5.5 (the total background expectation in October 1993), with a standard deviation of 0.55. Monte Carlo experiments were then performed that drew from that Gaussian distribution a value for the background expectation, and then took that value as the mean of a Poisson distribution, from which a value for the observed number of tags was drawn. These experiments were repeated a large number of times. Then, the fraction of experiments yielding 15 or more tags was calculated, and this was taken to be the probability of getting 15 or more tags in the absence of the top quark. The resulting value was 8×10^{-4} (CDF 1993c, 5).

What this earlier calculation omitted, however, was any consideration of those upward fluctuations in background sources that would increase the SVX and SLT totals simultaneously. In October 1993, the background to the combined experiment was taken simply to be the sum of the backgrounds to the individual experiments. However, for the backgrounds *shared* by the two *b*-tagging algorithms, the probability of a double-tagged event would be higher than the preceding method suggests. Also, an upward fluctuation in a background to the lepton + jets search in general, even if the tagging rates for the two algorithms were not correlated on the source of that fluctuation, would yield a larger number of lepton + jets events to which the two algorithms would subsequently be applied, hence raising the probability of

getting a higher number of tags. (Here CDF was abiding by the requirement that a significance estimate take into account all relevant aspects of
the data selection procedure; see Chapter 7.) The top group tackled such
problems by, as before, performing a large number of Monte Carlo "background experiments." However, in the improved version of the method, the
experiments treated each *type* of background separately, generating a set of
background events from specified sources, to which the two tagging algorithms were then applied in order to arrive at the number of tags for that
experiment. Again, a large number of such experiments were aggregated.
The fraction of them containing 15 or more tags was taken to be the probability of getting such a result in the absence of any top-quark decays. The
result of this calculation is $P_{combined} = 2.6 \times 10^{-3}$, slightly higher than the
number obtained in the October 1993 analysis. The authors note that "for a
Gaussian probability function, which we do not have here," this value would
correspond to a 2.8σ excess (Abe, Amidei, et al. 1994b, 3006).

Section six also includes a consideration of alternate hypotheses that
might be proposed to explain the data. First the authors discuss a fourth-
generation quark, the b'. They note that, depending on the coupling of
fourth- to second-generation quarks, there would either be a threefold increase in the production cross section for the signature or else there would
be no decays into the channels considered for the top quark. They conclude
that this hypothesis is not consistent with their results.

The other alternate hypothesis considered is that the data are the result
of a standard model Higgs boson with mass less than $150 \text{ GeV}/c^2$. But based
on standard model processes for producing the Higgs, only a "fraction of an
event" could be expected based on this hypothesis. The authors proceed to
note, however, that "this would become an interesting alternative if production mechanisms other than what is currently assumed were active" (Abe,
Amidei, et al. 1994b, 3008).

The authors note the excess of tagged Z + jets events already mentioned,
a hint that they may not perfectly understand the production of heavy flavors
in association with vector bosons. But, they note, similar tests on samples of
$W + 1$ and 2 jet events, which have higher statistics and should also have
an insignificant proportion of top content, do not exhibit this inconsistency
with expectations.

More statistics, the authors assert, will resolve such issues. They note that
present statistics are insufficient "to test the production and decay mechanisms in any detail" (Abe, Amidei, et al. 1994b, 3009).

The original plan for the Evidence paper did not include the discussion of
alternate hypotheses. Tony Liss recalled, "That got stuck in there as a result
of somebody's comments – it might have been one of the godparents," and
described the alternates considered as a "reasonably exhaustive list of the
obvious candidates" for alternative explanations of the group's findings (Liss
1998).

The paper concludes by stating that the data presented "give evidence for, but do not firmly establish the existence of, $t\bar{t}$ production in $\bar{p}p$ collisions at $\sqrt{s} = 1.8$ TeV," and the authors note that they expect a fourfold increase in data from the next run (Abe, Amidei, et al. 1994b, 3023).

11. RUN IA AT D-ZERO

Run Ia provided the first opportunity for the D-zero collaboration to do physics, and the search for the top quark headed their agenda. Although many of the same physics problems captured the interest of both collaborations, D-zero had built a detector very different from CDF's. Both were cylindrical detectors surrounding the collision region, with segmented calorimetry. However, the D-zero detector had neither a central magnetic field nor a silicon vertex detector. On the other hand, the calorimetry at D-zero was more finely segmented and provided better coverage, and D-zero's muon detection was more efficient (see Figure 4.4). As noted in the previous chapter,

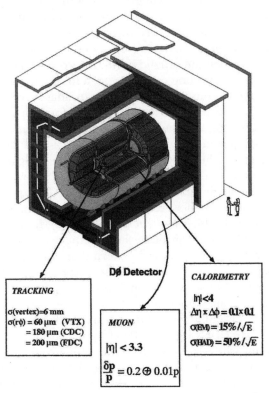

FIGURE 4.4. A cutaway schematic view of the D-zero detector. Courtesy of Fermi National Accelerator Laboratory.

such differences reflected distinct detector "philosophies" within the two collaborations, with CDF placing a stronger emphasis on tracking and D-zero stressing calorimetry. Mark Strovink, a Berkeley physicist working in D-zero, considered the two detectors to be "pretty much the direct descendants" of an earlier generation of detectors built at the Sp$\bar{p}$S collider at CERN. CDF followed approximately the strategy of the UA1 collaboration in building their detector, while D-zero patterned theirs after that built by the UA2 collaboration. "CDF and UA1 placed a great deal of emphasis, in the conception of the detector and in the money it took to construct the detector, on reconstructing individual charged particles," whereas "UA2 and D-zero concentrated on measuring the whole cluster of particles that appear when a quark is produced, the so-called jet particles, as a single fundamental object" (Grannis, Klima, et al. 1998).

These differences were a matter of degrees. D-zero did track individual particles, and CDF's detector had considerable calorimetry. However, D-zero's calorimeters yielded a finer spatial resolution than CDF's, and CDF's tracking chambers could do more than D-zero's, particularly with the help of a central magnetic field to distinguish different kinds of particles. Such differences translated into distinct strategies for finding the top quark.

The long delay that had preceded the beginning of run Ia was a blessing for D-zero, giving them a chance to roll an essentially complete detector into the beam line when the run did begin. D-zero came away from the run with two *PRL* papers reporting on their search for the top (Abachi, Abbott, et al. 1994; 1995a) and one remarkable top candidate event.

For the first search paper, D-zero looked in four top decay channels: the *e-e* and *e-μ* dilepton channels, and the $e +$ jets and $\mu +$ jets channels. CDF achieved background reduction in the lepton + jets channels primarily through *b*-tagging. In this paper, D-zero did not use *b*-tagging, but instead reduced background by requiring four jets (CDF had required at least three) and by using a kinematic quantity called *aplanarity*. This quantity is defined to be "proportional to the lowest eigenvalue of the momentum tensor for observed objects" (Abachi, Abbott, et al. 1994, 2141). In using this quantity, D-zero hoped to take advantage of the fact that the decay products of the top tend to emerge in a more spherical pattern than the background. D-zero found one $e +$ jets, one *e-e*, and one striking *e-μ* event. Assuming all of these to be signal events, they set a lower limit on the top mass of 131 GeV/c^2 at the 95% confidence level (ibid.).

D-zero's one *e-μ* event, recorded in January 1993, generated a great deal of excitement. Like CDF's DPF event, here was a striking event in the very clean *e-μ* channel, which came to be known by name. The D-zero physicists called it event 417 (see Figure 4.5) (Partridge 1995). The likelihood of the top decay hypothesis for this event is "at least ten times more than any of the known backgrounds," according to Boaz Klima, a Fermilab physicist who was D-zero's top group convener at the time. D-zero physicists scrutinized

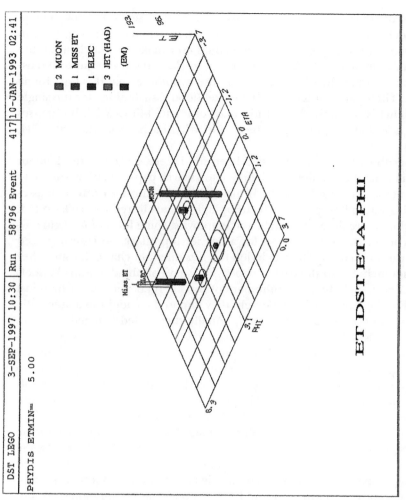

| DST LEGO | 3-SEP-1997 10:30 | Run | 58796 Event | 417 | 10-JAN-1993 02:41 |

PHYDIS ETMIN= 5.00

2 MUON
1 MISS ET
1 ELEC
3 JET (HAD)
(EM)

ET DST ETA-PHI

FIGURE 4.5. D-zero's event 417, shown in a Lego plot in the η-ϕ plane; note the prominent spikes identified as the muon and electron, as well as the neutrino, reconstructed from missing E_T; the three shorter towers are from jets. Courtesy of Fermi National Accelerator Laboratory.

this event at great length, and declared it a "perfect signature" for the top. They debated whether to publish a top discovery based on this single event, with some questioning whether one could ever make such a claim on the basis of a single event. They decided not to claim discovery but to discuss the event and analyze it for a mass on the hypothesis that it is top (Klima 1995). Their paper notes that "[t]he likelihood distribution [for the single e-μ event] is maximized for a top mass of about 145 GeV/c^2, but masses as high as 200 GeV/c^2 cannot be excluded" (Abachi, Abbott, et al. 1994, 2142).

After the 131-GeV/c^2 limit had been set, D-zero started to think about reoptimizing their analysis. In their first top search paper, they had sought to rule out top masses as low as the W mass, while remaining sensitive to masses at as high a limit as possible. By using tighter cuts to reduce background, including more decay channels, and employing some form of b-tagging, D-zero could extend the analysis for a heavier top quark. So D-zero revised their earlier analysis. They included the μ-μ dilepton channel, which they had not included in the earlier paper. They gave all the dilepton channels tighter cuts, including a requirement of two energetic jets rather than just one. For the lepton + jets channels, they imposed an additional kinematic requirement by placing a cut on H_T. Finally, they added b-tagged lepton + jets channels with looser cuts (e.g., only three or more jets were required, as opposed to four in the untagged lepton + jets channels). They subtracted background in these channels by means of soft-muon tagging similar to CDF's soft-lepton tagging. (D-zero placed their cut on the p_T of the soft muon at 4 rather than 2 GeV/c. However, D-zero's muon chambers were not efficient for muons with $p_T < 4$ GeV/c, so that a lower cut was not really a live option at D-zero.) In 13.5 pb^{-1} of data, D-zero found nine candidate events, with an expectation from background of 3.8 ± 0.9. One of these was the stunning e-μ dilepton event, and the other eight were distributed evenly amongst four different lepton + jets channels (e + 4 jets, μ + 4 jets, e + 3 jets with a soft-muon tag, and μ + 3 jets with a soft-muon tag). D-zero wrote up these results in a *PRL* paper titled "Search for High Mass Top Quark Production in $\bar{p}p$ Collisions at $\sqrt{s} = 1.8$ TeV" (Abachi, Abbott, et al. 1995b). They noted that, "In the absence of top, we calculate the probability of an upward fluctuation of the background to nine or more events to be 2.7%," and concluded that their experiment, "although consistent with the CDF result [the Evidence paper] and of comparable sensitivity, does not demonstrate the existence of the top quark" (Abachi, Abbott, et al. 1995b, 2426).

The attentive reader may have noticed that D-zero's conclusion implicitly criticizes CDF's conclusion. If D-zero's measurement was of "comparable sensitivity" to CDF's and yet in nearly the same amount of data did not "demonstrate the existence of the top quark," or produce evidence for the top quark, why did CDF's measurement yield a result that was claimed to be evidence? Many D-zero physicists felt that CDF had overplayed their hand in claiming that their Ia results constituted evidence for the top quark. According to Boaz Klima, the size of the effect reported in D-zero's paper on the search for a high-mass top quark was 1.9 standard deviations, compared to CDF's 2.8 standard deviations. Klima noted that "you can see many many two sigma effects in the data, all over the place, everyday. So you can't make a big deal out of that." Then he added, "It's also true for three sigma. You see quite a few of those." In presenting their result, D-zero "didn't even think about the word 'evidence.' That's something new that was introduced. Either you think you have it or you don't" (Klima 1995).[4]

Commenting on the claim that the CDF and D-zero searches on the Ia data had comparable sensitivity, one CDF physicist quipped, "Sure they're comparable – compared to CDF, D-zero's is much worse!" (Gerdes, personal communication). D-zero based their results on about 13.5 pb^{-1} of data, compared to the 19.3 pb^{-1} used for CDF's Evidence paper. Also, some search modes that had played important roles in CDF's results were unavailable to D-zero, such as the tagging of secondary vertices with the silicon vertex detector and the tagging of soft electrons. CDF physicists also criticized D-zero's greater dependence on kinematic features such as aplanarity and H_T to define candidate events. Taking such cuts to indicate top decays depends too much, they charge, on Monte Carlo comparisons of signal to background. D-zero physicists in turn deny that their analysis was more Monte Carlo–dependent than CDF's analysis. Paul Grannis, one of D-zero's spokespersons at the time, commented that "there was a misconception in CDF" concerning the role of Monte Carlo calculations in D-zero's analysis. Mark Strovink noted that D-zero had "two major backgrounds, one of which was determined in a straightforward way from data, the other which was determined by extrapolation from a region that we knew was dominated by background into the signal region" (Grannis, Klima, et al. 1998).

Such disagreements are small pieces of a complex relationship between the CDF and D-zero collaborations, in which rivalry is an important element. As Brenna Flaugher, a Fermilab physicist working on CDF who is also married to a member of the D-zero collaboration, put it, "Humans like to hate other humans . . . so you want to make the other experiment out to be the evil empire somehow, and I think both collaborations fell into that very easily [pause] very willingly [pause] enthusiastically!" (Flaugher 1998). This is not to make light of the substantive disagreements between members of the two collaborations. Such rivalries do, however, tend to generate more controversies than would otherwise occur.

In their second look at the Ia data, D-zero clearly did not yet think they could claim to have found evidence for the top quark. As members of the D-zero collaboration diverted their efforts into preparing for run Ib, they delayed their paper reporting the results of the search for a high-mass top quark based on the Ia data. The new run was well underway by the time they sent that paper to *PRL*, in November. By the time it was published, in late March 1995, much more exciting events had transpired.

12. SOME FEATURES OF THE PROCESS OF PRODUCING THE EVIDENCE PAPER

Although the Evidence paper is not at all typical of papers published in particle physics, the process of producing the results presented in that paper nonetheless displays some easily forgotten facets of the experimental enterprise in the physical sciences. I wish to focus in particular on what can be learned from this episode about the epistemic function of research reports in the social organization of science.

One reason for writing research reports intended for publication is to provide both a mechanism and a motive for critical deliberation over an experimental result. The publication of research reports is a fine example of a scientific institution kept in place, largely for methodological reasons, as a means of enforcing quality control over the output of a scientific community. Much of this work is accomplished before the paper appears. (Indeed, published papers probably do not do very much work *after* they are published anymore, in many fields of research, since most of the relevant audience for a paper will have already seen the work in preprint form, made available via the world-wide web.) This is an obvious point, but it is important enough to make explicit. Knowing that one's experimental claims will be scrutinized by other scientists after a paper is published provides a motive for its authors to scrutinize the result, and to ask questions, some of which will not have been anticipated by others in the group. In writing up their results for publication, experimenters seek to avoid being embarrassed.

By providing a textual focal point for argument and consensus building, the research paper also provides a means of organizing such deliberations. It both defines what any putative consensus will be *about* (imagine trying to forge a 450-person consensus about what everyone's personal degree of belief in a given empirical claim will be) and provides a practical context for debate. While disagreements over, for example, the best way to calculate the significance of a result could continue interminably, the need to make a practical decision concerning the paper – either agree on a statement that is "good enough" for almost everyone, or do not publish at all – makes debate manageable. The group's aim shifts from optimizing to satisficing. (As noted earlier, among those who did believe that CDF's results constituted good evidence for the top quark, beliefs varied widely with respect to the particular elements of the analysis that made that evidence claim sound.) At the same time, the process of coming to agreement on a single textual[5] summary forces advocates to make potentially relevant details – including those that count against the advocates' views – known to others, further facilitating criticism.

The outcome of an experiment is not always well defined prior to writing the paper. Although the result claimed in the Evidence paper did not shift significantly from that claimed in the October 1993 papers, the statistical significance did change as a result of improvements in the method of accounting for the effects of counting tags rather than events. Background estimates also changed slightly over the course of writing the Evidence paper (see Section 10 in this chapter; CDF 1993a; 1993b).

The degree of confidence experimenters have in an empirical claim can change during the process of writing the paper presenting that claim. Although the quantitative change in the top results between October 1993 and April 1994 was small, a much greater change took place in the opinions of CDF members regarding the soundness of the analysis that produced those numbers. In general, members of CDF were more convinced of their evidence claim at the end of writing the Evidence paper than they had been at the beginning of the process. (A few members of the collaboration did not become more convinced, but remained skeptical throughout.) Even those who were fairly well convinced in October 1993 insisted that by April 1994, they felt more comfortable with the analysis than they had six months earlier. Tony Liss had wanted to publish four *PRL* articles in October 1993. He later commented that members of the top group "didn't quite recognize" at that time just how far the analysis was from being ready for publication (Liss 1995). When asked if he felt more confident at the end of the writing process than at the beginning, he replied, "absolutely," and noted the advantages of having questions asked about the analysis by people who had not participated directly in developing it (Liss 1998). I asked numerous CDF physicists this same question. Nearly all of them responded in the same manner – they believed more strongly in the evidence claim at the end of the process than they did at the beginning. (The reverse also happened in at least one case: G. P. Yeh became less convinced of the evidence claim as he learned more about the SLT analysis.)

Mel Shochet, for example, noted that "even if I hadn't been paying attention to all of the details . . . I would have been more confident of [the evidence claim], just having sat in all of the meetings, in which so many people asked so many questions, demanding that they be answered at a quantitative level. . . . I was much more confident at the end." What Shochet found reassuring was that further work on the top results showed, in response to "hundreds of questions, detailed questions" that "the effect was not a systematic problem with the way the analysis was done, with the detector, with the way the sample had been selected" (Shochet 1998). Shochet himself contributed, producing in January 1994 a CDF note titled "How Robust Is Our Top Result?" that explored the effect of changing the cut on the transverse energy of jets on both background estimates and the number of candidate events (the importance of such investigations will be explored in chapter seven). Such prolonged, detailed questioning of the analysis served to eliminate the possible sources of error that initially concerned Shochet, and in doing so strengthened his conviction that the paper's central claim was correct.

Facts about the process of producing a paper can have a bearing on the evaluation of the result. Melissa Franklin also felt more comfortable with the physics in the Evidence paper at the end of the process than at the beginning, but she noted, "I felt more uncomfortable with my colleagues." Noting that the period of debate surrounding the Evidence paper was marked by "what I consider to be pretty bad behavior" such as defensive behavior and evasiveness on the part of some advocates of the evidence claim, she admitted that such behavior "makes you want to believe it less" (Franklin 1998). Likewise, Henry Frisch's recollection that deliberations over the Evidence paper had turned into a "non-convergent" process, and his partially enacted recommendations for reforming the conduct of meetings suggest that reaching consensus about what the result of an experiment is can require solving the social problem of how to create the conditions in which consensus is possible. Mel Shochet, meanwhile, was impressed by the severity of the process of exposing the top search results to the questions and criticisms of the collaboration. And Tony Liss commented that "having people who are not directly involved in the day-to-day stuff that you're doing look at it . . . carefully and critically, and say 'what about this' and 'what about that' . . . some of it you say, 'why are you bothering me with this?' but some fraction of them are really good questions, and you go off and you answer them and you feel much better about it" (Liss 1998).

Expertise was distributed widely within CDF. That is to say, members of the collaboration relied on their colleagues for knowledge of the details of the detector and the data it produced. Each person had a certain area of expertise, yet each typically placed great value on the independence of her own judgment. In this respect, the CDF collaboration stands as a microcosm of the scientific community in general. No one member of the collaboration knew everything about the work that they did, but every member felt himself to be capable of figuring out anything he should put his mind to. The top-quark analysis attracted much attention, but not so much that each member of the collaboration became an expert on it. Partly for this reason, some people – the godparents assigned to each prospective paper – were designated to serve as internal referees. Such independent criticism also addressed the collaboration's awareness of the potential for personal biases to influence a result. By having an internal review by physicists who were knowledgeable of the physics tackled in a given piece of analysis, but who remained independent of the process of developing that analysis, CDF exposed its own work to a level of criticism compared to which the criticism of external referees was, in Henry Frisch's words, "pretty mild" (Frisch 1995).

Accordingly, much of the harshest criticism of the Evidence paper was presented by group members who also felt that the social structures and institutions of CDF failed to function properly on this occasion. Such critics believed that the godparents for the Evidence paper were not sufficiently independent and not sufficiently critical of the work being done by the top group.

Accepting a claim regarding the outcome of an experiment involves accepting that certain questions that can be raised about such claims have been satisfactorily answered. The questions concern possible sources of error, and, to be satisfactory, the answers must show that those sources of error can be ruled out. Such questions can concern the calibration of devices, the contribution of systematic errors such as neglected sources of background, or the elimination of biases. Some of those biases can enter, however, as a result of decision-making procedures that may be difficult to verify as unbiased (such as when it is unclear whether colleagues chose data-selection criteria in a manner that was "blind to the data"). To accept an experimental claim means to have one's suspicions laid to rest. One cannot specify a priori the point at which raising more questions becomes unreasonable (for reasons that I will discuss in Chapter 6).

In a complex analysis carried out by many different people, biases can be introduced by overzealous, careless, or unscrupulous colleagues, or even by colleagues who have none of these flaws but who simply fail, on a particular occasion, to exercise sufficient diligence to *avoid* introducing biases. While physicists rely on additional empirical study to answer their questions about such biases, the point at which one decides that one has no more questions to ask can depend on one's opinions about colleagues and their behavior. A dedicated conspiracy theorist will never run out of questions to ask that may reveal a well-planned act of cheating. (Hence it is helpful if scientific work is organized in such a way as to make conspiracies difficult to execute.) Considering the independence and motives of a colleague has special importance when that colleague is acting as part of an institutional structure specifically designed to serve as a quality control filter for new work, as were the godparents.

K. Brad Wray has noted the importance of the fact that scientific work is done within institutional structures intentionally designed to promote the elimination of biases, not so much in individual scientists, as in the work that passes institutional muster (Wray 2000). Wray's point is that certain kinds of institutional practices employed within the sciences were both deliberately chosen to promote such epistemic aims as the elimination of biases, and figure importantly in the explanation of why science succeeds in achieving those aims: "The success of science is the result of the intentions of, not only those scientists working in labs, but also those scientists who have designed the institutions constitutive of science" (ibid., 170).[6]

Wray's point receives support here, insofar as suspicions that the institutions meant to act as a check on the ambitions of collaboration members had failed prompted CDF physicists to question the validity of the claims of the

Evidence paper. The godparent committee was not simply another group of interested CDF physicists, but an organ of the collaboration designed to function in a certain way, and a number of collaboration members worried that it had malfunctioned in the case of the Evidence paper. Consequently, they worried more about that paper than they might have otherwise.[7]

A recurrent phenomenon during the production of the top papers was the scrutiny exercised by collaboration members over the process of production itself. While concerns about the reliability of particular analyses arose primarily from factors that might be thought "internal" to those analyses (signal-to-background ratio in the SLT analysis, the reliability of Monte Carlo–dependent calculations, etc.), CDF physicists exhibited a great concern over how their colleagues conducted themselves and interacted with one another (the heated competition in the SVX b-tagging group, the role of the top godparents in the early stages of the Evidence paper, postponed decision making regarding the p_T cut in the SLT analysis). Thus, factors typically thought of as "external" also helped shape CDF members' epistemic attitudes toward the top-quark evidence claim.

13. THE RESEARCH PAPER: THE PURSUIT OF EXPERIMENT BY OTHER MEANS

Several authors have recently pointed out the importance in science of the published research report, not simply as a means of communicating the results of an experiment performed, but as part of the means of generating those results. Even in the narratives that describe the context of problems and related results in which work was done and that describe the methods used in carrying out that work, researchers engage in a reconstruction that distills a meaningful claim about the subject of investigation from a mess of contingent details. While some have decried such narratives as a "fraud" because of the very selective, even historically misleading ways in which scientific authors typically depict the history of their own disciplines (Medawar 1964; 1969), Thomas Nickles notes that even here "scientists are not writing *about* science; they are *doing* it" (Nickles 1992, 96). Argument and unavoidably whig-historical narrative are in this sense inseparably intertwined. Furthermore, the genre of the scientific research report has exhibited this mingling of narrative and argumentative tasks from its inception, as Frederic Holmes has argued (Holmes 1991).

Nickles focuses primarily on the discussion, in research reports, of the existing literature, and the ways in which scientists reconstruct the history of their discipline in relating their own work to the work that was done before. I wish here to extend that insight to the description by experimentalists of how their own work was done. The process of writing up the results of an experiment is a process, not of reporting, but of developing the very material that is to be reported on ("writing up" rather than "writing down" the results) and of probing the reported results for possible errors. Nickles describes the research process in general as "self-transforming" (Nickles 1992). I would add that the process of deliberation and writing that leads to the publication of a research paper is "self-correcting" as well – fallibly so, of course. Part of the discovery process involves taking a retrospective view of one's own work. This enables the experimenter to find in that work a meaning that she could not have found while engaged in it. The experimenter thus assures herself, by probing possible sources of error, of the correctness of the results that she will present in the publication.

Even such conventional aspects of the research report as the citation of related work can assist the author in making her case. Cited works are more than allies on the author's side in disputes between one network and another (cf. Latour 1987); they are evidence of the competence and thoroughness of the investigator. Philip Kitcher has noted the cognitive functions served by the rhetoric of a scientific publication (Kitcher 1995). In particular, numerous rhetorical aspects of scientific writing aid the reader in both deciding whether to accept the findings reported in the paper and determining the relationship between those findings and the reader's own endeavors. The first decision rests in part on a judgment of the competency of the authors of the report, and the research report provides both "external" clues for such a decision (institutional affiliations, the prestige of the publishing journal, etc.) and "internal" evidence of such competency. These internal signs of competence include the skillful handling of references, the appropriately expert use of specialized techniques, the anticipation of objections and possible systematic errors in the experiment, and so on.

I wish to add one more point regarding the scientific value of research reports. In addition to presenting a reconstruction of work already done and epistemically relevant rhetoric to convince readers of the result produced in that reconstruction, researchers, in writing a paper, engage in the final stages of deciding just what their result is and of convincing themselves, if they can, that it is correct. Where Nickles and Kitcher emphasize the work done by the finished paper as written, I wish to highlight the work done by striving to produce a paper at all.

The process by which CDF wrote the Evidence paper highlights the *methodological significance* of the scrutiny of experimental results that takes place during deliberations over what to put into the report, in deciding what kinds of checks on the result to consider, and what kind of language to use to describe the epistemic attitude of the experimenters toward their result. I also wish to note the role played by the need to reach agreement within the collaboration on the basis of that scrutiny. These processes leading up to the publication of a research report constitute an extension of the experimental testing procedures themselves in that they help to rule out errors and ensure the reliability of claims resulting from the procedure. These methodological aspects of the paper-writing process mesh smoothly with the methodological functions of the process of "pretrial" planning and the careful monitoring of the experiment in progress. They also reflect the epistemic functions of the process of scrutiny that takes place after the research report is completed, when it is circulated through paper and electronic preprints and through the journal in which it is published. They help experimenters test and refine their claims after the experiment proper has ended, in anticipation of the testing to come under the critical gaze of a broader audience.

5

Run Ib

"Observation" of the Top Quark and
Second Thoughts about "Evidence"

What we look forward to in science is further data, probably of a somewhat different kind, which may confirm or elaborate the conclusions we have drawn; but perhaps of the same kind, which may then be added to what we have already, to form an enlarged basis for induction.

> R. A. Fisher, "Statistical Methods and Scientific Induction" (1955)

This increasing precision is achieved . . . through the obligation to observe a series of additional facts at the same time that the principal fact is being observed, and through the necessity of submitting the raw data of experience to manipulations and transformations that are more and more numerous and delicate.

> Pierre Duhem, "Some Reflections on the Subject of Experimental Physics" ([1892] 1996, p. 93)

In March of 1995, CDF and D-zero simultaneously submitted papers to *Physical Review Letters* claiming the "observation" of the top quark (see Figure 5.1). In stating that they "observed" the top quark, they were not claiming that they had any kind of direct sensory experience of the top quark itself. Rather, they were claiming that the statistical evidence based on measurements of decay products was strong enough, in the words of CDF's article, to "establish the existence of the top quark" (Abe, Amidei, et al. 1995a, 2627).

In one sense, the findings presented in CDF's "Observation of Top Quark Production in $\bar{p}p$ Collisions with the Collider Detector at Fermilab" and D-zero's "Observation of the Top Quark" supported the claim of CDF's Evidence paper: Here was additional, stronger evidence of the top quark's existence. As we will see, however, in developing that stronger evidence, some CDF members believed that they uncovered something else: confirmation of their suspicions that the analysis used in the Evidence paper contained biases.

In this chapter, I will examine the developments that led to the publication of the two "Observation" papers, the changes in CDF's analysis as

FIGURE 5.1. Members of D-zero submitting their "Observation" paper, electronically, to *Physical Review*, February 24, 1995. Courtesy of Fermi National Accelerator Laboratory.

they searched for additional evidence of top-quark decays, and the ways in which additional data and the reanalysis of existing data can reinforce or cast doubt on evidence claims.

1. RUN IB AT CDF: A NEW DETECTOR AND A NEW ALGORITHM

Throughout CDF, physicists were aware of the dangers of making strong claims on the basis of a small number of events. They had accordingly adopted a cautious tone in the Evidence paper's conclusion: "The data presented here give evidence for, but do not firmly establish the existence of, $t\bar{t}$ production in $\bar{p}p$ collisions at $\sqrt{s}$ = 1.8 TeV" (Abe, Amidei, et al. 1994b, 3023). Any remaining uncertainties, they hoped, would be resolved with more statistics: "we anticipate an approximate fourfold increase in data from the 1994–1995 run [run Ib]" (ibid., 3023).

CDF made several detector improvements in preparation for run Ib. For example, they completed the central muon extension, which had been only partially operative during Ia. Fermilab upgraded the linear accelerator that functioned as an injector for the collider, as well. This upgrade was significant for CDF in that it would result in greater luminosity, so that data would accumulate more rapidly in run Ib than it had in run Ia.

The most significant hardware upgrade for the top results was the installation of a new silicon vertex detector dubbed SVX′. Even before run Ia began, a small group began planning the SVX′ as a replacement for the SVX. As expected, the performance of the SVX deteriorated significantly over the course of run Ia, as radiation produced in high energy particle collisions took its toll on the SVX readout chips. The SVX′ was to be equipped with new readout chips that were "radiation-hard." But when work began on the SVX′, the people who had worked on the original SVX were trying to get away from working on hardware and to get involved in analysis, which many saw as more advantageous to one's career as a physicist than hardware work.

Joe Incandela led the effort to build the SVX′. He had just been hired by Fermilab after working in the UA2 collaboration at CERN. Faced with the option of joining either CDF or D-zero, Incandela joined CDF because he thought it had more interesting data for analysis. He worried, however, that CDF did not need people as badly as D-zero did so that it would be harder for him to get substantially involved at CDF. Looking for a project to work on, Incandela noted that, while meetings concerning the SVX upgrade were very well attended, "there were [only] one or two people who were going to work on it." The densely packed meetings that had made it seem as though people were not needed very badly were packed with people who were interested in the project as spectators. Carl Haber of Berkeley and Nicola Bacchetta, then at the University of New Mexico, two physicists regarded as "detector types," were working on the SVX upgrade (Incandela 1995). Johns Hopkins University also had a postdoc named Rick Snider working on the SVX′, along with a couple of Hopkins graduate students (Barnett 1996).

Once committed to the SVX upgrade, Incandela found it difficult to extract advice and information from those who had built the original SVX. He discovered that "[m]ost of the people who worked on SVX would not even talk to me." These physicists, most of whom were in temporary or postdoc positions, needed to work on physics problems in order to advance their careers, and "they were very afraid of getting sucked into more hardware work." Incandela sent e-mail messages promising SVX veterans that he would not ask them to work on the upgrade, that he only wanted to interview them, and that the most he would ask of them would be that they help to select new people to work on the upgrade. Then he got responses, and he interviewed each member of the SVX team (Incandela 1995).

Incandela's original impression of a superabundance of physicists at CDF was completely reversed. "I had thought that they didn't need anyone, and I discovered that there was no one working on [the upgrade]." Understaffed, and with less than a year to achieve their goal, the SVX′ construction team began detector assembly in the early spring of 1993, with run Ia still under-way. The team worked 12-hour days and weekends. They were installing the SVX′ while the Evidence paper was being written (ibid.).

Along with the upgrade of the silicon vertex detector went an upgrade of the algorithm using that detector to tag *b* decays. When run Ib began, a group of physicists in the *b*-tagging group began a new effort to develop a more efficient algorithm than the jet-vertexing algorithm then in use.

Run Ib began in December 1993, when the first draft of the Evidence paper was just being written. People like Paul Tipton and Brian Winer who had developed the jet-vertexing algorithm were furiously at work answering questions about the evolving Evidence paper. At just that time Dave Gerdes, of the University of Michigan, and Joe Incandela became the new conveners of the *b*-tagging group. Alvin Tollestrup had persuaded Incandela to take a leadership position in the group, urging upon him the importance of having a single *b*-tagging algorithm, rather than three algorithms with their respective partisans competing for the favored position, as had happened during Ia (ibid.).

According to Dave Gerdes, "One of our first acts as *b*-tag conveners was to issue a ban on all talks involving tagger development." They wanted to take their time and begin with a "generator-level" Monte Carlo that simply models the physics processes of production and decay without attempting to model how the products of those processes propagate through the detec-tor. In a long CDF note, they outlined a formal program for developing a new algorithm, beginning with these Monte Carlo studies. By banning talks on tagger development, they conveyed to members of the group that they could take their time and do careful work on laying the foundations of an algorithm, and "no one's going to scoop you by showing the cooi plot next week" (Gerdes 1995).

Also important to the process of developing a single algorithm was the inclusion of people who had worked on the taggers the new algorithm was intended to replace, especially those who had worked on the jet-probability and *d*-ϕ algorithms. By contrast, people who had developed jet vertex-ing "did not get involved. . . . They were still working on the run Ia result [the Evidence paper]. . . . They didn't come to the meetings" (Incandela 1995).

The idea behind the new algorithm came to be known as "seed vertex-ing." Some members of the *b*-tagging group believed that jet vertexing was not as efficient as it could be and began searching for a more efficient way to tag secondary vertices. SVX *b*-tagging relied on identifying jets that ap-peared to originate at secondary vertices. A jet, though, is really a tightly clustered group of individual particles traveling in roughly the same direc-tion, and the silicon vertex detector allowed CDF to identify and measure tracks of individual particles within a jet, something that earlier generations of tracking chambers were unable to do. Jet vertexing required tagged jets to contain two intersecting SVX tracks and imposed strict cuts on the quality of those tracks to reduce background from "fake" tracks.

The basic idea behind seed vertexing was developed by Joe Incandela and Weiming Yao, a physicist at Berkeley who had entered CDF with an interest in *b* physics and had worked on the calibration of the first SVX (Yao 1995). Yao and Incandela proposed using two intersecting SVX tracks with somewhat looser cuts than jet vertexing imposed to define a "seed." After the seed had been identified, seed vertexing imposed an additional requirement that it be possible to fit a third track to the same vertex. The third track requirement would reduce the background, while a more generous allowance for the quality of individual tracks would keep the signal from being choked off. Incandela thought, "[I]f you started with two tracks... and require a third one for confirmation, there can't be very high backgrounds" (Incandela 1995).

Seed vertexing quickly won over most of the people working on SVX *b*-tagging. While the creators of the jet-vertexing algorithm were preoccupied with the Evidence paper, everyone else interested in secondary vertex tagging became involved in a single effort to develop a new algorithm based on seed vertexing, called SECVTX. With (almost) everyone working on a single algorithm, the group "felt like the new tagger was theirs, and they had some hand in it.... SECVTX was their algorithm" (ibid.).

Incandela and Gerdes were trying to organize the group to prevent the recurrence of problems that had appeared during run Ia. With three different algorithms, each with its own constituency, there had been "a tremendous amount of competition and fighting amongst the different tagger groups, and a lot of hostilities" (ibid.). The problem with such discord was not only that it was unpleasant but that the "free-for-all" atmosphere also threatened to undermine the quality of the physics being produced. The implicit message to those developing algorithms during Ia, according to Incandela, was "whoever comes in with the best signal, you're going to get the job" (ibid.). Consequently, as described in the previous chapter, members of each of the three groups were looking at the data and over their shoulders. They looked at the data to see how many events their own algorithm tagged. They looked over their shoulders to see how many events their competitors' algorithms tagged. All this looking caused concern for some of their colleagues about "tuning on the signal." The worry was that decisions about individual cuts might be made in such a way as to exaggerate the size of the signal being picked up, producing a very misleading appearance of a significant statistical excess over background (Incandela 1995).

As Dave Gerdes noted, looking at the data while working on the analysis to be applied to those data "is not the same as tuning on the signal, but it creates an environment where you have to be careful not to tune the cuts" (Gerdes 1995). In a mixed judgment, Incandela remarked that "I am sure in fact that there was probably some level of tuning to get the highest signal to noise in the data. . . . I don't think it was conscious. I don't think that they faked anything. But there was no way to control that. . . . I'm not saying they cheated. It looks like most of the events were real. Some of them I think were junk. I believe they got the right answer" (Incandela 1995).

Even some who suspected that biases may have entered into the run Ia SVX analysis as a result of the competitive atmosphere readily acknowledged that it would not have been easy, and perhaps not even possible, to develop an algorithm for tagging secondary vertices at that time without a certain amount of peeking at the data. The SVX was a new detector using technology never tried before at a hadron collider. Although a sure way to avoid tuning on the signal is simply to avoid looking at the data, such an approach would have been very risky. The *b*-tagging group faced the danger that, were they to develop a tagging algorithm based entirely on "fake" Monte Carlo data, and then look at real data only after months of work, they would find that their algorithm did not work on real data as they expected. Consequently, they could make a plausible argument that they *had* to run their algorithms on the data to be sure they were identifying the right kind of events. The price they paid was that, when they finalized their algorithm, some of their colleagues doubted the reliability of their statistical assessment of the excess they found.

Gerdes and Incandela wanted to avoid any such ambivalence regarding the new SECVTX algorithm. So in addition to an initial ban on tagger development talks and an effort to unify the group behind a single algorithm, they decided to impose a moratorium on analyzing the data. Joe Incandela sent out an e-mail message proposing this rather drastic measure. "I was very worried," he recalled, "that when I came in to the next *b*-tag group meeting people would react very violently." When he came into the next meeting of the group he was surprised to find that people were "so grateful that I had stopped this mad race, because all of them were exhausted. None of them wanted to do it either. . . . Everyone agreed to abide by that [decision]" (ibid.).

Once word of the moratorium circulated, some people outside the *b*-tagging group expressed displeasure. "A few bigwigs came and wanted to know what the hell I was doing," according to Incandela. As the run progressed, some CDF members who were not in the *b*-tagging group pressured the group leaders to break the moratorium and show tagging results (Incandela 1995). Gerdes recalled that they discussed looking at the data in June or July but did not decide that they were ready to do so until the fall of 1994.

2. ANOTHER BAD OCTOBER MEETING

The collaboration was due to meet in October of 1994. In the period before the meeting the leaders of the *b*-tagging group "were getting a little pressure from the people higher up" to show *b*-tag results (Gerdes 1995). In addition, Joe Incandela was getting nervous. Taking data "blind" like they were had its own risks. Incandela had memories of these dangers from his time as a member of the UA2 collaboration: they had run for three months with a faulty trigger, and had lost three months worth of data – a disaster (Incandela 1995). The *b*-tagging group had optimized SECVTX to match JETVTX's efficiency, but with a lower background. So, in the face of mounting pressures to show results, the SVX *b*-tag group ran their algorithm on the data. Dave Gerdes presented their findings at a collaboration meeting on October 13. "Things didn't look that good" (Gerdes 1995). In an amount of data equal to that produced during run Ia (19 pb^{-1}), there were only three tags in two events, with a background expectation of about 1.9. According to Gerdes, "We'd gotten unlucky," as some otherwise good events just failed one or another of the cuts" (ibid.).

As a check on the results from the Ia analysis, Gerdes also showed the results of running the same algorithm on the old Ia data. The search yielded five tags in four events. Those four events had all been tagged by the jet-vertexing algorithm used in the Evidence paper, but two other events that jet vertexing had tagged failed the cuts for SECVTX. Both of those events had SLT tags. One was the disputed pseudo *e-μ* event. Although Gerdes's presentation called these results "consistent with the previous result," he also noted that the SECVTX algorithm was about 10–15% more efficient at finding top events and had a 40% lower rate of mistags. This troubled some in the collaboration. If the original results were sound, then a more efficient algorithm should find more, not fewer, events. (We will return to this question in Section 7.1.)

The numbers were more encouraging when Gerdes showed results on the Ib data using the old algorithm: JETVTX tagged five events with a background expectation of approximately 2.3.

The results from the SLT search were even more disappointing than the SECVTX numbers. The SLT analysis had been extended to include previously excluded muons detected in the central muon extension. This made the SLT analysis more efficient, so it should select more top events than before. Instead, in an amount of run Ib data roughly equal to that used for the Evidence paper, the SLT algorithm found only four events, with a background expectation of 3.5. When the CMX data was included in the run Ia data, the SLT search found no additional candidate events beyond the seven previously selected, but the background expectation increased from 3.1 to 3.8.

The meager *b*-tagging results had implications for the group working on the mass estimate. In the analysis used in the Evidence paper, events used for purposes of estimating the top mass had to have been tagged by at least one of the *b*-tagging algorithms and have a fourth jet with $E_T > 8$ GeV (for the *b*-tagging algorithms themselves, only three jets were required, but all three were required to have $E_T > 15$ GeV). This had yielded seven events for mass fitting from the Ia data. When Berkeley physicist Lina Galtieri presented the findings of the mass and kinematics group at the October collaboration meeting, the Ib data included only two events that met those requirements.

The dilepton search found one additional candidate event in the Ib data beyond the two identified in the Ia data, while the background estimate roughly doubled. This yielded three candidates, all of them e-μ events, with a background expectation of approximately 1.1.

3. ANOTHER COMPETITION OVER SVX b-TAGGING

In the SVX b-tagging group, the meager results yielded by the new algorithm caused disappointment and renewed the commitment to the data moratorium. Tony Liss took the disappearance of the apparent signal as a "motivation for not looking at it again for a long time." Under such conditions, he noted, frequent examination of the data would create a considerable "danger of biasing yourself somehow if a signal that you had starts to go away, and then you try to bring it back" (Liss 1995).

The poor showing in the October meeting also prompted self-reflection in the SVX b-tag group. The background rejection obtained with SECVTX was in fact somewhat better than it needed to be. Working group members believed they might gain some advantage by loosening some of the cuts to make the algorithm even more efficient. Such a change let in a little more background but did not diminish the signal-to-background ratio much. The group did Monte Carlo studies on three different configurations of the algorithm, with varying degrees of looseness, to see which would have the highest probability of yielding a four–standard deviation excess in 50 pb^{-1} of data. This showed that the loosest of the three configurations was the most likely to find the top, so the group adopted that version of SECVTX as their tagger (Gerdes 1995; Incandela 1995).

The University of Rochester physicists who had worked on jet vertexing pursued a different course. Since the SECVTX was not after all giving results dramatically better than JETVTX, they argued that it would be better to maintain continuity with the Evidence paper analysis and retain JETVTX as the tagger of choice. The fact that JETVTX had tagged more events than SECVTX at the October collaboration meeting added to the older algorithm's appeal. The authors of SECVTX maintained that they had the superior tagger, but collaboration members were not happy with the signal that it was showing. Joe Incandela recalled, "A lot of people started leaning back toward the old tagger and saying, 'Why are you developing this new thing? The old one does fine. It gives us what we want.'" Encouraged by these shifting sentiments, the Rochester group made a strong effort between October and December to "defend their turf" (Incandela 1995).

The rest of the *b*-tagging group, who had by this time invested considerable effort in SECVTX, resisted this move. In December, the antagonists in the group came into open conflict. With another collaboration meeting scheduled for January, the SECVTX developers were ready to run their newly reoptimized algorithm on the data, but they ran into opposition from the Rochester group, who insisted that the algorithm was not yet ready to be run on the data. Some members of the *b*-tag group regarded this call to optimize the new algorithm as part of a Rochester plan to "kill" SECVTX, and as a "stalling tactic" intended to give the Rochester group time to reassert their dominance in the group after having been preoccupied with the Evidence paper during the early stages of run Ib.

On December 6, the SVX *b*-tag group met to prepare for the top group meeting the next day. Incandela proposed that they run the SECVTX algorithm on the data so that they could prepare presentations of the results for the January collaboration meeting. The Rochester group objected. Hesitating to strong-arm the group, Incandela left the issue unresolved until the next day's top meeting. There the top group voted on the issue and voted in favor of running SECVTX. "There was only one vote against it, and it was from Rochester" (ibid.).

Mark Timko was at that time a research associate at Tufts University and had worked on a kinematic analysis for top-quark evidence as well as on the Dalitz-Goldstein-Sliwa analysis. According to Timko, the JETVTX algorithm could not actually have been used in January 1995 to make a discovery claim regarding the top quark. Although it yielded about the same number of tags as the SECVTX algorithm did at that time, only four of those events had four jets meeting the requirements of the mass-fitting analysis. Furthermore, the mass estimates based on those few four-jet events tagged by JETVTX did not form a peak. Two of them yielded estimates below 160 GeV/c^2, and two yielded estimates above 180 GeV/c^2 (Timko 1998). This was more than just inconvenient; it suggested that the jet-vertexing algorithm was not tagging events of a single kind, that it was identifying events that could not be ·reconstructed as decays of just one kind of particle.

Another change in the SVX analysis deserves attention: in the Evidence paper, CDF estimated the SVX background using two methods. "Method 1" applied the tagging algorithm to a sample of 67,000 events that passed the 50-GeV jet trigger (these were known as inclusive jet events). They then measured the tagging rate for that control sample to estimate the probability of tagging for each of the $W+$ jets events to which the SVX *b*-tagging algorithm was applied (there were 52 such $W+$ jets candidates for the Evidence paper). This more conservative method, argued CDF, tended to overestimate background because of the sources of genuine heavy quarks (such as bs) that would be present in the inclusive jet sample but not in the $W+$ jets sample.

"Method 2" made more use of Monte Carlo calculations. For that method, CDF estimated the rate at which the tagging algorithm tagged events in the inclusive jet sample that did not have any heavy quark content at all (which they assumed would not be different for the sample of $W +$ jet events to which they applied the tagger). This estimate involved looking at an event parameter called L_{xy}, defined to be the projection onto the r-ϕ plane of the distance from the primary to secondary vertex. Roughly, CDF estimated the mistag rate for the SVX b-tagging algorithm by looking to see how often it tagged jets such that the vector giving the direction of L_{xy} (away from the primary vertex) did *not* point in the same approximate direction as the momenta of the particles in the jet that was tagged. Method 2 then estimated the background from tagged events containing Ws and $b\bar{b}$ or $c\bar{c}$ pairs separately, using Monte Carlo calculations. This yielded lower, and supposedly more accurate, background estimates, but depended on Monte Carlo calculations which not all collaboration members trusted. In the Evidence paper, both background estimates were given, but only the method 1 result was used for the "official" results and the significance calculation. By the time that the collaboration was developing the run Ib analysis, collaboration members generally felt more comfortable with the Monte Carlo routines that CDF was using, and more confident in their ability to simulate the operation of the silicon vertex detector, so the secondary vertex b-tagging group decided to adopt the method 2 background estimate, arguing that it gave a more accurate value.

4. THE JANUARY 1995 COLLABORATION MEETING AND THE OBSERVATION PAPER

The CDF collaboration meeting held on January 21, 1995, was well remembered by those who were there. In August 1993, the presentations that precipitated the Evidence paper were neither immediately nor unanimously seen as significant. But at the January 1995 meeting no one could doubt the importance of the SVX results that were shown.

Since October, the data set had grown from about 20 pb^{-1} to nearly 50 pb^{-1}. Using the optimized SECVTX algorithm, the SVX group ran on the data and produced a signal that "stuck out like a sore thumb" (Jensen, personal communication). Joe Incandela showed an overview of the SVX b-tag analysis, describing the performance of the SECVTX algorithm and the optimization process, and Weiming Yao showed the actual results. There were 18 tags in 14 events from the run Ib data, with an expected background of 5.6 ± 0.6. (Although changes in SECVTX had made it even more efficient since October, it continued to find only five tags in four events in the run Ia data.) The presentation of the mass distribution also made an impression. It showed a well-formed peak of just the sort that one would expect if the events being selected were top decays. When the mass peak was shown, "people's jaws hung open" (Incandela 1995).

The SLT results were not as clear-cut at the January meeting. The SLT developers continued to show results using both the 2- and 4-GeV/c choices for the soft-lepton p_T cut. Using the cut placed at 2 GeV/c, they found 13 tags in 12 events in the Ib data, but with a background expectation of 9.6. The background dropped to 5.6 when they moved the cut to 4 GeV/c, and the number of candidates dropped to 8 tags in 7 events. In their presentation, the SLT authors – Claudio Campagnari, David Kestenbaum, and Avi Yagil – expressed a preference for setting the cut at 4 GeV/c for the Ib analysis, although they maintained that the p_T cut had little effect on the expected significance of a top signal for top masses near that estimated in the Evidence paper.

The dilepton analysis had changed in minor ways. CDF had made small changes to all three algorithms in order to unify the particle identification cuts used in the three counting experiments. Since physicists working on the three searches had not at first collaborated on setting cuts, for the Ia top search the three groups had arrived at slightly different definitions of important elements in the analysis – such as what to count as an electron or muon. For the Ib analysis, they adopted a single set of cuts to define these elements, preferring slightly tighter cuts suited to a higher mass top quark. As of the January meeting, 3 dilepton candidate events had turned up in the run Ib data, with a background expectation of 1.06, yielding a total for both runs of five candidates with 1.60 expected from background.

Faced with such a large excess in the SVX search, CDF decided at the January meeting to write up their newest results, and to do so quickly. They wanted to be able to present "real results . . . for the winter conferences in March, so we wanted to pretty much be done by the end of January to February" (Gerdes 1995).

CDF collected more data, and the top group presented one last set of updated results to the collaboration at a meeting on February 9. The results did not change dramatically from what they had been in January (see Table 5.1). The SLT group continued to report results for both the 2- and 4-GeV/c choices for the p_T cut. (When the Observation results were finalized, the cut was left at 2 GeV/c.)

The paper itself came together quite quickly, and with very little friction, in strong contrast to the Evidence paper. "That was CDF's finest hour," said Paul Tipton. "I don't think we'll ever [again] get the lead out and have that little resistance to motion. . . . Maybe because we did really bleed through [the writing of the Evidence paper]" (Tipton 1995).

CDF had another reason to move quickly besides the upcoming winter conferences. D-zero had scheduled a talk at Wilson Hall for the end of February, and some CDF physicists speculated that this was going to be a "top talk." CDF wanted to be ready if this hunch was correct. (In fact, the

TABLE 5.1. Numbers of Candidate Events and Estimated Backgrounds, Runs Ia and Ib

Channel (SLT p_T cut)	Date / Data Sample	4/26/94 (Evidence) Ia	10/13/94 Ia	10/13/94 Ib	1/21/95 Ia	1/21/95 Ib	2/9/95[a] Ia	2/9/95[a] Ib
SVX	Candidates events (tags)	6(6)[b]	4(5)	2(3)	4(5)	14(18)	4(5)	16(22)
	Background (Ia: events/ Ib: tags)	2.3	1.9	1.9	2.7	5.6	2.53	4.92[c]
SLT (2)	Candidates events (tags)	7(7)	7(7)	4(4)	7(7)	12(13)	7(7)	16(17)
	Background events	3.1	3.8	3.5	3.8	9.6	3.8	11.6
SLT (4)	Candidates events (tags)	4(4)	4(4)	4(4)	4(4)	7(8)	4(4)	10(11)
	Background	1.7	2.1	1.8	2.06	5.6	2.1	6.6
DIL	Candidates	2	2	1	2	3	2	2
	Background[d]	0.56	0.56	0.56	0.54	1.06	0.25	0.59

[a] February 9, 1995, dilepton results use more stringent requirements than were used either in January, or in the Observation paper itself. Elsewhere dilepton candidates were required to have two jets with $E_T > 10$ GeV. Here they were required to have two jets with $E_T > 15$ GeV.

[b] For the Evidence paper, SVX results are based on use of JETVTX tagger. All other SVX numbers are from the application of SECVTX.

[c] Estimated by scaling total background to relative size of data set.

[d] For October 13, January 21, and February 9, dilepton background estimates were given for the combined Ia + Ib data. Separate Ia and Ib backgrounds are here estimated from total backgrounds by scaling the total to the relative size of each data set.

scheduled talk concerned *W* bosons, not the top quark, but was canceled at the last minute.) The two collaborations had an agreement with the lab administration that if either was to make a major announcement, they would give two weeks' advance warning to the other collaboration, giving the other collaboration time to prepare a response. In mid-February, CDF submitted a copy of their paper to lab director John Peoples (who had replaced Lederman in 1989), who notified the D-zero collaboration.

On March 3, 1995, CDF made another front-page announcement. The paper they submitted to *PRL* was titled "Observation of Top Quark Production in $\bar{p}p$ Collisions," and, to the surprise of CDF members, they were joined in this announcement by the D-zero collaboration submitting to *PRL* their paper, "Observation of the Top Quark."

5. OBSERVATION OF THE TOP QUARK BY D-ZERO

CDF had for some time been proceeding with one eye on their neighbors around the ring. D-zero's first top paper had reported a mass limit of 131 GeV/c^2 (Abachi, Abbott, et al. 1994). In their second search on the Ia data, D-zero had seen a slight excess of top candidate events but had not claimed it to be evidence for the top (Abachi, Abbott, et al. 1995a). For run Ib, they planned to wait until they had an amount of data roughly equal to the total for Ia, and then run a top analysis on it, with an eye to reporting the results at conferences, particularly the Aspen conference, in January 1995 (Narain, personal communication).

The top group at D-zero was trying to take advantage of their increasing understanding of their detector by developing better cuts for identifying event features such as leptons, but they also revised the cuts they were using to search for the top. Looking at the data, Boaz Klima recalled, "I realized that we simply don't see top which is 140 GeV." This realization was not based on a systematic study. Rather, looking at the top candidates selected using the cuts from the second Ia search paper, two facts struck Klima. First, the numbers of candidate events in the Ia data and the early Ib data in each decay channel individually were not as large as one would expect for a top with a mass near 140 GeV/c^2. It looked more like the low production cross section of a more massive top. Second, if the top had a mass around 140 GeV/c^2, then the reconstructed mass of the top candidates should show a peak near that value. Looking at the mass analysis for the lepton + jets top candidates in the Ia data, Klima saw "a big blob... [meaning] we have something else here" (Klima 1995). Rich Partridge confirmed that they "didn't see anything striking in our re-optimized cuts with the early Ib data," and that "[o]ur mass was coming out 200 for our events. Our best guess was that it was heavy" (Partridge 1995).

If the top was very heavy, then D-zero would have a better chance of finding it if they could make tighter cuts to reduce the background. If they were wrong about the mass and it was in fact lighter, they could still see it with the tighter cuts (because it would be produced at a faster rate), although they would not pick out the signal as efficiently. Using cuts suited for a lighter top, Klima concluded, they might miss entirely the heavy top they believed they were beginning to see in the Ia and early Ib data.

While Boaz Klima was contemplating a strategy for finding a heavier top, his colleagues Meenakshi Narain and Rich Partridge went to Aspen to present the latest D-zero results. They presented a mass analysis formalizing Klima's hunch by showing that the Ia candidate events chosen by the cuts used in the second D-zero search paper formed a mass peak at around 200 GeV/c^2. They also gave an estimate of the cross section for the top based on a combination of Ia and Ib data (Narain 1995).

With Narain and Partridge in Aspen presenting results based on the old cuts, Klima was organizing a few of his colleagues to revise those cuts. He sent e-mail messages to "very few, very senior" D-zero physicists proposing a strategy: "If you change too many things, people will be very suspicious, so ... let's keep the analysis as identical as possible to the old one with very few changes. And the change was to ... use the parameter called H_T." D-zero had in fact already used a cut on the quantity they called H_T in an earlier paper (Abachi, Abbott, et al. 1995a), where they defined H_T as the scalar sum of the transverse energies of the jets in an event. As mentioned previously (Section 7 in Chapter 4), some members of CDF had also advocated using H_T to discriminate top decays from background (although definitions of the quantity differed slightly).

Klima and his confidants tried out an increase in the cut on H_T. "We did it with one channel, and then we did it with a few, and then I said, 'We have a four sigma effect. This is top. We have it'" (Klima 1995). Having satisfied themselves on this point, it remained to go back and carry out the reoptimization systematically, making sure that each step could be justified.

Partridge and Narain both recalled being in Aspen and receiving excited e-mail messages from D-zero physicists back in Illinois reporting the results of these first attempts to look at the data with a revised H_T cut. Rich Partridge recounted getting messages reporting that by selecting "more stringent" cuts his collaborators were seeing "very significant results" (Partridge 1995).

D-zero's systematic revision of their cuts used Monte Carlo studies and background control samples to establish the optimal set of cuts for a top with a mass around 180 to 200 GeV/c^2. The primary change concerned the use of H_T. In the dilepton channels and the soft lepton–tagged lepton + jets channels, which had earlier not employed an H_T requirement, they imposed cuts requiring H_T greater than 100 to 140 GeV, depending on the channel. They raised the H_T cut from 140 to 200 GeV for the untagged lepton + jets channels (Abachi, Abbott, et al. 1995a, 2634; 1995b, 2424).

Collaboration members presented and discussed the results of the revised top analysis at a two-day marathon meeting, held in a crowded room at the D-zero site for fear that holding it at the high rise would tip off CDF that something important was happening. The comments from the collaboration were supportive of the work that had been done, although some asked for some additional checks to be carried out (Narain, personal communication).

The evening following the meeting, a group of D-zero physicists who had been working on the top analysis began, in the words of Rich Partridge, "rabble rousing" over dinner at a restaurant near Fermilab. The collaboration had been working on a long, detailed paper presenting their top analysis, but "it was clear that that was not going to converge in any short period of time." Why not, they asked, simply write a *PRL* presenting the results that they had? They could write it over the weekend. The next day this same group met with one of the D-zero spokespersons. They encountered some resistance to their idea at first but insisted that they could write the paper quickly, letting the collaboration decide whether to submit the paper. By the end of the meeting, they had produced a consensus that they would write the *PRL*. "By the end of that day, Friday, we heard from CDF that they . . . had a *PRL* written" (Partridge 1995).

D-zero's top group set out to produce a paper at breakneck speed. They began writing on a Friday evening, remembered Meena Narain, and by Monday morning the editorial board (roughly the D-zero equivalent of a CDF godparents' committee) had met twice to work on revisions. Fermilab physicist Herb Greenlee "was writing furiously" and received "a lot of help from others staying up all night" (Narain 1995). The first draft was done on Saturday, and on Sunday they produced at least one more draft. A copy of the paper went out to the collaboration on Monday (Partridge 1995). About 130 people responded to the paper with questions for the authors, and the authors answered all of them (Narain 1995), with the possible exception of some handwritten questions that came from D-zero representatives attending a conference in Rio De Janeiro, including one spokesperson. "I think those got lost in the shuffle. The requirement was that you had to use e-mail" (Partridge 1995).

By Thursday that same group of D-zero physicists was back at that same local restaurant, this time with one of the spokespersons, celebrating their completed *PRL*, and the next day both collaborations sent papers reporting the observation of the top quark to *Physical Review Letters* (ibid.).

6. TWO OBSERVATION PAPERS

CDF intended their analysis in "Observation of Top Quark Production" to follow up on that used in the Evidence paper. For run Ib the SVX detector was replaced with a more efficient version, a more efficient algorithm was employed for finding secondary vertices, and the fiducial region used in the identification of leptons was increased to reflect the increased coverage of the muon detectors. Other small changes made to simplify the presentation of the analysis arose from primarily aesthetic considerations, as they made for a simpler, clearer presentation. The search again consisted primarily of three counting experiments: dilepton, SVX, and SLT.

The results they reported were these: the SVX search identified 27 events with a method 2 background expectation of 6.7 ± 2.1 (more data had been added after the January 1995 collaboration meeting). Method 1 applied to the same data yielded a background estimate of 9.2 ± 1.2 (Yao, Bedeschi, et al. 1995, 15). However, CDF noted that, due to the smaller systematic uncertainty on the method 1 estimate, they arrive at approximately the same significance estimate using either method (Abe, Amidei, et al. 1995a, 2631).

The SLT search turned up 23 tagged events against an expected background of 15.4 ± 2.0. Thirteen events were tagged by both SVX and SLT. In the dilepton channel, they found six events with a background of 1.3 ± 0.3. CDF calculated the probabilities of finding as many or more candidate events in these three channels to be $P_{SVX} = 2 \times 10^{-5}$, $P_{SLT} = 6 \times 10^{-2}$, and $P_{DIL} = 3 \times 10^{-3}$, where these values are calculated in the same manner as the corresponding values in the Evidence paper. For the Observation paper, however, CDF combined the results of the counting experiments differently from the method used in the Evidence paper. Some collaboration members argued that simply adding together the numbers from the three counting experiments and then calculating a significance gave a misleading estimate because the three searches had quite different expected signal-to-background ratios. For example, the SVX and SLT tags in the Ia data would combine to give 13 tags on a combined background of about 5.4 (3.1 from SLT, 2.3 from SVX) whether there were 7 SLT tags and 6 SVX tags (as they in fact found) or 4 SLT tags and 9 SVX tags. But the latter combination would have constituted stronger evidence owing to the larger excess in the channel with the smaller background. Hence CDF adopted a method whereby they multiplied the significance estimates of the three counting experiments (P) and then calculated a probability, assuming only background sources were present, of arriving at results yielding "a value of P no larger than we observe" (ibid., 2629). This produced a significance estimate of 1×10^{-6}, corresponding to a 4.8σ effect for a Gaussian distribution. Furthermore, an extrapolation from the SVX excess to the other searches on the assumption that the SVX excess is the result of the presence of some top decays yielded expectations of excess events consistent with those found in the data (ibid., 2629–30).

Another independent source of evidence for the top quark presented in the Observation paper was the peak in the reconstructed mass distribution. CDF noted that after including systematic effects in the background prediction, the probability of finding such a mass peak simply from background is 2×10^{-2} (ibid., 2630). Although their official result was the significance estimate of 1×10^{-6} based on the counting experiments alone, the Observation paper noted that the significance level of results based on combining the counting experiment and mass peak significances is 3.7×10^{-7}, which is equivalent to 5.0σ for a Gaussian distribution (ibid., 2631). Also, the excess of tagged $Z +$ jets events that had worried CDF in the Evidence paper had disappeared (ibid., 2630).

The authors concluded that their additional data "confirm the top quark evidence" presented in the Evidence paper (ibid., 2630). Noting the significance of the counting experiments, the sharpness of the mass peak, and the fact that a "substantial fraction" of the jets in the dilepton events were tagged as b jets, they concluded, "This establishes the existence of the top quark" (ibid., 2631). Based on a sample of events with b tags (SVX or SLT) and four or more jets, they estimated the mass of the top to be $176 \pm 8 \pm 10$ GeV/c^2, where the first error is statistical and the second is systematic (ibid., 2630).[1] They estimated the production cross section for the top to be $\sigma_{t\bar{t}} = 6.8^{+3.6}_{-2.4}$ pb, considerably lower than the value of $\sigma_{t\bar{t}} = 13.9^{+6.1}_{-4.8}$ pb reported in the Evidence paper (ibid., 2627, 2629).

The analysis used in D-zero's Observation paper departed more significantly from what had been used before. As noted previously, the main difference in D-zero's Observation paper analysis and that used in their earlier "Search for High Mass Top Quark" paper was the expanded use of cuts on the quantity H_T and the movement of cuts on that quantity to higher values. For the $e\mu +$ jets and $ee +$ jets channels they redefined the quantity as the scalar sum of the transverse energies of the jets *and* the leading (most energetic) electron (only the jet energies were included for the other channels) (Abachi, Abbott, et al. 1995b, 2634). This quantity, the authors pointed out, "is a powerful discriminator between background and high-mass top quark production" (ibid., 2634).

D-zero applied this revised analysis to data from an integrated luminosity of approximately 50 pb^{-1}. In this data, they found 17 candidate events, with an expected background of 3.8 ± 0.6 events, spread somewhat evenly through six of the seven search channels that they defined (see Table 5.2). The probability of such an excess (or more) given only background sources they estimated to be 2×10^{-6}, equivalent to a 4.6σ effect for a Gaussian distribution (ibid., 2635). Using a looser set of cuts, they selected events from which they derived a top mass estimate of $m_{\text{top}} = 199^{+19}_{-21} \pm 22$ GeV/c^2, where the first error is statistical and the second is systematic. They concluded that the production cross section is $\sigma_{t\bar{t}} = 6.4 \pm 2.2$ pb (ibid., 2636).

7. DISCUSSION: DETECTING BIAS WITH MORE DATA AND WITH DIFFERENT TESTS

As CDF moved on to gather more data during run Ib and develop new means of analyzing that data, they learned some things about their earlier Ia results. In the Observation paper, CDF claimed that "additional data confirm the

TABLE 5.2. D-zero's Candidate Events in Various Search Channels from Their Observation Paper

m_t (GeV/c^2)	$e\mu$ + jets	ee + jets	$\mu\mu$ + jets	e + jets	μ + jets	e + jets/μ	μ + jets/μ	All
140 $\varepsilon \times \mathcal{B}$ (%)	0.17 ± 0.02	0.11 ± 0.02	0.06 ± 0.01	0.50 ± 0.10	0.33 ± 0.08	0.36 ± 0.07	0.20 ± 0.05	
$\langle N \rangle$	1.36 ± 0.21	1.04 ± 0.19	0.46 ± 0.08	4.05 ± 0.94	2.47 ± 0.68	2.93 ± 0.68	1.48 ± 0.42	13.80 ± 2.07
160 $\varepsilon \times \mathcal{B}$ (%)	0.24 ± 0.02	0.15 ± 0.02	0.09 ± 0.02	0.80 ± 0.10	0.57 ± 0.13	0.50 ± 0.08	0.25 ± 0.06	
$\langle N \rangle$	0.94 ± 0.13	0.69 ± 0.12	0.34 ± 0.07	3.13 ± 0.54	2.04 ± 0.53	1.95 ± 0.39	0.92 ± 0.24	10.01 ± 1.41
180 $\varepsilon \times \mathcal{B}$ (%)	0.28 ± 0.02	0.17 ± 0.02	0.10 ± 0.02	1.20 ± 0.30	0.76 ± 0.17	0.56 ± 0.09	0.35 ± 0.08	
$\langle N \rangle$	0.57 ± 0.07	0.40 ± 0.07	0.19 ± 0.04	2.42 ± 0.67	1.41 ± 0.36	1.14 ± 0.22	0.64 ± 0.16	6.77 ± 1.09
200 $\varepsilon \times \mathcal{B}$ (%)	0.31 ± 0.02	0.20 ± 0.03	0.11 ± 0.02	1.70 ± 0.20	0.96 ± 0.21	0.74 ± 0.11	0.41 ± 0.08	
$\langle N \rangle$	0.34 ± 0.04	0.25 ± 0.05	0.11 ± 0.02	1.84 ± 0.31	0.95 ± 0.24	0.81 ± 0.16	0.41 ± 0.10	4.71 ± 0.66
Background	0.12 ± 0.03	0.28 ± 0.14	0.25 ± 0.04	1.22 ± 0.42	0.71 ± 0.28	0.85 ± 0.14	0.36 ± 0.08	3.79 ± 0.55
$\int \mathcal{L}\,dt$ (pb^{-1})	47.9 ± 5.7	55.7 ± 6.7	44.2 ± 5.3	47.9 ± 5.7	44.2 ± 5.3	47.9 ± 5.7	44.2 ± 5.3	
Data	2	0	1	5	3	3	3	17

Source: Abachi, Abbott, et al. (1995b, 2635).

top quark evidence presented in [the Evidence paper]" (Abe, Amidei, et al. 1995a, 2630). This was correct, insofar as the Observation paper constituted a stronger confirmation of the claim that the top quark exists and that its mass is approximately what they had initially claimed. At the same time, the new data and new analysis employed in the Observation paper hinted at biases in the tests used in the Evidence paper. Some collaboration members took these hints as confirmation of their suspicions regarding the earlier analysis.

These after-the-fact assessments were based on two different ways of subjecting evidence claims to empirical test: (1) the application of a different testing procedure to the same data on which the initial evidence claim was based and (2) the collection of additional data subjected to the same analysis as that on which the initial evidence claim was based.[2]

7.1. The SVX Analysis: Finding Bias with a New Test on the Same Data

Some CDF members suspected that biases had crept into the algorithm used for SVX *b*-tagging during run Ia. The fact that the algorithm had been run on the data numerous times while physicists were developing it raised the fear that jet vertexing may have been tuned on the signal. Additionally, some feared that the selection process in the SVX *b*-tagging group had introduced a bias. They worried that yielding the largest number of candidate events was in fact a criterion for selection as the tagger of choice, although those making that choice may not have consciously relied on that criterion.

But such evidence of bias was, as one group member acknowledged, "circumstantial." Group members might have run their algorithm on the data without the results influencing their subsequent work on that algorithm. Furthermore, even if the tagger chosen from the three candidate algorithms yielded the most significant result, the top group might have chosen the same tagger anyway. (Members of the *b*-tagging group insisted that the choice was directed by quite different considerations. In any case, they pointed out, jet vertexing did not yield the most candidates until after they chose it.)

If an analysis of experimental data does detect a real effect, however, then the effect should be detected by other, independent analyses. An effect that survives such further testing is *robust*. As William Wimsatt has described such robustness analyses, they involve four procedures: (1) an analysis of "a *variety* of *independent* derivation, identification, or measurement processes"; (2) a search for results of those processes that are "*invariant*... or *identical*"; (3) a determination of the "*scope* of the processes" across which those results are invariant or identical and "the *conditions* on which their invariance depends"; and (4) an analysis and explanation of "*failures of invariance*" (Wimsatt 1981, 126, emphasis in original; for a closer analysis of the evidential value of robustness, see Staley 2003b).

Such robustness tests appear wherever experimenters seek to avoid inter-preting an artifact of their instrument or analysis as an independent phe-nomenon. Ian Hacking famously drew philosophers' attention to the prac-tice with the example of a microbiologist examining red blood platelets with an electron microscope, noticing what appear to be dense bodies within the cell, and then checking on this apparent finding by applying a fluorescent stain to the platelets and viewing them through a fluorescence microscope. The latter operates on quite distinct principles from those of an electron microscope. Hacking calls them "unrelated chunks of physics" (Hacking 1983, 200–1).

The CDF top search incorporated some elements of a robustness analysis throughout. For example, the use of three different counting experiments, with each one finding an excess over background, indicates that the excess was robust. That support would be weakened, however, if one or more of those counting experiments were tuned on the signal, since the tuned algo-rithms would not then constitute *independent* tests. Being tuned on a single set of data, the tuned algorithms would tend to agree with excesses reported in the other searches whether they were due to top decays or not. Wimsatt identifies such failures of independence as a major source of error in the use of robustness analyses (Wimsatt 1980, 306–15; 1981, 156–9).

When the *b*-tagging group applied the SECVTX algorithm to the Ia data and compared the results to those from the jet-vertexing algorithm JETVTX, they were carrying out a modest piece of robustness analysis: subjecting the jet-vertexing excess to an independent testing process (the SECVTX algo-rithm) and looking to see whether the excess remained invariant under that new process. One additional feature of this instance of testing for robustness is noteworthy, however. Because the SECVTX algorithm was known to be *more efficient* than JETVTX (about 25% more efficient by the time that they finalized their results for the Observation paper), a truly satisfactory result would have been, not merely an excess of the same size, but a *larger* excess. Applying SECVTX to the Ia data, CDF had every reason to expect to find an additional event or two, if the jet-vertexing excess was an entirely real effect and had no artifactual enhancement owing to bias.

But that was not what they found. As noted earlier, SECVTX tagged two *fewer* events than JETVTX had when applied to the Ia data. Although that is not compelling evidence that the developers of JETVTX had tuned their algorithm on the signal, it suggested that the size of the effect detected using JETVTX was exaggerated. The *b*-tagging group developed SECVTX using "fake" Monte Carlo data, whereas work on JETVTX had continued well after members of the *b*-tagging group had looked at the data to which the algorithm was to be applied. The smaller numbers yielded by SECVTX raised suspicions that some of the Ia SVX events would not have been tagged if the developers of JETVTX had not been aware of the details of those particular events.

CDF's use of a robustness analysis that ought to have yielded an enhancement of the effect suggests adding a nuance to Wimsatt's account of robustness, insofar as such testing may aim at something other than strict invariance. A sign of robustness might be a particular *change* in a result, when background knowledge indicates that such a change is to be expected as a reliable discriminator between the "ontologically and epistemologically trustworthy" and the "unreliable, ungeneralizable, worthless, and fleeting" (Wimsatt 1981, 128). What is sought in such an instance of robustness analysis is an indication that the result *responds* to independent processes of generation in a way that it would be unlikely to do if it were not sound. In such cases, invariance is sought at the level of the phenomenal claim being tested (e.g., "the results indicate the occurrence of top quark decays"), but the invariance of this result is sought by means of looking for *changes* of a specific sort in one's experimental outcomes (e.g., the measured value of a test statistic). Thus, the term "convergent" (as was used by Donald Campbell and Donald Fiske, 1959) rather than "invariant" might better express the somewhat more flexible criterion needed to characterize robustness.

7.2. The SLT Analysis: Finding Bias with the Same Test on New Data

One of the themes that I will explore in the next chapter is how experimenters can evaluate evidence by the way it develops over time. This *dynamic* evaluation of evidence is one aspect of the susceptibility of evidence claims to empirical testing. One way to test the soundness of an evidence claim is to see how a result responds to changes in various components of the experimental test. In the last section, I described how CDF physicists changed the algorithm used to identify SVX candidates and watched to see how the statistical excess in that counting experiment responded to that change. The SLT algorithm, on the other hand, did not change in any dramatic ways from run Ia to run Ib, apart from incorporating data from the central muon extension. This increased both the expected signal and the expected background by about 20%. CDF members could, however, employ another method of testing the SLT algorithm for biases (thus testing any evidence claim based on the application of that algorithm as well): look to see what happens to the excess over expected background as more data accumulates. If an effect is real, then as more data is added, the excess should gain in statistical significance.

Such testing occurred in stages. In October 1994, when the top group showed the first top search results from run Ib to the collaboration, the presentation included SLT results for both Ia and Ib data. Even the Ia results, however, amounted to a test by adding data, since the incorporation of data from the central muon extension amounted to applying the same algorithm to data that had not been available before. The results were neither encouraging nor damning. The background estimate increased from 3.1 to 3.8, and no new candidate events were selected. More disturbing were the early results from the Ib data: the SLT search found virtually no excess in the new data that had accumulated since the beginning of run Ib. CDF had only four SLT candidates, with 3.5 expected from background. The SVX results in October were disappointing as well, however, and a depression in *both* channels would be consistent with a statistical lull in the signal. Thus, one could plausibly regard the dearth of top candidate events as the result of an unlucky downward fluctuation in the production of real top quarks.

At the January collaboration meeting, the SLT group showed 12 candidate events from the Ib data, with a background expectation of 9.6, and by February 9, the number of candidate events had increased to 16, while the background expectation rose to 11.6.

What do these numbers tell us? By themselves, perhaps not very much. However, when looked at in terms of statistical significance, we can learn more from them. In looking at the significance estimates based on these numbers, it is important to keep in mind that not all results that yield the same significance estimate are equally impressive as evidence.[3]

Consider a simple, rather artificial example. Imagine that we have three coins that we wish to test for fairness. Suppose that coin 1 is loaded such that the probability of its landing heads when it is tossed is 0.8. Coin 2 is also loaded, but not as heavily. The probability of coin 2 landing heads when tossed is 0.6. Coin 3 is a fair coin. Suppose that we toss each coin ten times, and each coin lands heads eight of the ten times it is tossed: coins 2 and 3 are subject to a statistical fluctuation during the first ten times they are tossed. We proceed to collect additional data, stopping every ten tosses to calculate the significance of the excess of heads over what we would expect for a fair coin, up to 40 tosses. During the remaining 30 tosses, coin 1 yields 8 heads out of every 10 tosses, coin 2 yields 6 heads out of every 10 tosses, and coin 3 yields 5 heads out of every 10 tosses.

Consider what happens to the significance estimates based on these results (see Figure 5.2).[4] For coin 1, the results gain steadily in significance as data accumulates. For coin 2, however, there is a drop in significance from the first set of results to the second, followed by slight gains. This reflects the fact that the initial set of results gave a misleading impression resulting from the upward fluctuation in the number of heads in those results. When the pattern subsequently normalizes, the effect of that fluctuation is diluted. (However, a coin that is loaded in the way that coin 2 is would eventually yield the kind of steadily rising significance shown for coin 1, but more slowly. Continuing in the same pattern, coin 2 would reach a significance level after 120 tosses (76 heads) comparable to that yielded by 20 tosses of coin 1.) Meanwhile, the initial appearance of a significant deviation from what one would expect for a fair coin quickly disappears with coin 3, as the significance of its results declines with accumulating data.

(a) **significance of coin tossing results, coin 1**

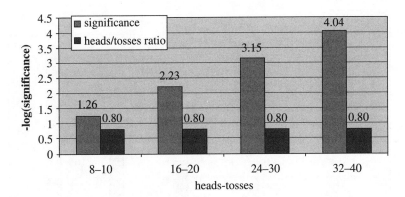

(b) **significance of coin tossing results, coin 2**

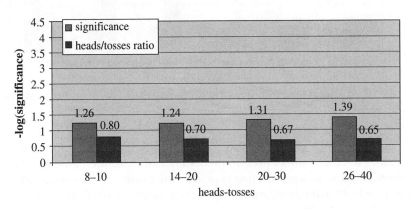

(c) **significance of coin tossing results, coin 3**

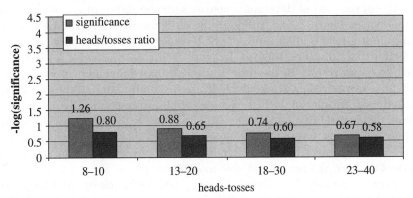

FIGURE 5.2. Results of coin tossing experiment: (a) coin 1, loaded: p(heads) = 0.8; (b) coin 2, loaded: p(heads) = 0.6; (c) coin 3, fair: p(heads) = 0.5.

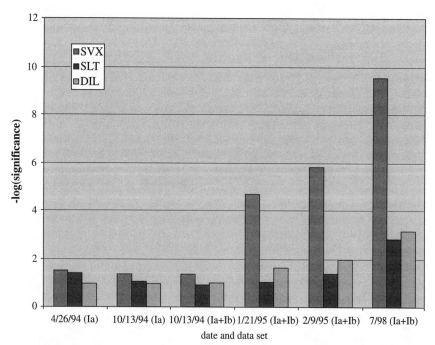

FIGURE 5.3. Statistical significance of CDF's counting experiment results, April 1994 to July 1998.

The three coins exemplify three quite different experimental possibilities: an effect that shows itself immediately and continues; an effect that, while real, is not as large as at first appears; and an effect that appears briefly to be real, but turns out to be an illusion of statistics, resulting from an initial fluctuation in the data.

Likewise, when comparing the course of events during run Ib, we see different patterns of developing statistics for the different counting experiments (see Figure 5.3).[5] Although the early months of run Ib yielded disappointing results all around, the SLT results in particular show an initial decline in significance, followed by a modest increase just prior to the Observation paper results. By the time of the February 9 meeting, approximately 65 pb^{-1} worth of data was available from the combination of runs Ia and Ib, but the significance of the SLT excess was only just regaining the significance level that it had initially attained with just 19 pb^{-1}. To critics, this vindicated their suspicions that the SLT analysis had been skewed by the inclusion of a few extra events from the Ia data that greatly exaggerated the size of the effect that the SLT search was able to detect.

Even in January 1995, when Campagnari, Kestenbaum, and Yagil presented the SLT results to the collaboration, they noted that, although they expected to find a 5σ effect in 100 pb^{-1} if the production cross section for

$t\bar{t}$ was as reported in the Evidence paper, they could only expect to see a 2σ effect in that same amount of data if the $t\bar{t}$ production cross section was equal to that predicted for a 170 GeV/c^2 top quark based on theory (see Laenen, Smith, and van Neerven 1994). The cross section reported in the Evidence paper was widely suspected to be too high even at the time that it was published and had quite significant uncertainties attached to it. More recent estimates have brought it down from the $\sigma_{t\bar{t}} = 13.9^{+6.1}_{-4.8}$ reported in the Evidence paper to $\sigma_{t\bar{t}} = 6.5^{+1.7}_{-1.4}$ pb (Affolder, Akimoto, et al. 2001b).[6]

This analysis is rough. I have offered it only to show how views within CDF regarding their own analysis evolved and to illustrate how scientists use additional data to test their own analyses and evidence claims. I do not intend my own analysis to have any demonstrative force regarding CDF's actual results. With that caveat in mind, it was just such indications that led some CDF members to conclude that the initial SLT results gave a misleading impression of the strength of the evidence for the top quark based on that part of the search. Yet no particular *cause* of that excess can be inferred from these indications. Although the pattern is what one would expect from a biased algorithm (setting aside the question of the cause of any such bias), it is also consistent with bad luck (a misleading upward fluctuation of the signal) of a kind that is not extraordinarily rare. Here, as in the case of the jet-vertexing analysis, one can at most say that the conditions under which the algorithms were produced prompted a question about the possibility of bias, and that subsequent reanalysis (with a different algorithm or with more data) failed, in the short term, to answer those questions in a reassuring way.

The relevance of these unanswered questions for CDF's evidence claim is this: To accept the significance estimates of the counting experiment results as reported requires that one accept the underlying assumptions of the analysis, namely, that the algorithms used in producing those estimates were not biased toward a favorable result on the data at hand. When questions are raised about such underlying assumptions, experimenters try to test those assumptions, just as they test the primary claims they are making. If the assumptions pass those tests, and the tests are severe (in a sense to be articulated in the next chapter), then the assumptions themselves receive support, and a possible source of error has been ruled out. If the assumptions fail such a test (as here), it does not prove that they are false, or even necessarily give evidence that they are false. It does, however, mean that a plausible source of error remains unresolved, casting doubt on the primary evidence claim.

Many CDF physicists who worked on the Evidence paper insisted that any bias or statistical fluctuations that may have influenced that paper's results would have been more than compensated for by other "conservative" choices regarding the analysis. These choices, such as the use of the method 1 technique for estimating background, were indeed conservative insofar as they made it less probable that the top-quark hypothesis would pass the test to which it was being subjected if it were false and they were only looking at background. That is to say, such conservative choices did make for a more *severe test* of that hypothesis. However, the precise *extent* to which these choices would compensate for possible biases elsewhere in the analysis is difficult to quantify.

Hence, CDF member Morris Binkley concluded that the attempt to "boil [the Evidence paper results] down to one number" was misguided. Such an attempt at precision in the face of unquantifiable corrections for unknown potential biases, Binkley alleged, called to mind his fellow Baltimore native H. L. Mencken's description of theology as "the attempt to put the unknowable in terms of the not-worth-knowing" (Binkley 1995).

This, however, is not to say that significance estimates are not useful. In the next chapter, I wish to explore a view of experimental inference that will shed light on the uses of significance estimates and bring out, through the background noise of the busy life of an experimental collaboration, a clear image of the structure of some parts of CDF's argument for the top-quark hypothesis.

6

A Model of the Experiment

Error-Statistical Evidence and the Top Quark

> [I]n the mind of the physicist there are constantly present two sorts of
> apparatus: one is the concrete apparatus in glass and metal, manipulated by
> him, the other is the schematic and abstract apparatus which theory substitutes
> for the concrete apparatus and on which the physicist does his reasoning.
> Pierre Duhem, *The Aim and Structure of Physical Theory* ([1914]1954, 182–3)

The word "evidence" appears prominently in the title of CDF's first paper
supporting the existence of the top quark. Furthermore, the choice of that
word in the title was a matter of debate turning primarily on how compelling
an argument CDF could make for the top-quark hypothesis on the basis of
their data. Yet, the collaboration did not attempt to articulate an explicit *theory of evidence*. Neither did they seek any guidance from theories of evidence
offered by philosophers of science.[1]

In this chapter and the next, I wish to discuss a theory of evidence – the
error-statistical theory – that could shed light on the methodological difficulties encountered by the CDF physicists in their search for the top quark. For
a philosopher to attempt to dictate practices on the lab floor from his desk
chair would of course be an act of unpardonable hubris. I hope that I will
not be accused of such a vice if, however, I propose that a correct theory
of evidence may assist the deliberations of experimenters in tackling the
problems that they do encounter. Indeed, a theory of evidence that cannot
serve this function does not deserve the allegiance of either scientists or
philosophers.

To understand experimental evidence, however, we need to understand
how evidential arguments relate to experimental practice. The first task for
this chapter is therefore to consider two distinct perspectives on experiment:
the *procedural* perspective and the *representational* perspective (Sections 1
and 2). This will help to set the stage for a theory of evidence that portrays
evidential relationships as empirical and determined by objective features

201

of experimental procedures, but also revealed and justified by means of explicit representations of such procedures, the data they produce, and the hypotheses they seek to test. Representations of the elements of experimental inquiry will here be treated in terms of a *hierarchy of models* (Section 3). The relationship of such models to the material procedures of experimentation will be explored in Section 4. With this notion of a hierarchy of models in hand, I will present in Section 5 a first-order approximation to a model of CDF's counting experiment test of the top quark hypothesis based on their data from run Ia.

In Section 6, I articulate the error-statistical theory of evidence, distinguishing it from other interpretations of testing procedures and their outcomes and explicating the precise sense in which the concept of evidence at the center of the error statistical theory is objective.

1. PROCEDURE

Experimentation comprises a range of activities. Consider the kinds of activities engaged in by the physicists at CDF: stringing cables, calibrating phototubes, scheduling meetings, mapping magnetic fields, ordering parts, writing tagging algorithms, creating plots, running Monte Carlo programs, sitting up all night with the detector as, for example, particles become pulses of light in scintillating material, which become electrical charges, which become data. . . . Which of these activities are to be thought of as part of the experiment being performed? What do they contribute to that experiment? These questions arise when attending to the procedural aspect of experimentation.

But the kind of experiment we are examining may also be considered, not as a collection of activities but as a representation of such activities and their relationship to certain facts and to certain theoretical claims of interest to the experimenters. There is a hypothesis to be tested, there are data that are to serve as the basis for that test, there are probability distributions for possible outcomes, and so on. All of these aspects of the experiment can be presented more or less abstractly, without mention of concrete details of the activities experimenters actually engage in. One might call this the representational aspect of experimentation.

What is the relationship between these two perspectives?

In his 1990 book, David Gooding proposes a "procedural" philosophy of experiment that focuses on "human agency" and "practice." One of his theses is that "human agency is essential to both exploratory observation and experimental testing" (ibid., 10). To the extent that this amounts to saying that in order to do an experiment you have to *do* something, it is, of course, a truism. But Gooding emphasizes the point because he believes that it is something that philosophers of science often forget.

Whether or not that is the case, it is helpful to keep the truism in mind. The context in which one is liable to forget the truism is itself important, however: the context in which an experiment is depicted and evaluated by means of mathematical models that facilitate the application of the probability calculus to the subtleties of empirical investigation.

There is a danger in fixation on these models. The danger exists both for philosophers studying an episode of experimental investigation and for experimenters themselves. The danger is greater, however, for philosophers who never actually do any experiments and typically lack the training that experimenters receive. The danger is that in simply looking at data as a finished product, waiting to be related to the hypothesis that is to be tested, one might lose sight of an important fact: the data, and the ways in which experimenters relate the data to the hypothesis being tested, might have turned out to be quite different. The data and their interpretation might have turned out differently not just if the systems under study behaved differently but also if the experimenter had responded to such behavior differently – had reacted differently to events involving data, instruments, and other experimenters. This may be the best way to understand Gooding's claim that "[o]bservers' experience of the world is construed, that is, mediated by their exploratory behaviour, their instruments and by their interactions with other observers" (Gooding 1990, 76).

Prior to and during the collection of data relevant to the top results, CDF members made many decisions that could affect both the data and the choice of analysis for using those data to evaluate the top-quark hypothesis. Even after they stopped collecting data, serious discussions continued regarding the proper means of analyzing the data. Some measurements, such as the energies of jets measured by the hadron calorimeters, were known to be systematically in error and had to be "corrected." Whether even to *use* some data called for deliberation, as in the case of data from the silicon vertex detector. CDF had to decide whether the detector had suffered too much radiation damage to be producing usable data. During and after run Ia, and in the midst of run Ib, disagreements arose over which algorithm to use to identify events with secondary vertices resulting from *b* decays. Decisions such as these, along with thousands of day-to-day decisions made while monitoring the incredibly complex CDF detector, had effects – some small, some profound – on the data produced and the way CDF used it to test the top hypothesis. One of the skills of a gifted experimenter is the ability to keep in mind the many ways in which the *humanity* of experimenters can be relevant to assessing the results of the experiments they do.

Just as important, however, is the recognition that experimenters use abstract models of their experiments precisely to address the relevance of such issues. Having used their knowledge and experience to determine in what ways the experiment could go *wrong*, they employ such models as tools to help them avoid making those errors. One of the purposes of models, particularly statistical models, is to help experimenters address the question: what might have happened that did not? (The importance of this question is a matter of philosophical controversy. I will return to it in Chapter 7.)

2. REPRESENTATION: MODELS

In 1960, at the first International Congress for Logic, Methodology and Philosophy of Science, Patrick Suppes presented a paper titled "Models of Data" (Suppes 1962). In this paper Suppes proposes understanding the relationship between experimental events and theories under test by means of a "hierarchy of models." He presents this idea in a very brief exposition, which more recently has been taken up and elaborated by Deborah Mayo (Mayo 1983; 1996). In what follows I will draw on the expositions of both Mayo and Suppes.

For Suppes, a model of a theory is "a possible realization in which all valid sentences of the theory are satisfied," where a possible realization is "an entity of the appropriate set-theoretical structure" (Suppes 1962, 252). Suppes contends that "exact analysis of the relation between empirical theories and relevant data calls for a hierarchy of models of different logical type" (ibid., 253).

In her elaboration of Suppes's account, Mayo employs a looser model concept, claiming that investigation of the methodology of experimentation "requires going beyond the set-theoretical relationships between models" (Mayo 1996, 131). The exact nature of the concept of model that Mayo does employ, however, remains unspecified. At times, she appears to use the term synonymously with "representation."

Here I wish to steer a middle course. I do not claim to be able to give an analysis of the case at hand in strictly set-theoretical terms. However, I wish to delineate a notion of model that may serve the purposes of understanding experimentation. The resulting conception, while still not as precise as I would like, will make for a realistic explication of actual experimental practice while retaining the two most important foci of Suppes's discussion: (1) the importance of abstract or mathematical representations of the elements of experimental inquiry and (2) the hierarchy, or series, of such representations that mediates the relationship between experimentally produced data and the theories that experiments (often) test.

R. I. G. Hughes has suggested a promising account of models as they are employed in physical theory (Hughes 1997) that emphasizes three components of theoretical representation: denotation, demonstration, and interpretation. Hughes calls it the "DDI" account of theoretical representation. I will give just a brief sketch of DDI here. My aim is not to argue that this theory is an adequate account of the uses to which models are put in the sciences generally. Indeed, so many different kinds of models are used in so many different ways that it seems likely no such general account will succeed, as Peter Achinstein has already argued (Achinstein 1968, chs. 7–8).

My aim is only to apply the DDI account of modeling to a "hierarchy of models" of the sort proposed by Suppes. Such an application bears out Hughes's claim that through an analysis of the use of models that keeps "these three activities in mind, we shall achieve some insight into the kind of representation" involved in such uses. "Furthermore, we shall rarely be led to assert things that are false" (Hughes 1997, S329) – at least for the type of case under discussion here.

Let me first sketch the DDI account.

Denotation. On the DDI account, a theoretical model denotes in that it acts as a symbol. Such models can serve to denote either particulars or types of physical systems. (A "global theory" such as quantum theory, Hughes notes, will describe a class of theoretical models, which in turn will denote specific types of systems (ibid., S330; see also Cartwright 1983).) Hughes invokes Nelson Goodman: "[D]enotation is the core of representation and is independent of resemblance" (Goodman 1968, 5).

Demonstration. In speaking of the use of theoretical models for purposes of demonstration, Hughes fully intends the double entendre of that term: using a mathematical model one can demonstrate (i.e., prove) that a certain theorem or result is true of a system of the sort denoted; using a "material model" (a "slinky" toy as a model of a medium transmitting sound waves, for example) one can demonstrate (i.e., display) the occurrence of a phenomenon that occurs in the system denoted. (What happens when a sound wave strikes a hard surface or enters a denser medium? The answer can be "demonstrated" using the slinky model by fixing one end of the slinky, or by connecting one end to another slinky that is more rigid.)[2] By supporting demonstrations models become predictive. Hughes comments, "From the behavior of the model we can draw hypothetical conclusions about the world over and above the data we started with" (Hughes 1997, S331). While this should come as no surprise when speaking of general theories like quantum electrodynamics, I will claim that models of *experiments* also have such predictive content, and that experimenters take advantage of this in order to test the adequacy of those models.

Interpretation. Finally, the results of such demonstrations must be interpreted. This step leads back to the system denoted by the model. Having produced a result from demonstration, one ought to be able, if the model is a good one, to interpret that result in the model as corresponding to or denoting some real phenomenon in the system denoted by the model. (Note that this is a necessary but not sufficient condition for a good model.) Ideally, the model that denotes the system is produced by an isomorphism from some aspects of the system onto some aspects of the model, so that, having determined a result in the model (through demonstration), the interpretive step simply consists in taking the inverse of that isomorphism for whatever element in the model results from the demonstrative step. More generally, as Hughes notes, requiring empirical adequacy amounts to requiring "that interpretation is the inverse of denotation" (ibid., S333).

However, because not every aspect of the model will denote a real characteristic of the system being modeled (e.g., the color of the slinky will not denote any property of the sound-transmitting medium that it is supposed to denote), no model can be evaluated without some understanding of which of its aspects are meant to denote. Hence, a full specification of a model should also indicate which of its features are relevant for its functioning as a model (although in practice the relevance specification is often understood implicitly).

A wide variety of models can support the functions of denotation, demonstration, and interpretation. We can see this in the diversity of the models (models of the hypothesis, of the experiment, and of the data) that serve to relate experimental data to theories. I will next describe Suppes's hierarchy of models and then illustrate and clarify that description with an example.

3. A HIERARCHY OF MODELS

A model of what Deborah Mayo calls the "primary hypothesis" (i.e., the hypothesis that the experiment serves to test) occupies one end of the hierarchy. This model may be a structure representing certain parameters of a population, for example a probability distribution or class of probability distributions for a measurable characteristic of members of some population, or a region or class of regions of some phase space representing the states of a physical system considered possible under a given hypothesis. One can often depict this part of the hierarchy in Suppes's terms: a model of a theory is a set-theoretical structure that satisfies the axioms of that theory. On the other hand, in accord with the DDI approach to theoretical models, one can think of such a model, not as a model of the theory, but as a model of the physical system (as a type or as a population) that the theory describes. Thus a model of a population of coin-toss experiments involving a fair coin might be thought of as a structure including among its elements the set of all possible outcomes (rendered as the relative frequency of heads) of experiments of a given length, along with a probability measure on that set, constituting a normal distribution.

Below (or beside) this model of the primary hypothesis is the model of the experiment, which "provides a kind of experimental analog of the salient features of the primary model" and specifies "analytical techniques for linking experimental data to the questions of the experimental model" (Mayo 1996, 133–4). In Suppes's account, the theory of the experiment specifies possible realizations of the theory that are "the first step down from the abstract level of the . . . theory itself" (Suppes 1962, 255). Specifications of the particular sample to be collected in the experiment serve as the basis for a structure that is the analog of the model of the primary hypothesis for that sample. A testing rule may be specified as part of the model of the experiment that maps possible outcomes in the model of the data to regions in the parameter space employed in the model of the hypothesis. Much of the difficulty and nuance that preoccupy the skilled experimenter arise from the need to determine an adequate model of the experiment.

At this level, the notion of model taken over from mathematical logic does not sit as comfortably. Although Suppes speaks at this level of the "*theory of the experiment*" (ibid., 255), which is the theory whose axioms are satisfied by the model of the experiment, this will typically not be a theory that anyone can formulate. Nor, for most experimental situations, will there be any need to do so. In fact, as we will see in the case of CDF's top search during run Ia, the model of the experiment that CDF did use was generally believed to be a distortion, so that any theory of the experiment of which it could be said to be a model would have been believed to be false. Rather than regarding this as a flaw in their model of the experiment, CDF found the inaccuracy of their model (because of the kind of inaccuracy involved) to be a strong argument in *favor* of using that model.

At the third level of the hierarchy is the model of the data, which puts the information contained in the data into a canonical form for purposes of comparison, via a statistical measure, with the model of the experiment (possibly as an input to the mapping function specified in the testing rule specified at the previous level). The model of the data is "designed to incorporate all the information about the experiment which can be used in statistical tests of the adequacy of the theory" (ibid., 258). Here again Suppes insists that a model of the data must, for complete specificity, invoke a "theory of the data" (ibid., 253). I will only insist that the model of the data represent a set of measurements actually carried out during the experiment.

We next have the level at which problems of experimental design arise. Here we find "all of the considerations in the data generation that relate explicitly to . . . some feature of the data models" (Mayo 1996, 139), such as the calibration of instruments, the randomization procedures necessary to select a sample, and so on. Such considerations do not directly show up in comparisons of data to theory but are relevant to the adequacy of the model of the data and can be investigated using the same kind of mathematical models employed at the level of the data model, even if no one bothers to do so. Finally, there are the ceteris paribus conditions, including "every intuitive consideration of experimental design that involves no formal statistics. Control of loud noises, bad odors, wrong times of day or season go here" (Suppes 1962, 258).[3] Such considerations are potentially relevant to the assessment of a model of the data, yet typically resist treatment by means of formal statistical measures.

To illustrate this framework, consider the following simple experiment: A manufacturer of bowling balls wishes to test a new ball (called "Hume") that is larger in diameter but less dense than conventional bowling balls, so that a bowler can hurl a fatter ball at the pins without having to manage a heavier ball. The company wants to know whether bowlers might perform better using this new kind of bowling ball. How might the company test their product?

The hypothesis in question needs to be put into the form of a statistical model. The relevant population here is the class of ordered pairs of bowlers using "Hume," and their individual performances in particular games. Let us call such ordered pairs *tenpinomos*.[4] We next define a function $f(t)$ on the class of tenpinomos, such that for each tenpinomo t, $f(t) = X$, where X is the score obtained by that particular bowler as a result of that particular game. This quantity X can be regarded as having a probability distribution $p(X|\cdot)$, which can be characterized in terms of certain population parameters, such as the average value of X, denoted by θ, and the variance, σ^2. The set of possible values of θ is known as the parameter space Ω, some subset of which is associated with each statistical hypothesis. Representing the statistical hypotheses by a model $M(\theta)$, where $M(\theta) = [p(X|\theta), \theta \in \Omega]$, the hypothesis under test concerns which distribution $p(X|\theta)$ in $M(\theta)$ describes the population in question. Specifically, supposing that we know that the average score for bowlers using conventional bowling balls is θ_0, the hypothesis we are investigating, H′, will state that the distribution of X for the population of bowlers is to be described by a function $p(X|\theta)$, where $\theta > \theta_0$, and the null hypothesis H_0 will state that the distribution of scores among bowlers is to be described by the function $p(X|\theta_0)$.

As called for by the DDI account, each probability distribution in M is potentially a probability model of the population of tenpinomos that *denotes* a particular feature of that population. Furthermore, the model supports *demonstrations*. For example, one might use a particular probability model to determine values of X that will be upper and lower limits between which 95% of the area under the curve describing the probability distribution will be contained. If the probability model used for such a demonstration is in fact an empirically adequate model of the performances of tenpinomos, then one will be able to *interpret* the results achieved correctly for that determination as upper and lower limits between which 95% of all tenpinomo scores will fall.

Suppes notes "two obvious respects in which a possible realization of the theory cannot be a possible realization of experimental data" (Suppes 1962, 254). First, a possible realization of the theory H′ will include an infinite sequence of pairings between bowlers and scores, which no actual experiment can include. Also, the probability distributions in the model of the hypotheses are continuous functions, where the data are discrete, and the parameter θ that characterizes that distribution in the population is not experimentally observable but must be estimated. Relating the experimental data to the theory requires an intermediate level.

This is the job of the model of the experimental test. The experimental model also serves as the primary representation of the procedural aspects of the experiment emphasized by Gooding. According to Suppes, to be a model of the experiment, a structure must be a possible realization of the theory that is "the first step down from the abstract level of the... theory itself" (ibid., 255). Suppose we set out to test our hypothesis by using the members of a particular bowling league, with, say, 200 members. Each member of the league is given a "Hume" ball of the same weight as the conventional ball that bowler ordinarily uses, and with identically drilled holes. Each bowler is then asked to bowl three games with that ball, and a score s is recorded for each game played by each bowler. This amounts to recording values for 600 independent random variables $S_1, S_2, \ldots, S_{600}$. An experimental outcome will be represented as a 600-tuple of values for those 600 random variables: $s_1, s_2, \ldots, s_{600}$, which can be abbreviated as $<s_{600}>$. (Alternatively, one could represent it as a 200-tuple of triples.) Just as at the level of the hypothesis a parameter space was specified, here a sample space, ϑ, is given, which is the set of possible values of $<s_{600}>$. As Mayo notes, "On the basis of the population distribution $M(\theta)$ it is possible to derive, by means of probability theory, the probability of the possible experimental outcomes" (Mayo 1983, 301), here $p(<s_{600}>|\theta)$, for any $<s_{600}> \in \vartheta$, and any $\theta \in \Omega$. Also at this level one may specify a rule T that maps particular outcomes $<s_{600}>$ in ϑ onto particular members of the set of probability distributions (statistical hypotheses) in $M(\theta)$. This yields an experimental testing model characterized as ET $(\vartheta) = [\vartheta, p(<s_{600}>|\theta), T]$.

The model of the experiment thus described is not a model of simply one performance of an experiment, but of a *population* of experiments of the same kind as that under consideration, for an assumed value of the population parameter of interest. Many of the difficulties of producing reliable evidence center on the need to have an adequate model of the experiment that one is performing – that is, to ensure that the probability distributions that one uses in comparing the data model to the primary model are not biased so as to undermine any evidential claims arising from that comparison.

Here again, the DDI account may help to shed light on the relation between the experimental model and the population it represents. The experimental model can not only denote but also serve as the basis of demonstrations regarding that population. Because the actual value of the population parameter (here θ) is unknown, the model of the experiment must be an adequate model of what the make-up of that population *would* be for any given value of that parameter. Thus, if the model of the experiment is such that, for given θ, the probability distribution integrated over a certain subspace of the sample space ϑ contains 95% of the area under p for ϑ as a whole, then 95% of the outcomes of the population of repetitions of the experiment should be such as are denoted by the elements in that region of the sample space, *if in fact the population is correctly described by the assumed value of θ*.

In our example, the mapping rule T makes use of a test statistic Ψ, which is the mean of the scores in $<s_{600}>$. This statistic Ψ is a function of the random variables $S_1, S_2, \ldots, S_{600}$ defined on ϑ, and it is possible, for a given value of θ, to calculate its probability distribution. The mapping rule serves to stipulate which subset of ϑ is to be associated with H′, and which, with H₀.

At the level of the data, the observation of the bowlers' performance is modeled as an element of ϑ (i.e., as a 600-tuple of individual scores). Each score represents the result of a certain measurement on a certain tenpinomo, and the array of scores constitutes a model that denotes the particular tenpinomos measured during this experiment (or, more specifically, as a model of that property of each such tenpinomo that is measured by means of its score). Here again, the model can be employed for purposes of demonstration. In fact, in the present example (as in most experiments), the model consisting of the results of individual measurements only becomes useful after it has served as a basis for a demonstration regarding some property of the sample it represents that was not directly measured. In this case, the primary feature of the sample to be demonstrated is precisely that denoted by the test statistic: the mean score Ψ. It is to this quantity that the testing rule specified in the model of the experiment will be applied. This data model can then be represented as $D = [<s_{600}>, \Psi(<s_{600}>)]$.

In short, the levels of experimental inquiry work as follows:

1. The data model yields a test statistic.
2. The model of the experiment stipulates a testing rule and a family of probability distributions defined on the sample space.
3. The model of the hypothesis specifies a parameter space, different regions of which are denoted by the hypotheses under test.

The model of the experiment links the models of data and hypothesis insofar as the testing rule associates different possible values of the test statistic with different regions of the parameter space. The model of the experiment also provides the probability distributions that yield the *error probabilities* of the test with respect to the hypotheses in question, the importance of which I will address shortly.

Thus far I have not said much about the two remaining levels of Suppes's hierarchy, the levels of experimental design and ceteris paribus conditions. Suppes's own remarks here are rather brief, noting that "[t]he analysis of the relation between theory and experiment must proceed at every level of the hierarchy. . . . Difficulties encountered at all but the top level reflect weaknesses in the experiment, not in the . . . theory" (Suppes 1962, 259).

Suppes notes that considerations that enter at the level of experimental design "can be formalized, and their relation to models of the data . . . can be made explicit" (ibid., 258–9). In our bowling example, we might ask, at this level, whether the bowling league we chose for our study was itself a source of error, such that the statistic Ψ could not be related in the way we thought to the experimental model. Perhaps it was a league with an inordinately high number either of very good or of very poor bowlers, or a league that served more as an excuse to drink beer than as a genuine sporting institution. The number of games to be played may turn out to be too small because it takes more time for the bowlers to become accustomed to the new ball. Here one can take advantage of the fact that the data model lends itself to diverse demonstrations regarding facts about the sample. To see whether initially depressed scores were rising with experience, one might restructure the array of 600 tenpinomo scores as a 200-tuple of triples, such that the first, second, and third element in each triplet is the first, second, and third score respectively, of a given bowler. One can then take the averages of the first, second, and third members of the triples separately. A problem would be indicated by average scores that rise steadily and significantly.

Suppes readily admits that considerations at the level of ceteris paribus conditions will typically resist formalization – he refers to the "seemingly endless number of unstated ceteris paribus conditions," which cover "every intuitive consideration of experimental design that involves no formal statistics" (Suppes 1962, 258–9). At this level, we might ask whether the bowlers in question were bowling under unusual conditions of lighting, noise, and the like; whether unusual environmental factors (the passage of a new law against serving beer in bowling alleys, or the repeal of an old law prohibiting it) might be impairing or enhancing their performance; and so on.

Mayo combines the levels of experimental design and ceteris paribus conditions into one, noting that "features assumed to be irrelevant or controlled may at a later stage turn out to require explicit scrutiny" (Mayo 1996, 139).

Nothing very weighty would seem to depend on whether questions of experimental design and ceteris paribus conditions form one level or two. I do wish to emphasize not only that aspects of the experimental context initially judged to be harmless or irrelevant may later demand attention but also that a rather surprising range of types of fact can, as a result, become relevant. This point is important for understanding the relationship between experimental procedures and formal models of experiments.

4. RECONCILING THE TWO PERSPECTIVES

What does this framework of models do for our understanding of experimentation? I believe that it conceptualizes the ways in which arguments are drawn out of experimental activities via abstraction. In working with probability distributions and data structures, one is working with materials that have been drained of the blood and guts, the materiality and agency of experimental activity. Such abstraction is simultaneously useful and dangerous.

On the surface, Suppes's formal approach to representing experiments would seem to be open to Gooding's charge against philosophical studies of experimentation, that they abstract out the procedural aspect of experimentation and the fact of agency (i.e., the fact that human beings actually set out to *do* things in experimental activities, and that experimental outcomes are contingent upon such actions as performed during an experiment). According to Gooding, the "received view" among philosophers depicts "the relationship between theory and experiment as a logical relationship between propositions" (Gooding 1990, 9). It is not clear who Gooding finds guilty of these errors. However, at least in the case of Suppes, though he is an arch-formalist, the charge seems misdirected.[5]

Suppes certainly is interested in formal depictions of the relationships between models of the theory, the experiment, and the data. But (even if we take the term "logical" in a very broad sense to refer to formally specifiable relations and replace Gooding's "propositions" with set-theoretical structures) Suppes's description of the hierarchy of models as an account of the relationship between theories and experimental data agrees with Gooding's description of the received view only if we understand Suppes to be claiming that the hierarchy of models *is* or exhaustively describes the experiment, and nowhere does he make any such claim. The purpose of the hierarchy of models is to *model* the experiment and the theory in such a way that the two *can* be related to one another in an argument leading to a conclusion about the theory, premised on what is learned in the experiment. The formal relationships are within, and with, the model of the experiment, which is necessarily abstracted from the actual performance of the experiment. Suppes does not set out to describe a logical relationship between the theory and the experiment itself, in the sense of a set of procedures carried out by human agents. Indeed, it is hard to understand what the latter kind of relationship could be.

Perhaps Gooding's complaint against formalist philosophical accounts of experimental evidence is that they deemphasize the very features that interest Gooding. It is certainly true that Suppes, for example, in detailing the formal relationships between the parts of experimental inquiry, does not say very much about the grubby details of what experimenters *do*. In part this is because of his emphasis on the first three levels of the hierarchy. While the procedures of the experiment are reflected to some degree at every level of the hierarchy except the model of the hypothesis, they are most apparent at the levels of experimental design and ceteris paribus conditions, about which Suppes has less to say.

More important, however, is an apparent difference in the questions that these two authors are trying to address. Gooding proposes a framework for describing experimental activity as a phenomenon of interest for its own sake, whereas Suppes emphasizes drawing out from that activity representational structures in order to arrive at results. From this perspective, the *epistemic* role of experimental procedures is of primary interest.

It would not be surprising to find a similar divide among experimentalists themselves. It may be that some experimentalists find experimentation primarily interesting for its own sake but recognize that they also are expected to produce theoretically interesting results. (Certainly we saw in Chapter 3 that there are "hardware types," such as Aldo Menzione, who take special pleasure in the design and construction of novel, but also theoretically useful, instruments.) Other experimenters may pursue experimentation primarily for the chance to address what they regard as exciting theoretical questions.

Whatever their preferences, experimenters cannot afford to ignore either aspect of experiment. Procedures such as the selection of a sample, the recording of data, and the control of complicating factors, can often be modeled within the hierarchy. But to a large extent the design of a good experiment is a problem of making procedures (and agency) *irrelevant* to the conclusions to be drawn from the experiment. One wishes to avoid making relevant things that are difficult to model because then it would be difficult to assess their effect on the reliability of the argument being presented.[6] To be sure, it will generally take a great deal of purposive planning, practical acquisition of skills, and sensitive response to the vicissitudes of life in the laboratory to be able to produce an experiment in which all of those facts about who did what and why can be *left out* of the list of relevant factors in the modeling of the experiment. But when such measures are carried out successfully, the arguments produced will be able safely (i.e., without becoming less reliable) to ignore such features of the experiment.

Typically, the interests and abilities of the particular person who performed a certain action during the experiment is deemed irrelevant because of considerations at the level of ceteris paribus conditions. For example, in a good experiment, one avoids setting people to tasks that they are incompetent to carry out or, in a case where a person has a strong reason to favor one outcome over another, tasks the outcomes of which are sensitive to manipulation. However, sometimes it becomes clear that such facts about "agency" cannot be ignored in evaluating the results of an experimental test. In such a case, the consideration becomes explicit and must be addressed by means of either formal or informal statistical considerations.

This brings us back to Mayo's point that "features assumed to be irrelevant . . . may at a later stage turn out to require explicit scrutiny." The models employed in experimental reasoning – the model of the theory, the model of the experiment, and the model of the data – are designed to contain all of the information that must be *explicitly* attended to in reaching an experimental conclusion. Yet the potential relevance of considerations of experimental design and ceteris paribus conditions seems to be open-ended. A consideration that might not seem important at first may turn out to require careful study later.

In the statistical assessment of their experimental results, scientists frequently cite error probabilities that allegedly represent the relative frequency (in the long run) of a certain type of outcome, for a sequence of repetitions of the experiment at hand. These probabilities will vary, however, depending on the specification of the experiment. The probability you cite, in other words, depends on your choice of reference class. Here arises a philosophical problem: At what level of description should one specify the class of experiments thus referred to in these statistical assessments? It would seem natural to say that one should describe the experiment so as not to exclude any *statistically relevant* factors, but such factors could include all manner of facts about the experimenters performing the experiment. Such could include, for example, facts about their mental dispositions at the time of the experiment, their willingness to believe the theory being tested, the strength of their motivation to achieve a particular result, and the availability to them of means for producing the appearance of having achieved that result.

The broader philosophical difficulty, then, becomes this: How could such statistical evaluations of experimental results be objective, when they might need to take all of this into account? Our canonical formulations of the notion of objectivity suggest a difficulty. For knowledge to be objective is, according to those canonical formulations, for it to be independent of the thoughts and wishes of epistemic agents.

In the remainder of this chapter, and into the next, I will argue for a way of understanding experimental evidence that *is* objective in the very face of these considerations and that retains its objectivity precisely by means of the use of probabilistic models. Those models are not enough by themselves, however. For them to yield objective assessments of evidence, the way must be prepared for them. Still, if you fail to secure the conditions for the appropriate use of a statistical tool, a different statistical tool might very well help you to find that out. In some cases, you may even be able to use a different method to give an objective, if approximate, evaluation of the evidence in spite of such failure.

5. A MODEL OF CDF'S IA TOP SEARCH COUNTING EXPERIMENT AS A STATISTICAL TEST

In this section I wish to show, in a highly simplified sketch, how the framework just described could be applied to the case of CDF's Evidence paper results (limited, however, to the results of the counting experiments). Because of the complexities of the experiment as a whole, the following should be regarded as a first-order approximation to a full model of the ways in which CDF used their run Ia data to develop an evidential argument for the existence of the top quark.

5.1. The Primary Hypothesis

In theoretical terms, the primary hypothesis of interest could be expressed as follows:

Θ : $t\bar{t}$ pairs are produced in $\bar{p}p$ collisions at $\sqrt{s} = 1.8$ TeV.

So that this hypothesis can be related to the results of experiment, however, a model of it must be specified. More specifically, we seek models representing the population studied by CDF, that will distinguish scenarios in which Θ is true from scenarios in which it is false. The population of interest is a population of proton-antiproton collision events of the sort produced in Fermilab's Collider. We first specify a model of the "null" hypothesis, characterizing that population under the conditions in which there is no top quark.

The production of top decay candidate events (whether signal or background) from $\bar{p}p$ events can be regarded as a random process with two possible outcomes: either the event yields measurements that pass the candidate event selection cuts ("success"), or it yields measurements that fail those cuts ("failure"). Such random processes with exactly two possible outcomes are known as *Bernoulli* processes. Assuming that the probability of success in each Bernoulli trial is p, the probability of getting s successes in N trials is given by the *binomial distribution*

$$f(s, N, p) = \frac{N!}{s!(N-s)!} p^s (1-p)^{N-s}.$$

A limiting case of the binomial distribution as $N \to \infty$, $p \to 0$ is the *Poisson distribution*, which is determined by a single parameter $\lambda = Np$. This parameter represents the average number of successes in N trials, and the Poisson distribution gives the probability of finding exactly s successes in N trials as

$$f(s, \lambda) = \frac{\lambda^s e^{-\lambda}}{s!}, \text{for } s = 0, 1, 2, \dots.$$

The Poisson distribution is a good approximation of the distribution for Bernoulli processes that have a very low probability of success, when the number of trials involved is very large (hence such processes are called Poisson processes). The production of top-quark candidates from high energy $\bar{p}p$ collisions is an excellent example of such a process.

Now we are in a position to present a mathematical model of the null hypothesis, according to which only background sources produce top-quark candidate events. That scenario is represented by means of a Poisson distribution:

$$H: f(s, \lambda) = \frac{\lambda^s e^{-\lambda}}{s!}, \text{ for } s = 0, 1, 2, \ldots, \text{with } \lambda = \lambda_0.$$

Here λ_0 is the expectation value for s (the number of candidate events) for a number of trials equal to that in the data set at hand. In order to represent the null hypothesis, we will set the value of λ_0 equal to the number of background events expected, on average, in a series of repetitions of such an experiment, generating the same size data set under the same conditions.

We also seek a representation of the other possibility: the case where the top quark does exist. Here enters an asymmetry, however, for although Θ may be either true or false, there are many more ways that it can be true than there are ways that it can be false, for the purpose of the modeling of hypotheses. If the top quark does exist, then it is produced and decays by a process that we can represent as a Poisson process with some probability p_t of yielding a positive count for the experiment, but the actual value of the parameter p_t (and hence of $\lambda_t = Np_t$) depends on the mass of the top quark, unknown to CDF at the time that they began their experiment. In addition, in the case where the top quark does exist, candidate events will still be produced from background sources according to the same probability distribution already used to represent the null hypothesis. So s, the number of candidate events, will reflect the contributions of two independent Poisson processes, governed by the parameters λ_0 and λ_t. The distribution for s will therefore be a Poisson distribution with the parameter $\lambda = \lambda_0 + \lambda_t$. However, no single Poisson distribution can represent the alternative hypothesis because it comprises a range of possible values of λ_t. Rather we must employ a family of distributions:

$$J: \{ f(s, \lambda) = \frac{\lambda^s e^{-\lambda}}{s!}, \text{ for } s = 0, 1, 2, \ldots : \lambda > \lambda_0 \}.$$

In other words, we represent the alternative hypothesis with a *set* of distributions for the probability of finding s candidate events in a sample of N events drawn from the population of similarly produced $\bar{p}p$ collisions. Each member of the set is a Poisson distribution $f(s, \lambda)$ with expectation $\lambda > \lambda_0$.

Thus, the model of the primary hypothesis utilizes a conceptual apparatus representing distinct physical scenarios by means of a distinction between different members of a family of probability distributions M: $[f(s, \lambda), \lambda \in \Omega]$, where Ω denotes the *parameter space*. Within Ω, we distinguish two subspaces: Ω_H, which contains the single element λ_0, and Ω_J, defined as the set $\{\lambda: \lambda > \lambda_0\}$. To complete the specification, we can note that for CDF's experiment reported in the Evidence paper, $\lambda_0 = 5.96^{+0.49}_{-0.44}$.

Here it is worth observing that this model of the hypothesis uses the "method 1" background estimate (see Section 4 in Chapter 5), which CDF physicists regarded as an *over*estimate of the backgrounds in the lepton + jets counting experiments. Hence, one of the virtues of this model was not its perfect accuracy, but the nature of its expected inaccuracy. By overestimating the background, CDF essentially "raised the bar" on their own results. With a higher background estimate, a larger number of candidate events would be required to achieve an impressive statistical significance level.[7]

5.2. The Experiment

There are three components to the model of the experiment: $\mathrm{ET}(\lambda) = [\vartheta, f(s, \lambda), T]$. The first element is the *sample space*, the second is a *probability distribution* for the test statistic, and the third is the *testing rule*.

The sample space ϑ represents the set of all possible experimental outcomes. Consider a family of probability distributions defined over ϑ. We might imagine these to be probability distributions defined over a space of very large dimensionality, yielding probabilities of getting any particular set of data. Practically speaking, however, we require only probability distributions defined over the set of all possible values of the test statistic s. Different probability distributions will be generated by different assumed values for λ.

We next define a testing rule T that maps from different regions of the sample space onto different members of M (the model of the primary hypothesis). In particular, we can decide which outcomes we will choose to map onto the subspace of M indicated by J: those distributions with $\lambda > \lambda_0$. This defines a critical region CR, a subset of ϑ. All other outcomes map onto the H distribution. We can then ask about the error characteristics of this testing rule. We can add a layer to this structure by partitioning M into two regions. We will label the distribution $f(s, \lambda_0)$ the "accept H" subset of M, and the set of all distributions $f(s, \lambda)$ with $\lambda > \lambda_0$ the "reject H" subset of M. Then we can regard T as a map from ϑ to {reject H, accept H} that maps all members of CR to "reject H," and all members of the complement ϑ − CR to "accept H." To reflect CDF's run Ia top search experiment, we can define CR as $\{s: s \geq 15\}$.

The testing rule T and the probability distributions defined over ϑ combined give us the crucial error characteristics of the test (or at least, in this case, one of them):

$$p(T \text{ rejects } H/H \text{ is true}) = p(s \in \mathrm{CR}/\lambda \in \Omega_\mathrm{H}) \leq \alpha.$$

CDF's calculations indicate that $\alpha = 2.6 \times 10^{-3}$ (see Section 10 in Chapter 4). However, if this is taken as the probability of getting a test result that rejects the null hypothesis when in fact there is no top quark, then it is professed by many CDF members themselves as probably inaccurate. This is, however, alleged to be an epistemically virtuous kind of inaccuracy. Because the model of the null hypothesis itself is an overestimate (as mentioned in the previous section), this significance estimate is (ceteris paribus) an overestimate of the probability of getting such a result that rejects the null hypothesis when in fact there is no top-quark content in the data. Many CDF members, consequently, felt that the actual significance level of their results was smaller than the quoted value.

5.3. The Data

For all the hundreds of thousands of channels of information gathered by CDF during run Ia, and the many millions of collisions they recorded, a small handful of numbers summarizing those measurements became the focus of all the debates over evidence. For our present, first-order reconstruction, one single number is of paramount importance: the value of the test statistic s discussed in the model of the experiment. This test statistic is the distillation, produced by the algorithmic selection of candidate events, of all the data insofar as it can be brought to bear on the hypothesis, by means of the model of the experiment just described. Hence the model of the data may be regarded as including both an immense array consisting of the results of each measurement made on each collision event, Ξ, and an operation $\mathbf{S}$ defined on Ξ such that $\mathbf{S}(\Xi) = s$. That is to say, $\mathbf{S}$ applied to any given data set simply yields the number of "successes" recorded as outcomes of the set of trials recorded in that data set. Thus, the data model is specified as D: $[\Xi, \mathbf{S}(\Xi)]$. As it happened, from the run Ia data, $\mathbf{S}(\Xi_{Ia}) = s_0 = 15$. This nice, tidy model is too simple. In one sense, it is much too simple; in another sense, it is only a little bit too simple.

The model is only a little bit too simple in that CDF really did attempt to combine all their counting experiments' results into a single statistic that would yield a single significance calculation. However, the actual definition of that statistic was not simply "number of candidate events" (because they counted tags rather than events for the SLT and SVX searches). Furthermore, they did not employ a simple Poisson distribution (they used a Monte Carlo routine that generated Poisson-distributed results, but with parameters that were themselves drawn from a Gaussian distribution).[8]

At the same time, the simplified model captures none of the complex and varied ways in which CDF used their own data to reassure themselves (with mixed results) of the accuracy of that primary statistical assessment. The statistical assessment of the evidence for the primary hypothesis based on the significance calculation just mentioned could only be accepted on the conditions that the model of the primary hypothesis represented the distinct possible states of the population correctly, the probability model used to represent the experiment was adequate, and the data satisfied the assumptions underlying the use of the corresponding probability distributions in the model of the experiment. Establishing good reasons to believe that these conditions were satisfied was no trivial task. Furthermore, it was a task carried out largely by the use of a wide variety of models developed from the very same body of data.

CDF's Evidence paper is littered with examples of such "cross-checks" that seek to verify in detail the satisfaction of the conditions for the soundness of their evidence claim. A single example will suffice for illustration.

5.4. Checking the *b*-Tag Distribution

Looking at the model of the primary hypothesis described in the preceding section, one peculiarity stands out. The population that is represented by means of the family of distributions is represented in terms that are highly specific to CDF's experimental setting. The parameter that governs those distributions, λ, is defined as the product of the number of trials N (the number of collision events recorded) and the probability p of "success" for each of those trials. Note, however, that "success" here is defined entirely in terms of the specific algorithm used by CDF to select top candidates, an algorithm that made sense only in the context of CDF's detector and its specific features. Although the general characteristics of the decays of top quarks do not depend on the instrument used to detect them, a different detector would have called for a very different algorithm at the level of choosing particular cuts or even particular search strategies. (D-zero, for example, could not feasibly try to tag *b* events by looking for secondary vertices because they lacked a high-resolution vertex detector and central magnetic field during runs Ia and Ib.)

This means, however, that for the model of the primary hypothesis to be a reasonable choice for representing the distinct theoretical possibilities that CDF wished to discriminate experimentally, the algorithm needed to make the kinds of physical discriminations that it was intended to make. CDF needed to do more than establish that they had recorded a statistical excess of *something*, they needed to show that the excess was of *the right sort*. But for this they could use additional, often qualitative, statistical assessments drawn from the very same body of data from which they drew their primary assessment of the top-quark evidence.

As mentioned in Chapter 3, the jet-vertexing algorithm, which was de-signed to tag events with secondary vertices resulting from particles carrying *b* quarks, tagged events in part based on a measurement of the distance L_{xy}, the projection onto the r-ϕ plane of the path from the primary to the secondary vertex. This same quantity could be used to estimate, for each tagged event, the *effective proper decay length,* $c\tau_{\text{eff}}$. CDF took advantage of this in order to test their assumption that jet vertexing was identifying events with genuine *b* quarks. This amounted to a test of the adequacy of their model of the primary hypothesis in one respect: part of the definition of "success" involved in specifying that model was that an event resulted in a success if it passed the cuts for a $W+$ jets event, and then was also tagged by the jet-vertexing algorithm. CDF imposed this requirement to distinguish $t\bar{t}$ pairs decaying to two W bosons and two b quarks from other similar events involving Ws and jets by means of the separate identification of a b quark. If jet vertexing was *not* identifying events with b quarks, then the parameter λ in the model of the primary hypothesis would have little connection with the presence or absence of top quarks in the $\bar{p}p$ collisions under study.

To test their tagger, then, and the adequacy of their model of the primary hypothesis, CDF applied the tagger to two sets of "data." One set was real data: a sample of "inclusive electron events." This sample contained a large number of events sharing one feature: they each passed the electron identification cuts with an electron having $P_T \geq 10$ GeV/c in the region $|\eta| < 1.0$ (in other words, an electron with relatively high momentum, in the central region of the detector). The other set of events to which CDF applied their b tagger was generated by computer using a Monte Carlo routine named HERWIG (for Hadron Emission Reactions With Interfering Gluons) (Marchesini and Webber 1988). CDF used HERWIG to simulate $\bar{p}p$ collisions producing events with b quarks in them, then ran the results through their own computer simulation of the CDF detector. To that set of simulated data, they applied the jet-vertexing algorithm. For both the genuine and simulated data, CDF calculated $c\tau_{\text{eff}}$ for each tagged "event." They then compared the results for the two sets in a plot (see Figure 3.8).

By doing this, CDF essentially employed a model (the HERWIG Monte Carlo model) that they *knew* to be a good model of a hypothesis intimately related to the primary top-quark hypothesis, and one that they considered to be already well established by experimental evidence: the standard model account of the production and decay of b quarks. In running HERWIG, they essentially "sampled" from that model of the b quark.[9] This gave them a model that could be brought into direct comparison with CDF data. Those data could be compared to data, not from genuine b quarks, but from the simulated analogs to b quarks. If indeed their choice of parameter λ was sensible for characterizing the top-quark hypothesis, then, since part of what determined the definition of λ was the use of jet vertexing to identify b quarks, jet vertexing should identify the analogs of b quarks correctly and should yield measurements in b events in the data close to those from b analog events from HERWIG. As can be seen in Figure 3.8, the plots of the two sets of measurements do in fact match very closely.

This comparison tests the hypothesis that the parameter λ correctly models the primary hypothesis about the top quark insofar as the models thus constructed represent a particular aspect of the population of $\bar{p}p$ collisions producing $t\bar{t}$ pairs: their subsequent decay to b quarks. Put another way, this test hypothesis states that the population of $\bar{p}p$ events that would be selected by the jet-vertexing algorithm consists, for the most part, of events that do include the production and decay of *b* quarks. Because the hypothesis is described qualitatively in this way, no actual distributions are specified for it. However, CDF does produce a model that functions like Suppes's model of the experiment, enabling a confrontation between abstract hypothesis and discrete, particular data. That model is the ordering of measurements on the Monte Carlo "data" that are displayed in the plot in Figure 3.8. The element that functions as a model of the data for this test is the corresponding ordering of measurements on the inclusive electron events selected by jet vertexing that are also displayed in that plot. Here the assessment of the "fit" between the two models is strictly qualitative. In a section of the Evidence paper in which CDF describes a number of cross-checks on the jet-vertexing algorithm, they refer to Figure 3.8, and note simply, "The data agree well with the Monte Carlo simulation of *b* decay" (Abe, Amidei, et al. 1994b, 2987).

6. ERROR-STATISTICAL EVIDENCE

Thus far I have described how the hierarchy of models can represent those features of experimental procedure that are relevant to drawing reliable scientific conclusions from experimental tests of hypotheses. Next I want to discuss how such conclusions relate to experimental procedures and the models that represent those procedures.

The kind of test I have described in Sections 5.1–5.3 of this chapter is an example of a Neyman-Pearson Theory (NPT) significance test. In some respects, CDF did not conform to orthodox NPT practice. They did not specify a "stopping rule" in advance, indicating at what point they would declare a complete data set and report the results. Indeed, at the outset of run Ia, they were not even certain whether they would report their results as a significance test or as a confidence limit on the mass of the top quark.[10] They did not set a significance level in advance and report their results simply as significant or not significant at that level. Instead, they took the number of candidate events they actually recorded and reported a kind of "observed" significance level. In the next chapter, I will examine some of these departures from orthodox NPT practice and discuss their relevance to the assessment of CDF's evidential claims. In any case, there is no question that the basic methodology of CDF's test of the top-quark hypothesis was that of an NPT significance test. They tested a null hypothesis against a composite alternative. They estimated the probability of their test resulting in a "type I" error (erroneously rejecting the null hypothesis) and interpreted that estimate as relevant to their assessment of the results of their experiment. I take these two features to be central to the NPT tradition as distinct from alternative approaches such as likelihood or Bayesian analyses.

Deborah Mayo has analyzed the various approaches to interpreting NPT tests in terms of the criteria of a "good test" that different interpretive models would deploy. She has advocated her own "error-statistical model" for interpreting NPT tests that is intended to avoid the difficulties of the alternatives. The error-statistical model of evidential reasoning can be explained in terms of the *severity requirement* (SR) and the *severity criterion* (SC).

> SR: An experimental result E constitutes evidence in support of a hypothesis H just to the extent that:
>
> 1. E fits H, and
> 2. H passes a severe test with E.

We can further specify the severe test requirement (SR2) by means of the Severity Criterion:

> SC: A hypothesis H passes a severe test T with outcome E just in case the probability of H passing T with an outcome such as E (i.e., one that fits H as well as E does), given that hypothesis H is false, is very low.

In what follows I will examine the relationship between this error-statistical model and the behavioral and evidential alternatives against which it is opposed, explaining how it avoids problems that the alternatives face. I will argue that the error-statistical model does not constitute a complete rejection of an evidential interpretation of statistical tests. Indeed, in my own discussion I intend to treat it – even more explicitly than does Mayo herself – as principally a *theory of evidence*. Rather, it constitutes a rejection of the literal interpretation of the accept/reject terminology and of the idea that error probabilities can be translated into a global measure of evidential strength. In so doing, error-statistical theory offers an approach to evaluating evidential claims based on tests of hypotheses grounded in two of the most important aims of scientific research:

1. Avoid erroneous conclusions and
2. Discover the reasons for believing specific empirical claims.

I do not claim that this amounts to an exhaustive methodology with respect to all the aims of science (scientists aim also at developing theories that exemplify such virtues as explanatory power and simplicity, for example, and also strive for such "nontheoretical" utilities as marketable technologies and career advancement). I do claim that error-statistical methods prove useful for pursuing these two central aims of the experimental sciences, and that these aims make scientific inquiry an evidence-driven enterprise.

The error-statistical concept of evidence also purports to be objective. Employing some helpful distinctions drawn by Peter Achinstein, I will attempt to specify the precise sense in which error-statistical evidence might be said to be objective.

6.1. The Error Statistical Model as an Alternative to Other Views

In two seminal articles published in 1928 and 1933, Jerzy Neyman and Egon Pearson invoked the fact that the methods they proposed enable one to limit the *rate* of one's errors in making decisions as the rationale for using NPT tests. They rejected any notion of relating outcomes of such tests to assessments of evidential strength:

> We are inclined to think that as far as a particular hypothesis is concerned, no test based upon the theory of probability can by itself provide any valuable evidence of the truth or falsehood of that hypothesis. . . . Without hoping to know whether each separate hypothesis is true or false, we may search for rules to govern our behaviour with regard to them, in following which we insure that, in the long run of experience, we shall not be too often wrong. (Neyman and Pearson 1933, 290–1)

In following the NPT method, one selects an accept/reject rule to meet one's requirements regarding the rate at which one is willing to make errors, but

> [s]uch a rule tells us nothing as to whether in a particular case *H* is true . . . or false. . . . But it may often be proved that if we behave according to such a rule, then in the long run we shall reject *H* when it is true not more, say, than once in a hundred times, and in addition we may have evidence that we shall reject *H* sufficiently often when it is false. (Neyman and Pearson 1933, 291)

This approach may be termed the *behaviorist* interpretation of statistical tests. Mayo states the criterion for a good test on the behaviorist model as follows:

> A *good test* is one that has an appropriately small frequency of rejecting *H* erroneously, and at the same time erroneously accepts *H* sufficiently infrequently (in a given sequence of applications of the rule). (Mayo 1985, 501)

It is certainly true that Neyman-Pearson significance tests can be good tests in this behaviorist sense. Hence, the behaviorist interpretation provides a convincing rationale for some uses of such methods, such as to reduce the frequency with which a manufacturer ships out defective merchandise. It also indicates the usefulness of such tests for promoting scientific aim (1) mentioned earlier in this section, provided that we interpret that aim to be the avoidance of errors *in the long run*. A scientific researcher might find this way of thinking about statistical tests to be remote from his concerns, however. Scientists aim not only to avoid a high rate of errors over the long run. They also seek to rule out specific types of errors in individual cases, and this is the sense of aim (1) that best reflects the aims of scientific inquiry. In addition, one aim of experimental science is to gain knowledge of particular experimental phenomena (aim 2 from earlier in this section). A test that is good in the behaviorist sense may not be good for either of these two aims.

Scientists seek to find evidence for empirical claims of interest to them. Consequently, one might seek to derive from the outcome of a statistical test a measure of the strength of the evidence for a hypothesis that is contained in a given body of data. Many criticisms of NPT testing have assumed that statistical tests ought to provide such a measure, and have charged NPT tests with failing to provide it. Alan Birnbaum proposed an interpretation of NPT intended to support conclusions regarding the strength of the evidence in support of a given hypothesis (Birnbaum 1977). This approach might be called *evidentialist.* Mayo regards the evidentialist criterion for a good test to be the following:

A good test rejects *H* (accepts *J*) if observed data ... provides [sic] appropriately strong evidence against *H* and in favor of alternative *J.* (Mayo 1985, 503)

Critics of NPT have been quick to point out difficulties in formulating a measure of evidential strength based on the outcome of NPT tests. They note that, as pointed out originally by David Lindley (1957), for any given level of significance, it is possible, given a sufficiently large sample, for a test to reject a hypothesis at that level, when intuitively we would say that the evidence favors that hypothesis over the alternative.

Colin Howson and Peter Urbach have illustrated the problem nicely with the following example. A purchaser of tulip bulbs is unable to recall whether he ordered a shipment of 40% red-blooming bulbs and 60% yellow-blooming bulbs (H), or a shipment of 60% red and 40% yellow (J). One way to decide the issue would be to select a sample at random from the shipment and plant them to see what blooms. Suppose that the bulb merchant selects a sample of ten bulbs, with the intention to reject the null hypothesis (40% red, 60% yellow) only if the results come out significant at a level of 0.05. In that case the bulb merchant would reject the null hypothesis only if seven or more of the bulbs bloomed red, and the power of the test against J would be 0.37 (i.e., the probability of a type II error of mistakenly rejecting the alternative hypothesis would be 0.63). Lindley, Howson, and Urbach, among other critics of NPT, point out that as the sample size increases, the proportion of bulbs observed to bloom red that would lead to a rejection of the null hypothesis at a significance level of 0.05 decreases, gradually approaching 0.4. For a sample size of 100,000, only 40.26% of the bulbs need bloom red for the null hypothesis to be rejected, and the power of the test against J would be nearly 1.0 (i.e., there would be almost no chance of accepting the null hypothesis when the alternative J is in fact true). Given that only two hypotheses are considered possible, however, and that the alternate hypothesis is that 60% of the bulbs are red, a result such as finding that 40.26% of the bulbs bloom red would seem to be much stronger evidence for the null hypothesis than for the alternate hypothesis. Howson and Urbach conclude: "The thesis implicit in the current [NPT] approach, that a hypothesis may be rejected with increasing confidence or reasonableness as the power of the test increases, is not borne out in the example, which signals the reverse trend" (Howson and Urbach 1993, 209).

Mayo argues that both the behaviorist and evidentialist models, and the criticisms directed at NPT based on them, are misguided and proposes an alternative approach to interpreting statistical tests, the error statistical approach. Quoting some suggestive comments made by Egon Pearson, she asserts that

> error frequencies are important, not because one is concerned simply with low error-rates in the long run; but because they provide "that clarity of view needed for sound judgment." ... Tests accomplish this learning function by providing tools for *detecting* certain *discrepancies* between the (approximately) correct parameter values (within a statistically modeled problem) and the hypothesized ones. (Mayo 1985, 507, emphasis in original)[11]

Consider again the model of CDF's counting experiment that I provided previously. The value of the parameter λ under the null hypothesis is λ_0. We might also consider whether λ exceeds some value λ', where $\lambda' > \lambda_0$. We can define D_{obs} to be the difference between the observed value of the test statistic s_0 and the hypothesized value λ_0 of the parameter: $D_{\text{obs}} = s_0 - \lambda_0$. Mayo claims that the question, what has been learned from this test?, is to be answered by means of the following principle:

> IND: (i) D_{obs} is a *good* indicator that λ exceeds λ' only if (and only to the extent that) λ' *infrequently gives rise to so large a difference*. (ibid., 510, emphasis in original)

That is to say, such a discrepancy is a good indicator that λ exceeds λ' only to the extent that the hypothesis that λ exceeds λ' passes a severe test with the result D_{obs}. Mayo continues:

> (ii) D_{obs} is a *poor* indicator that λ exceeds λ' to the extent that λ' *frequently gives rise to such a large difference*. (ibid., 510, emphasis in original)[12]

In other words, if the excess of the observed value for the test statistic is only slightly higher than the average expected assuming $\lambda = \lambda'$ ("slightly higher" in the sense that the excess would frequently occur in that case), then the hypothesis that λ exceeds λ' does not pass a severe test with the result D_{obs}, and hence the discrepancy does not indicate that hypothesis.

Critics of NPT conclude from examples such as the tulip case cited earlier that the outcome of a test might be such that, while the test is very powerful and rejects a hypothesis with a statistical significance that can be cited at a very low number, the data might nevertheless provide stronger evidence for the rejected hypothesis than for the alternative.

Mayo argues for distinguishing the statistical conclusion (i.e., the accept/reject output of the test) from the scientific conclusion that one draws based on one's knowledge of the characteristics of the statistical testing situation. In light of this distinction, she proposes two distinct criteria:

(i) A *statistical testing procedure* is good iff one is able to objectively evaluate what has and has not been learned from a statistical conclusion (reject or accept H).
(ii) A *statistical test conclusion*... is [poor] good for learning about a given discrepancy between λ and λ' to the extent that it is a [poor] *good indicator* that λ exceeds λ' [in the sense of IND above]. (ibid., 512)

Mayo proposes that evaluating the outcome of a statistical test requires that one consider the frequencies with which the test result occurs on the basis of a variety of values for λ'. Thus, in the tulip example, supposing λ represents the percentage of red tulips in the sample, one would not be satisfied with simply rejecting the null hypothesis at a given significance level regardless of sample size. One would want to consider the frequency with which the observed proportion of red bulbs in the sample would occur not only on the basis of the null hypothesis value of 40% red bulbs, but also on other possible proportions of red bulbs, particularly the possibility of there being 60% red bulbs. Naturally, one would find that 40,260 red plants out of 100,000 would be even less likely given a population of 60% red plants than given a population of 40% red plants. If we take λ' to be a value just slightly higher than 40% (such as 40.26%), then the observed difference between the value of the test statistic and the value hypothesized by the null hypothesis would in fact arise rather frequently. Consequently, for $\lambda' = 40.26\%$, the test conclusion "reject H at a significance level of 0.05" would be a *poor indicator* that the value of λ exceeds λ' (and consequently an even worse indicator that $\lambda = 60\%$). Hence such a test outcome would not constitute evidence that $\lambda = 60\%$.

Such an analysis of the tulip example reflects a point with which users of statistical procedures such as the CDF physicists are quite familiar, and which I employed in my discussion of the search for biases in the top experiment (see Section 7.2 in Chapter 5): If the effect that you are detecting is genuine, then, as the size of your sample increases, the significance level at which the null hypothesis is rejected should decrease. For this reason, test results that appear favorable at one stage can be defeated by the kind of diachronic assessment I described in the previous chapter. In using such tests, one does not consider the outcome of a statistical test as a static result, but as part of the dynamic process of developing evidence for or against a hypothesis. Good statistical method requires that one observe (where possible) how the outcome of the test varies as one changes parts of the analysis (as in the assessment of the jet-vertexing algorithm in Section 7.1 in Chapter 5), or as one collects more data (as in the assessment of the SLT analysis in Section 7.2 in Chapter 5).

The relevance of the diachronic behavior of statistical measures in evaluating evidence arises from two distinct factors. The first constitutes the core of the response to the tulip example: As sample size increases, the experimental (sampling) probability distribution becomes narrower. Consequently, when the sample size is large, a result that has a very low probability on the null distribution might fit very well (i.e., have a much higher probability) on the basis of a hypothesis that departs from the null hypothesis to a very small degree. Tracking the statistical significance of the result as sample size grows gives one, in effect, an additional means of ruling out errors due to statistical fluctuations. One might get a result that allows one to reject the null hypothesis at the 0.05 significance level for a given size sample. Of course, the null hypothesis could nevertheless be true, and the result might be one of those relatively infrequent outcomes that occur five times in a hundred. But if the significance level at which the null may be rejected becomes smaller and smaller as the sample size grows, then the behavior of this statistical measure can itself be regarded as highly improbable, if indeed the null hypothesis were true. Thus, the severity of the test passed by the alternative hypothesis is enhanced, and one has stronger evidence for that alternative than one would have based on a single statistical result.

The second reason why good experimenters are interested in the diachronic behavior of statistical measures arises from the possibility of errors that arise not merely from statistical fluctuations but from biases in the testing procedure. An error-statistical evaluation of evidence relies on the experimental distribution. It is on the basis of this distribution that both fit and severity are estimated. However, the error probabilities of one's testing procedure are themselves empirical facts. Although ideally one arranges the experiment so that these facts can be determined independently of the data themselves, where doubts arise as to the accuracy of the experimental model, the behavior of one's results over time can yield relevant information. In particular, as discussed in the previous chapter, a testing procedure biased by an attempt to squeeze out a certain result will typically not continue to yield that same result as additional data are added.

Both of these factors are at work in CDF's response to the data they collected during run Ib. The disappointment that CDF members felt when they learned of the October 13, 1994, results from the early Ib data was not caused by an increase in the probability of a type I error. The cause of the disappointment was the *failure* of that probability to *decrease*, as would be expected for a genuine effect. That failure suggested (though it certainly did not establish) that the initial excess over estimated background resulted at least partially from biases in the initial test. Those worries were mitigated by the subsequent performance of the revised algorithm, which delivered, from January 1995 onward, results that behaved according to the expected pattern: As more data were added, the probability of a type I error went down. (However, since the algorithm *had* been revised, this did not completely erase the suspicions of some collaboration members regarding the original algorithms from run Ia.) CDF regarded the dynamics of the statistics over time as an important piece of information in deciding whether they had indeed had evidence in their Ia data. If the effect had not become significant at a smaller level with the addition of more data, CDF would have been forced to admit that they did not understand the nature of their findings.

Hence, the thesis that Howson and Urbach claim is "implicit" in the NPT approach to statistical testing is in fact not implicit in the actual use of such tests. Indeed, the error-statistical model explicitly rejects that thesis: One does not assume that a hypothesis may be rejected with increasing reasonableness as the power of the test increases. Rather, a statistical test is "a standard tool" that allows one to represent both the hypotheses being investigated and the experimental test of those hypotheses. Specifically, those tools permit one to represent the susceptibility of the test to certain kinds of errors with respect to those hypotheses by means of probability models. In so doing, they permit one to establish that particular experimental results are reliable indicators of both discrepancies and agreements between hypothetical models and experimental results represented as models of data (see Mayo 1996, 162–4).

6.2. Error-Statistical Evidence as Objective Evidence

Although the error-statistical model is meant to serve as an alternative to evidentialist interpretations of statistical testing such as Birnbaum's, it clearly does not entirely separate statistical tests from judgments regarding the strength of experimental evidence for or against a particular hypothesis. In this section, I wish to clarify the relationship between the statistical characteristics of a test and evidential relationships based on the results of that test.

It is important first to distinguish different ways of speaking about evidence. Peter Achinstein argued that it is essential to a concept of evidence of central importance in the sciences that evidence for a hypothesis constitutes a good reason to believe that the hypothesis is true (1983). More recently, Achinstein has distinguished four different concepts of evidence (Achinstein 2001).

E-S Evidence. One way of speaking about evidence would be to say that having evidence of the truth of a hypothesis is a matter of having information that, relative to an *epistemic situation* (a situation wherein one believes or knows certain propositions to be true, but not others, and knows how to reason about the truth of that hypothesis on the basis of the information in question), *justifies* one in believing the hypothesis to be true. This "E-S evidence" concept is objective insofar as whether a person in a given epistemic situation would be justified in believing a given proposition is an objective matter and does not depend on whether anyone actually is in such an epistemic situation.

Subjective Evidence. A distinct concept of evidence is tied to specific persons or groups of persons and their beliefs. Certain information may be regarded by a person as evidence for a hypothesis, such that the information in question is that person's reason for believing that the hypothesis is true. We could then say that this information is that person's "subjective evidence" for that hypothesis. Where E-S evidence is objective in being relativized only to a epistemic situation *type*, subjective evidence obtains only when some person or group is in fact in a specific epistemic situation. Furthermore, a person might have subjective evidence that constitutes *her* reason for believing a hypothesis yet fails to *justify* that belief.

Veridical Evidence. We can also note the use of a concept of evidence that is quite strongly objective: A proposition might describe evidence for some hypothesis by providing a good reason to believe the hypothesis is true, although no one knows or even believes the truth of that proposition, the hypothesis, or that the proposition expresses evidence for the hypothesis. This type of evidential relationship holds without relativization to either a person or group, or even to an epistemic situation type. If, in addition, the hypothesis in question is in fact true, then this is a case of "veridical evidence." Achinstein points out that with veridical evidence there is always the potential for "empirical incompleteness." That is, whether a given fact is veridical evidence for a certain hypothesis may depend on empirical information beyond what is expressed in the statement of the fact and the statement of the hypothesis.

Potential Evidence. It is also possible that certain information might constitute a good reason to believe that a hypothesis is true, although that hypothesis is in fact false. We might speak of evidence in a way that imposes the requirements of objectivity and nonrelativization, as in the case of veridical evidence, but without veridical evidence's requirement that the hypothesis is in fact true. Then we would be speaking in terms of "potential evidence." If some fact is potential evidence for a hypothesis, and the hypothesis is furthermore true, then that fact is also veridical evidence for that hypothesis. Potential evidence, like veridical evidence, is potentially empirically incomplete.

Although different concepts of evidence can be discussed within an error-statistical framework, I wish to argue that error statistics, through the severity requirement and severity criterion, primarily explicates *potential evidence.*

We can find a hint in Mayo's own statement of the severe test requirement, which generalizes the principle IND that articulated the concept of a *good indicator*: "Passing a test *T* (with *e*) counts as a good test of or good evidence for *H* just to the extent that *H* fits *e* and *T* is a *severe test* of *H*" (Mayo 1996, 180), where the criterion for the severity of a test should be understood in the sense (SC) specified earlier.

This correspondence between "good indicator" and "good evidence" is natural for a nonrelativized concept of evidence such as Achinstein's potential and veridical concepts of evidence. Achinstein compares his concept of a "good reason to believe" to concepts such as "sign" and "symptom." For example, the general flattening of the landscape as one travels east across northern Arkansas is a *sign* that you are approaching the Mississippi, and this is true whether or not anyone believes that to be the case. It would be the case even if no one had ever made the trip. Likewise, the probabilities involved in an assessment of error-statistical evidence are objective in that they do not depend on whether anyone believes those probabilities to have any particular values, or even whether they think about probabilities of outcomes at all.

In taking this stance, however, it might seem that the error-statistical account of evidence becomes committed after all to the kind of "evidential relation" interpretation of statistical tests from which it attempts to distinguish itself. If evidential relationships are objective because the error probabilities of tests are objective, then would not any two tests with the same error probabilities yield equally strong evidence, meaning that error statistics is after all saddled with the "Lindley paradox" discussed in the example of the tulip bulbs? To answer this objection requires distinguishing different bases of evidential assessments.

The rejection of evidentialist interpretations of statistical tests is a rejection of the notion that any judgment regarding strength of evidence can simply be "read off" of the outcome of a single statistical test, in the sense that a hypothesis passing a test with given error characteristics is confirmed just as much as any other hypothesis passing a test with identical error characteristics. The error-statistical account denies such a direct correspondence between error characteristics and evidential strength. Nevertheless, one can, on the basis of statistical test results, evaluate evidence on the basis of *comparative* evaluations of error probabilities with respect to various hypotheses. How can this be done?

Consider a comparison between the results of two tests that are significant at approximately the same level but based on different sample sizes. Two counting experiments similar to the sort performed by CDF can illustrate the point. We seek evidence for a new particle phenomenon by looking for an excess of events with a certain decay signature beyond what is expected from background. The first counting experiment is based on a sample of 10,000 particle events. According to the null hypothesis, depicting the case where all candidate events are from background, the probability of getting a candidate event in a single trial is $p = 5 \times 10^{-4}$. Hence the expectation from background in the first experiment is 5 candidate events. Suppose that our testing rule for this experiment directs us to do the following: (1) examine 10,000 particle events, (2) note the discrepancy between the observed number of candidate events and the null expectation, and then (3) reject the null hypothesis just in case the observed number exceeds the background expectation by an amount that would occur less than 3% of the time, assuming the null hypothesis is true. Finding more than 10 candidate events, then, would lead to a rejection of the null hypothesis. In a second experiment, 200,000 particle events are recorded. Now the background expectation, assuming the same null hypothesis, is 100 events. The rule for this experiment is the same except that it directs us to consider 20 times as many events in our sample. Observing more than 119 candidate events would yield a rejection of the null hypothesis in that case.

So here we have a case where two experiments have roughly the same error characteristics, at least as far as significance levels are concerned. However, error-statistical considerations dictate asking, not merely about significance levels but also about the frequency with which one would obtain such a result for various models of the population being studied that are closely related to the model of the hypothesis directly under test. Suppose that we ask, for example, which values of p would *infrequently* yield results such as found in the two experiments. We might specify that we mean by "infrequently" that such results would not occur more than, say, 10% of the time. It turns out that even if the value of p were to be as high as 7×10^{-4}, you would get more than 10 candidate events less than 10% of the time, in a sample of 10,000 events. However, in a sample of 200,000 events, you would find more than 119 candidate events less than 10% of the time only for values of p less than 5.27×10^{-4}. In other words, by error-statistical principles, if the two sets of data yield results with the same significance level against the original null hypothesis, the result based on the second set of data does not indicate a difference from the null hypothesis that is very large, while the first result, based on a smaller set of data, is much stronger evidence of a significant departure from the null hypothesis. The results from the larger sample indicate a much smaller deviation – one we may not even consider worthy of our attention.[13]

Now we are in a position to reply to the objection that if error-statistical evidence is objectively based on error rates of tests, then it is subject to the Lindley paradox. The evidential relationship supervenes on facts about the error rates of testing procedures. Although such facts about error rates are objective, the error rates with respect to a *single* hypothesis are generally not sufficient to fix the strength of that relationship. Instead, the results indicate a function of the rate at which the relevant test would erroneously accept or reject *various* alternative hypotheses within a certain family. (In terms of a hierarchy of models, they are all competing hypotheses at the same *level* in that they refer to a single parameter or characteristic of a population.) The evidential relationship obtains whether or not anyone actually considers such other hypotheses or notices their relevance. Hence no reference need be made to any person's beliefs, interests, or epistemic situation. The error-statistical concept of evidence is objective in a sense similar to the objectivity of Achinstein's potential and veridical evidence.[14] However, it is clearly not veridical evidence because it is not a requirement of this model of evidence that the supported hypothesis is true. (Of course, adding the requirement that H be true to SR would easily give us an error-statistical concept of veridical evidence.)

This analysis, however, generates a worry: In *The Logical Foundations of Probability*, Rudolf Carnap distinguished three concepts of confirmation (Carnap 1962): *classificatory, comparative*, and *quantitative*. Does the error-statistical account yield only a comparative concept of evidence? If so, one might suspect that it is incapable of meeting the Achinsteinian requirement of providing good reasons for believing hypotheses.[15]

Fortunately, we need not worry, for error-statistical assessments *are* capable of supporting both classificatory and comparative evidence claims. (As we will see, no quantitative concept of evidence can be found in the error-statistical account, and those who will settle for nothing less must look elsewhere.) To see this point requires distinguishing two different ways of talking about evidence and clarifying the nature of the hypotheses that are thus supported.

A comparative concept of evidence allows one to answer such questions as "are results e better evidence for hypothesis h than for hypothesis h'?" and "are results e better evidence for hypothesis h than results e' would be?" On the error-statistical account, such a comparative concept will typically be appropriate for evaluating a statistical "point hypothesis" (i.e., a hypothesis that specifies a precise value for a parameter describing some population). Point hypotheses are especially important for purposes of the precise evaluation of evidence, as only they can be represented with a single probability distribution.

Using this comparative concept of evidence, however, one can say things such as "these data constitute better evidence for h than for h', and better evidence for h' than for h'', even though hypotheses h, h', and h'' are incompatible with one another. Consequently, this comparative notion of evidence cannot meet the requirement that it only applies where there is reason to believe the hypothesis in question because plausibly an experimental result cannot simultaneously be a reason to believe both a hypothesis and its negation. (Nevertheless, one might still wish to make this comparative notion a threshold concept (Achinstein 2001). Thus, a result that does not fit h at all would not merely be regarded as worse evidence for h than for some other h', or worse evidence for h than some other result would be, but as no evidence whatsoever for h.)

Such a comparative concept of evidence can be useful in exploring the implications of a given experimental result for a class of point hypotheses regarding a single parameter. However, the classificatory concept is no less important. Indeed, it is the classificatory notion of evidence that has taken center stage in the story of the top-quark search. The title of CDF's paper was, after all, "Evidence for Top Quark Production . . ." rather than "Stronger Evidence than D-Zero Has for Top Quark Production" or "Stronger Evidence for Top Quark Production than for Fourth-Generation b' Production." (And this is not only because of the howls of protest that would have come from the D-zero collaboration if they had used the former title!)

To understand the relationship between the classificatory and comparative notions, we should start by scrutinizing the hypothesis for which CDF claimed to have evidence: that there is a top quark with $M_{\text{top}} = 174 \pm 10^{+13}_{-12}$. This is not a point hypothesis, but a *compound hypothesis* (i.e. a disjunction of different point hypotheses corresponding to distinct, but closely related populations, none of which is strictly compatible with the others). Each possible top-quark mass within the range specified by the errors on CDF's mass estimate corresponds to a different value of the parameter under investigation in CDF's experiment (the parameter λ in the model described in Section 5). Within that range, CDF's data did not yield classificatory evidence for believing one hypothesis over all the others.

Thus, we can say that the results of CDF's testing procedure yielded classificatory evidence for H (the hypothesis that there is a top quark whose mass lies within the range just specified), where H comprises numerous point hypotheses (h, h', h'', etc.), each strictly incompatible with one another, and each represented in the theoretical model by a distinct value of λ.[16] From the error-statistical perspective, this means that CDF's data fit some of the hypotheses comprised by H in a manner extremely improbable if the true value of λ were outside the range of H. In other words, a well-designed test (one that will be severe with respect to hypotheses of interest that pass that test) is a reliable indicator as to whether the true value of the parameter under investigation lies inside or outside a certain range. Because our reliable indicator indicates that the true value of the parameter lies inside a particular range when a certain compound hypothesis passes such a severe test, it provides a good reason to believe that it does.

This classificatory concept of evidence, unlike the comparative concept described earlier, does not permit application simultaneously to incompatible hypotheses, although it may apply to distinct but compatible hypotheses. In order to constitute evidence for the hypothesis

H: There is a top quark with a mass in the range from 165 to 185 GeV/c^2, our experimental results must be highly improbable unless that hypothesis is true. In other words, they must be highly improbable for any top-quark mass outside that range. That condition cannot simultaneously be met by H and the following hypothesis:
J: There is a top quark with a mass in the range from 140 to 164 GeV/c^2, or by H and the hypothesis:
K: There is no top quark.

Note, however, that the condition can potentially be met simultaneously by
H and the following:

L: There is a top quark with a mass in the range from 160 to 180 GeV/c^2.
M: There is a top quark with a mass in the range from 140 to 200 GeV/c^2.
N: There is a top quark.

None of these hypotheses is incompatible with H. Indeed, experimental
results that are highly improbable unless H is true *may* be highly improbable
unless L is true, and *will* be highly improbable unless M is true, and unless
N is true.

Based on what has been said thus far, one might be tempted to draw a simple distinction: To compound hypotheses only the classificatory concept of
evidence applies; to point hypotheses only the comparative concept applies.
However, such a claim would be mistaken.

It is easy to see in the present example that comparative judgments of
evidence can be drawn for compound as well as point hypotheses. A result
that constitutes evidence for H will constitute even stronger evidence for
the weaker hypothesis M, because such a result would fit each of the point
hypotheses comprised by M in a way that would be even less probable if the
true value of λ lay outside the range comprised by M. (To get to possible
values of λ outside the range of M, we would have to go even further away
from values that fit the data best, pushing our result even further out onto
the "tails" of the distributions comprised by the alternative to the hypothesis
(here M) being supported.) This illustrates how, in such cases where the
classificatory concept of evidence applies, one can still make comparative
judgments of stronger and weaker evidence, but now without necessarily
losing the ability to say that the evidence in question provides a reason to
believe the hypothesis supported (since the hypotheses being compared
need not be incompatible, as they are in the case of point hypotheses).

Perhaps less obviously, one can have classificatory evidence for a point
hypothesis. We can see this in the tulip example itself. Such evidence is
not provided by the results of the planting experiment alone, however. In
setting out the tulip example, Howson and Urbach take it to be background
knowledge that the batch of bulbs that was ordered was either 40% red
and 60% yellow, or 40% yellow and 60% red. If our concern were with
epistemic-relative evidence, we might say that relative to an epistemic state
that included that statement as background knowledge, planting 100,000
bulbs and having 40,260 come up red is evidence that the order was 40%
red and 60% yellow.

However, we can go further, and speak of classificatory, nonrelative evidence for a point hypothesis. In the tulip example, we might ask what the
evidence is that the order that was placed was one of just those two possibilities. This would presumably be something such as a copy of an order form
from the bulb company listing only those two red-yellow tulip combinations,

but on which one can no longer read which of the two combinations was checked off. Here we are not likely to be able to give a precise statistical characterization of the evidence, yet we can still ask, how likely is it that I would have a copy of the order form that looks like *this*, unless I ordered either 40% red and 60% yellow tulips, or vice versa? Then, supposing that such data do constitute evidence that the order was either 40% red and 60% yellow or 60% red and 40% yellow, I have evidence that the order was 40% red and 60% yellow (a point hypothesis) when I combine my two sources of evidence. One source of evidence (the planting experiment) distinguishes two point hypotheses at the comparative level, the other source (the order form) then rules out all the intermediate possibilities. Taken together, they constitute classificatory evidence for a single point hypothesis.

I have thus far argued that error-statistical evidence supervenes on objective facts about the error characteristics of tests with respect to a family of hypotheses. It is consequently an objective matter, for a particular set of data from a particular test, whether or not those data constitute evidence for a given hypothesis, belonging to a family of alternative hypotheses (the classificatory concept of evidence), and also whether, within that family the data constitute stronger or weaker evidence for one hypothesis as opposed to another (the comparative concept of evidence). Using the comparative concept, we can sometimes, on the basis of error-statistical considerations, compare the extent to which e supports h to the extent to which e' supports h'. But note that it is not possible, on this model, to make such a comparative evaluation for *every* possible combination of evidence statement e and hypothesis h. In the imaginary counting experiment just described, for example, we can say that the results that lead to rejection of the null hypothesis on the basis of 10,000 events are better evidence for a nonbackground particle process than the results rejecting the null from 200,000 events, although both have the same significance level, because we can compare their frequency characteristics, not only under the hypothesis that $p = 5 \times 10^{-4}$ but also under other hypothesized values of p. That is to say, a comparative question about the relative strength of evidence based on a given outcome e for various hypotheses h_1, h_2, ... has a determinate answer when h_1, h_2, ... constitute a family of competing hypotheses that differ with respect to the value of some parameter, such that the frequency with which e would occur on the basis of models of those competing hypotheses, for a given experimental test, is determinate for each hypothesis h_1, h_2, ... Likewise, objective comparisons of relative evidential strength are possible for a given hypothesis h for various possible outcomes of a given test, provided that the probabilities of those various outcomes are determinate and can be compared.[17]

However, the error-statistical concept of evidence does not yield a *quantitative* measure of evidential strength. If it did, then we should be able to define on its basis a confirmation function c such that, for any pair of hypotheses h and h' and any pair of experimental outcomes e and e', we could meaningfully ask whether $c\,(h,\,e) > c\,(h',\,e')$. In other words, a quantitative concept of evidence would entail the possibility of *global* comparative assessments of evidential strength. The error-statistical concept of evidence does not lend itself to such global comparisons; thus, it does not yield a quantitative measure of evidential strength. The comparisons of evidential strength that are meaningful, to which objective facts correspond, are local comparisons. For some statements e and e' and some hypotheses h and h', the question whether e is stronger evidence for h than e' is for h' will not have a determinate answer. (For comparison, we can safely say that the Beatles had a greater impact on pop music than, say, Paul Revere and the Raiders, and we can safely say that Charlie Parker had a greater impact on jazz than Sonny Stitt – not to malign the talented Mr. Stitt – but it is far from obvious that the question "did the Beatles have a greater impact on pop than Charlie Parker had on jazz?" has any determinate right answer.)

One type of situation in which such comparisons will not generally be possible is the comparison of test outcomes for tests directed at answering completely unrelated questions. Suppose that the results from the test of the experimental bowling ball described in Section 4 show that in 600 games the average score allows for a rejection of the null hypothesis (that games bowled with the new ball yield the same average score as those with a conventional ball) at a significance level of 0.05. Is that stronger or weaker evidence of the effectiveness of the experimental bowling ball than finding 10 candidate events when only 5 are expected in a sample of 10,000 is evidence of the existence of a new particle decay phenomenon? There is no common family of hypotheses within which we can regard these two cases as competing alternatives, nor can we regard the two experimental findings as outcomes the relative frequency of which can be evaluated under a single hypothetical model. While it is true that we can compare the frequency characteristics of these two experimental tests by simply looking at the significance levels (0.05 and 0.03, respectively), we cannot compare the strength of the evidence provided by those tests for the alternative hypotheses in question without being able to compare their ability to detect discrepancies of various relative sizes. And because these tests concern themselves with hypotheses about unrelated populations, characterized by unrelated parameters, there can be no direct comparison of their ability to do so. The relevant discrepancies in the two cases are entirely different, and so there is no common standard for comparing the evidential strength of these two outcomes on the error-statistical model.

On the error-statistical interpretation, Neyman-Pearson tests are not meant to provide a measure of the strength of the evidence produced by an experiment for a given hypothesis according to some global scale of confirmational strength. Nor do they aim to determine the probability of a hypothesis in light of all the evidence at hand. Hence it is a mistake to criticize these methods as unjustified for their flaws in providing such information. Instead, these methods are to be judged for their ability to indicate, in Mayo's phrase, "what has been learned" from a given experiment. In very many cases, as in the top search at CDF, the aim is to determine whether a given effect is real or not. For answering such questions, Neyman-Pearson methods of hypothesis testing are quite useful indeed because they direct our attention to the facts on which objective evidential relationships depend.

7. CONCLUSION

In *Representing and Intervening*, Ian Hacking issued a challenge to philosophers of science to abandon a theory-dominated concern with representation and to pay more attention to the ways in which investigators engage the material world through experimental intervention. Here I have suggested that representing and intervening are equally important components of experimental activity for the purposes of creating scientific knowledge.[18] Drawing on the error-statistical theory of evidence, I have shown how an appropriate representation of experimental procedures, and in particular the error probabilities of those procedures, can be crucial to the determination of the evidential status of the results of experiment.

I have also aimed, in the present chapter, to clarify certain aspects of the error-statistical concept of evidence. The error-statistical theory describes evidential relationships as supervening on probabilistic features of testing procedures. However, from this it does not follow that statistical measures such as significance or power are sufficient to determine a measure of confirmational potency or evidential strength. To the contrary, I have argued that error-statistical considerations support two distinct evidential concepts: *comparative* and *classificatory*. Based on the error characteristics of a testing procedure one can make "local" comparisons of evidential strength, as between the evidence provided by a given result of a given procedure for various members of a family of hypotheses, or between the evidential bearing on a given hypothesis of various possible outcomes of a given testing procedure. Furthermore, the error-statistical analysis yields a classificatory concept of evidence, enabling one to say, on the basis of the error characteristics of a testing procedure with respect to various hypotheses of interest, that one does (or does not) have evidence for a hypothesis (where often this hypothesis will in fact comprise a number of similar but strictly incompatible possibilities between which the test is unable to discriminate). The error-statistical theory of evidence does not, however, underwrite a global measure of evidential strength such that for any pair whatsoever of results e and e', and any pair whatsoever of hypotheses h and h', one can determine whether e is better evidence for h than e' is for h'.

Nevertheless, error-statistical evidence is *objective* in a sense similar to that articulated by Peter Achinstein in his characterization of "potential evidence": If a result is error-statistical evidence for a certain hypothesis, it is so independently of whether any individual believes the result, the hypothesis, or that the result is evidence for the hypothesis; furthermore, the evidential relationship is not relative to either particular subjects or to particular non-ideal epistemic situations (see note 14 to this chapter). In addition, the evidential status of an experimental result, on the error-statistical theory may (and indeed typically will) depend on empirical facts other than those expressed in a statement of the result and a statement of the hypothesis in question. In other words, error-statistical evidence is potentially empirically incomplete.

An adequate theory of experimental inference ought to yield insights into the practices of working experimentalists. My aim in the next chapter is to see whether the error-statistical account meets this requirement. In particular, I wish to take a close look at biases that can arise in statistical testing procedures by such violations of good experimental practice as "tuning on the signal." I will be looking to establish to what extent the error-statistical theory of evidence can diagnose with precision the nature of the difficulties thus created, and to what extent that theory can help to underwrite solutions to those difficulties.

7

Bias, Uncertainty, and Evidence

A card being drawn at random from a piquet pack, the chance is one-eighth that it is an ace, if we have no other knowledge of it. But after we have looked at the card, we can no longer reason in that way. That the conclusion must be drawn in advance of any other knowledge on the subject is a rule that, however elementary, will be found in the sequel to have great importance.

    C. S. Peirce (1931–1958, vol. 2, paragraph 696; 1982– , vol. 4, 411 [1883])[1]

Benutzt man den Wahrscheinlichkeitsbegriff und Formeln der Wahrschein-lichkeitsrechnung, ohne sich über die auftretenden Kollektivs und ihre gegen-seitigen Beziehungen klar zu sein, so gerät man leicht in die Irre. [If one uses the concept of probability and the formulas of probability theory, without un-derstanding the relevant collectives and their relations with one another, one will easily be misled.]

    Richard von Mises (1972, 196)

1. INTRODUCTION: TYPES OF EXPERIMENTAL SUCCESS

Philosophers see hypothesis testing as a logical activity. Yet such testing is carried out in contexts that sometimes complicate the logical schemata of philosophers. Scientists cannot afford simply to abstract the schema from the context, but must confront the context head-on. To do so, they some-times invoke methodological constraints that appear to have surprisingly psychological or social aspects. In this chapter, I will examine just such a constraint invoked in the debates surrounding CDF's top results. My aim is to try to understand the epistemological function of this constraint, which relates to some long-standing disputes among philosophers of science over confirmation.

    I wish to begin by distinguishing different perspectives for the assessment of an experimental outcome. More specifically, I wish to focus on those aims of experimenters relevant to the *epistemology* of science, and in light of those aims to distinguish three distinct types of epistemological success in experimental undertakings.

In the previous chapter, I described a theory of evidence that purported to be objective in the sense that, if some test procedure yields evidence for a hypothesis, then it yields a reason to believe that hypothesis, regardless of whether anyone does in fact believe the hypothesis for that reason (and hence regardless of whether anyone takes the test procedure to yield a reason to believe the hypothesis). From the perspective of such a theory of evidence, one can perfectly reasonably speak of an experiment yielding (potential or veridical) evidence without being committed to the claim that the experimenter or anyone else knows, or even believes, that the results of the experiment are evidence for a particular hypothesis.

Hence one type of epistemological success for experimenters can be described as follows: An experimenter (or experimental collaboration) executes a certain testing procedure. A particular hypothesis passes (or fails to pass) a severe test with the outcome of that procedure. If the hypothesis does pass such a severe test with the outcome of that test, then the result of the experiment is a reason to believe that the hypothesis is true, and we can say that the experiment is an *evidential success* with respect to that hypothesis. It is the job of the experimenter(s) to find out whether this is so. More precisely, I will say that an experiment is *evidentially successful* with respect to hypothesis *h* if and only if the experiment yields reasons for believing, of *h*, that it is true. (Whether there is any interest in a particular *h* for which an experiment is evidentially successful is obviously not independent of the opinions and interests of the experimenters. In what follows, I will sometimes use the term "evidential success" without relativizing it explicitly to a particular hypothesis, if it is clear from the context which hypothesis is of interest in assessing the evidential success of that experiment.)

Evidential success is merely a starting point, however. Although an experiment is evidentially successful for some hypothesis, those conducting the experiment might be unaware of this fact. They might believe that their experiment produced no evidence for any hypothesis in which they have any interest. Or they might believe that it yielded evidence for a hypothesis entirely different from that with respect to which it was in fact evidentially successful. Certainly no experimenter will be satisfied with this kind of success!

Let us then say that an experiment is *epistemically successful* for *h* if and only if: (1) it yields reasons for believing, of *h*, that it is true (i.e., it is an evidential success); and (2) the experimenters know, or are justified in believing, that (1) obtains for the experiment at hand.[2]

An experiment will be an evidential failure for a hypothesis, then, just in case it fails to produce reasons for believing that hypothesis. An experiment will be an epistemic failure for *h* either if it is an evidential failure for *h*, or if it is an evidential success for *h*, but the experimenters do not know (or are not justified in believing) that it is an evidential success for *h*.

Epistemic success as I have just defined it, however, is strictly a matter of first-person perspective. Thus, it cannot capture what is perhaps the most important epistemological aim of experimenters, which is to make their results, and the broader significance of their results, known to their colleagues. We need a third type of success to capture this aspect of the experimental enterprise. Given that nothing short of this would satisfy most experimenters, let's call this *experimenters' success*. An experiment is an experimenters' success with respect to *h* if and only if the following conditions are satisfied:

1. It is an epistemic success (and hence an evidential success) with respect to *h*; and
2. the experimenters who performed the experiment are able to provide the appropriate scientific community with a justification for believing that (or to provide them with the knowledge that) the experiment is an epistemic success with respect to *h*.

Certainly the notion of an "appropriate" scientific community is both vague and contextual. Did CDF wish to provide a justification to the community of high energy–experimental particle physicists? physicists working on theories of elementary fields and particles generally? the entire physics community? the "educated" public? Closely related will be distinctions between different types of justifications that are appropriate to these various communities. For the educated public, an appeal to authority might be all that one can reasonably expect, but for the community of high energy–physics experimenters, a much more detailed argument will have to be provided – hence the extraordinary length and intricacies of CDF's paper detailing their evidence for the top quark.

To further complicate things, the distinction between first-person (epistemic success) and third-person (experimenters' success) perspectives is not sharp in the context of collaboration work. CDF members working directly on particular pieces of the top analysis had to justify claims about the evidential status of the results of their work to other CDF members working on other parts of the top analysis. All the CDF members working on the top analysis (and "membership" in this subgroup of the collaboration would have to be regarded as a matter of degree) endeavored to justify their evidential claims to the collaboration as a whole. One could certainly say that for Tony Liss, John Yoh, and Claudio Campagnari (for example), the "appropriate scientific community" was, first, the other members of the small (yet still numerous) group of CDF members who knew the top search intimately, and, second, the CDF collaboration as a whole.

There are other kinds of experimental success besides those I have discussed. Demonstrating that one's experimental apparatus functions correctly is an important type of success and was the main goal, for example, of the 1985 "engineering run" at CDF. Experiments that yield similarly strong evidence may not be equally successful when judged by other standards. Actually finding evidence of the top quark's existence was in some sense more of a success than finding a new mass limit would have been. The experiments that formed the basis of CDF's 91 GeV/c^2 mass limit were (approximately) no less evidentially successful with respect to the hypothesis that the top was at least that heavy (if it existed) as their later experiment was with respect to the hypothesis that the top did indeed exist. Yet there can be no doubt that CDF valued more highly success with respect to the stronger (existence) claim. But it is important not to confuse "null" results, where one fails to find evidence of the hypothesis of immediate interest, with evidential failure. As long as the experiment is sufficiently well run, so that the data are not rendered evidentially useless, one can always ask, "What *did* we learn from this experiment?"

In terms of the epistemological function of the test of a hypothesis, however, success is to be understood with indifference as to whether the truth or falsehood of the hypothesis is indicated, independently of whether the hypothesis about which one has learned is exciting or dull, whether the result gives you a shot at a Nobel Prize, and so on. The aim of an experimental test, or at least the aim that gives experiment its central role in the sciences, is to produce knowledge. For this reason, I will focus on these three types of epistemological success. I wish to consider what conditions must be met before success is possible, and what factors can prevent such success, by looking at how CDF sought to avoid evidential, epistemic, and experimenters' failure.

Epistemological success can elude the experimenter. Much of the training of experimentalists helps them to avoid pitfalls leading to the failure of an experiment. Being a good experimenter is not so much a matter of making "big discoveries" as it is a matter of performing experiments that one can learn from, even if one fails to get the desired exciting result. What is learned from one experiment can be used in the next.

Note how different evidential and epistemic failures are, then, from errors. From errors – a previously accepted hypothesis or experimental assumption fails subsequent testing – one can hope to learn something about the phenomenon of interest. From complete evidential and epistemic failures, one learns nothing. Perhaps one learns about the inadequacy of one's apparatus, or about one's own weaknesses, but not about the inadequacy or weakness of any hypothesis. Even when an experiment is an evidential and epistemic success, if one lacks the means to justify what one knows to one's colleagues, then what one has learned from the experiment cannot contribute directly to the ongoing pursuit of scientific knowledge.

With these experimental aims in mind, then, I wish to consider the potential causes of epistemological failure. I am interested, in this chapter, in a particular source of failure, which has haunted the story of CDF's first top-quark evidence thus far: "tuning on the signal."[3]

We have seen how the question of bias arose with respect to some aspects of the experimental test reported in the Evidence paper. I now wish to probe more deeply the issues that arose during those debates. I am going to examine in detail the kind of evidence that counts for and against the accusation of bias owing to tuning on the signal. My aim in pursuing this investigation is not to reach a final verdict on whether these searches were biased but to try to understand what it means to say that a test is biased in this way, how such a claim is vindicated, and the nature of the failure that such bias yields. I also hope, by examining the social context in which these controversies arose and were resolved (to the extent that they were resolved), to learn something about how experimental reasoning works when carried out in a collaboration, point out some of the distinctive methodological problems raised by collaborative experimentation, and suggest some positive approaches to responding to those problems.

2. PREDESIGNATION AND TUNING ON THE SIGNAL

When the SLAC-LBL collaboration led by Burton Richter codiscovered the J/ψ in 1974, they were using the SPEAR electron-positron collider. The collaboration was looking at the products of e^+-e^- collisions and scanning for "resonances" – bumps in the energy spectrum of collision products that would suggest that a new particle was being produced when electrons and positrons were annihilating one another at a specific energy. The SLAC-LBL collaboration discovered the J/ψ as a result of looking at the cross section for the decay of e^+-e^- pairs into various final states. What they found was a peak in the cross section at an energy of approximately 3.1 GeV. In their publication announcing the discovery they noted, "Our attention was first drawn to the possibility of structure in the $e^+e^- \rightarrow$ hadron cross section during a scan of the cross section carried out in 200-MeV steps. A 30% (6nb) enhancement was observed at a c.m. [center-of-mass] energy of 3.2 GeV. Subsequently, we repeated the measurement at 3.2 GeV and also made measurements at 3.1 and 3.3 GeV" (Augustin, Boyarski, et al. 1974, 1406–7). These repeated measurements were not conclusive, so they went back and collected more data, "using much finer energy steps." The resulting data revealed a striking peak at 3,105 MeV (see Figure 7.1).

Having found one resonance, dubbed $\psi(3105)$, they began a search for other resonances in the same region. They instituted a "search mode" at SPEAR, in which the energy of the storage ring (twice the energy of an individual beam) was increased by a 1-MeV step every three minutes. In the process of this search, they found another resonance, this one at 3,695 MeV.

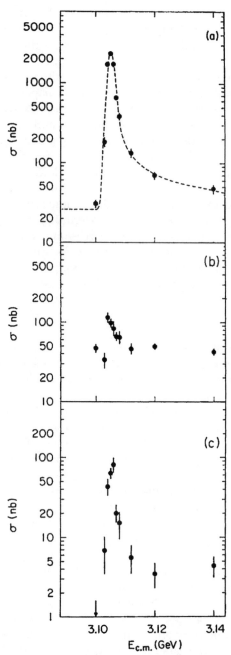

FIGURE 7.1. A "bump" in the energy spectrum. Shown are cross section versus energy plots for (a) multihadron final states, (b) e^+e^- final states, and (c) $\mu^+\mu^-$ final states (Augustin, Boyarski, et al. 1974, 1407).

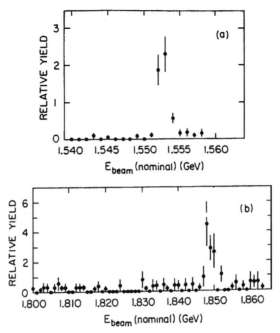

FIGURE 7.2. Results from data collected during "search mode" at SPEAR, showing (a) the previously identified $\psi(3105)$ and (b) a new resonance, $\psi(3695)$ (Abrams, Briggs, et al. 1974, 1454).

They named the new discovery $\psi(3695)$. Their paper displayed the results of the search both in the region of the previously announced $\psi(3105)$ (Figure 7.2a) and in the newly explored area between 3.6 and 3.71 GeV (Figure 7.2b). On the basis of the latter data, they comment, "A clear indication of a narrow resonance with a mass of about 3.70 GeV is seen. It should be emphasized that we have not yet scanned any mass region other than that between 3.6 and 3.71 GeV" (Abrams, Briggs, et al. 1974, 1454). (It may not be obvious *why* this should be emphasized, but there are good reasons for doing so, to which we will return.) The collaboration then collected more data in the region of that suggestive peak, using longer data-taking times, and smaller changes in beam energy. They obtained, again, an obvious peak at 3,695 MeV (Figure 7.3).

Consider now an imaginary experiment. You are given your own personal magnetic detector, like that used by the SLAC-LBL group, situated at an electron-positron collider. You are looking for resonances at a slightly higher energy, in the area around 9,000 MeV. And you find an excess of events beyond what you expect from background at around 9,010 MeV (see Figure 7.4). Specifically, you find 13 events, where you only expect 6 events based on the background distribution (which I have unrealistically

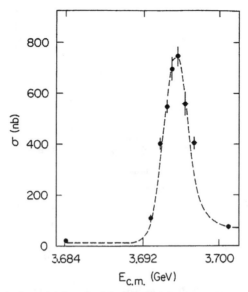

FIGURE 7.3. A closer look at ψ (3695) (Abrams, Briggs, et al. 1974, 1454).

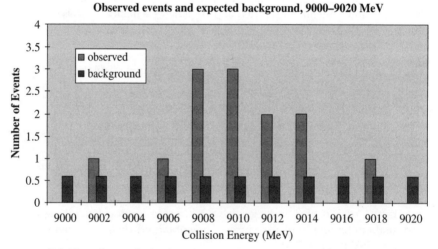

FIGURE 7.4. Experiment A. An imaginary experiment at an electron-positron collider, selected data.

represented as being absolutely flat in this region, to keep things simple). This is not by any means the sharp, narrow peak of the J/ψ discovery, but it has a statistical significance of 0.0088, which might be just enough for you to say you have evidence of something new.[4]

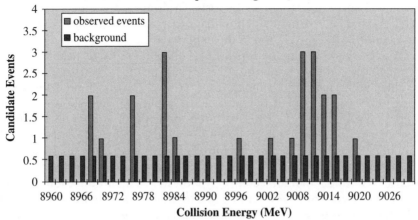

FIGURE 7.5. Experiment B. An imaginary experiment at an electron-positron collider, with results shown from all data collected.

The data given in Figure 7.4 suggest that you set your machine to produce collisions in the energy region between 9,000 and 9,020 MeV, and this statistical excess just turned up (we'll call that Experiment A). Suppose, however, that you *actually* scanned a larger range of energies, from 8,960 to 9,030 MeV. In that much larger energy range, you found 24 events, where the background distribution (still unrealistically flat) would lead you to expect 21 (Figure 7.5). This is an entirely different experiment, which we'll call Experiment B. The results of Experiment B seem much less impressive than those of Experiment A. The statistical significance for the Experiment B results is merely 0.28. That means that in a long run of repetitions of the experiment that was performed, excesses over background as large as this or larger would occur more than one time out of four, even if only background sources were present in this energy range. By looking in a larger energy range, you have increased the probability of a random upward fluctuation in background sources *somewhere* in that range.

Consider a rough characterization of this situation. If you actually performed Experiment B (i.e., you collected data over the entire range of center-of-mass energies from 8,960 to 9,030 MeV), then the statistical significance of the excess between 9,000 and 9,020 MeV is not 0.0088, and you should never have reported that it was. If you suspect that the slight excess in the 9,000–9,020 MeV range might be the first indication of a real resonance, then no matter how strong your hunch is, you ought to go back and collect more data in that area before announcing an important discovery. That is what the SLAC-LBL group did when they saw a slight "enhancement" above background expectations at approximately 3.2 GeV. Simply reporting the data from the region where the excess appeared misrepresents the experiment, and hence the statistical significance calculation. Since statistical significance is calculated on the basis of a given model of the experiment, its calculated value depends on how that experiment is described and is only as accurate as the description of the experiment.

We can see now why the SLAC-LBL collaboration went to the trouble to emphasize, in their paper on the $\psi(3695)$, that the data they presented in their initial scan of the range from 3.6 to 3.71 GeV showed the *entire* range they had looked at. They wished to avoid any suggestion that they were misrepresenting their results (even though they quoted no statistical significance for their result, and even though they went back and gathered additional data in the area of the apparent resonance, just to make sure).

The practices of the SLAC-LBL group reflect a widely accepted rule of statistical testing: when subjecting a hypothesis to a significance test, it is important to specify one's test statistic *in advance* of collecting the data against which the hypothesis will be tested. This *predesignation* rule has a venerable history in scientific methodology. A particularly clear-sighted early advocate of the predesignation rule was Charles Sanders Peirce:

> *The inference that a previously designated character has nearly the same frequency of occurrence in the whole of a class that it has in a sample drawn at random out of that class is induction.* If the character be not previously designated then a sample in which it is found to be prevalent can only serve to suggest that it *may be* prevalent in the whole class. . . . The induction only has its full force when the character concerned has been designated before examining the sample. (*CP* 6.409, *CE* 3, 313–16 [1878], emphasis in original)

Accordingly, when SLAC-LBL noticed a higher than expected frequency of hadron events at about 3.1 GeV, they took this only to suggest a possible effect at that energy. Then, in effect predesignating that same characteristic for their search in further data collection, they performed a new test that allowed them to claim justifiedly that they had uncovered evidence for a new phenomenon.

The predesignation rule, as applied to the kind of counting experiment discussed here, serves to eliminate tuning on the signal. However, if we take it seriously, it prohibits *more* than just tuning on the signal, for that practice is understood to involve somehow *using* one's knowledge of the data in order to tailor event selection cuts in favor of a particular result. This description is admittedly imprecise, but clearly it refers to something more specific than simply choosing cuts *after* having knowledge of the data. (The distinction resembles that between "use novelty" and "temporal novelty," which I will discuss later.)

As a general requirement, the predesignation rule requires the specification of all features of the test procedure that will make a difference to the probabilistic model of the experiment (the *sampling distribution*). This includes the definition of the quantity to be measured (the *test statistic*), the criteria for data selection (the *cuts*), and the size of the sample (the *stopping rule*).

I wish to acknowledge here at the outset a puzzlement that some feel over the predesignation requirement. Some will ask: "What difference could it make to the strength of an evidence claim whether the particular feature of interest to the experimenter was identified before or after the data was collected or examined? If I am told what the results of an experiment are, must I determine the epistemic state of each experimenter involved at each moment of the experiment in order to judge the worthiness of those results as evidence for some hypothesis? But surely such matters are completely irrelevant! Surely scientists do not respond to experimental results by asking who knew what and when did they know it?"

That scientists regard such information as evidentially irrelevant is asserted as fact, for example, by Colin Howson and Peter Urbach in their influential text on Bayesian inference. They charge that the Neyman-Pearson theory of inference has "subjective elements" arising from the fact that Neyman-Pearson statistical calculations depend on an experimenter's personal choices that affect the outcome space (their example concerns the choice of stopping rule, but they could just as well refer to the choice of test statistic): "A significance-test inference, therefore, depends not only on the outcome that a trial produced, but also on the outcomes that it could have produced but did not. And the latter are determined in part by certain private intentions of the experimenters . . . scientists would not normally regard such personal intentions as proper influences on the support which data give to a hypothesis" (Howson and Urbach 1993, 212).

Yet, in the controversies over the evidence for the top quark at CDF, we find just such a concern with who knew what, when they knew it, and what aims they had in making decisions based on that knowledge. Nor surprisingly, in such a vast collaboration, the answers to such questions are exceedingly difficult to find.

In what follows, I aim to explicate why such facts are relevant to evidential judgments from the error-statistical standpoint and also to discuss how experimenters can address situations in which these facts cannot be determined with confidence. In the process, I will show how, notwithstanding the relevance of "personal intentions," the error-statistical concept of evidence is objective in a very strong sense.

In part, my argument will follow the path laid out by previous defenders of predesignation in statistical inference, such as Giere (1969), Armitage and Barnard (see the discussion in Savage 1962), and Mayo (1996, esp. ch. 9), who have defended the rule not as a methodological absolute but as a rule the violation of which can lead to erroneous conclusions in certain circumstances. However, I wish to focus in particular on the requirement of predesignating the criteria for data selection. I seek to explicate precisely why predesignation facilitates an error-statistical evaluation of evidence, when predesignation is optional, and what can be done to evaluate evidence in the face of a failure to predesignate.

What follows thus constitutes an exercise in metamethodological criticism from within the error-statistical theory of evidence. In engaging in such criticism, I have two goals: (1) to demonstrate the ability of the error-statistical theory to explicate in a coherent way a common methodological intuition, to clarify why that intuition seems reasonable, and to identify its limitations; (2) to indicate the methodological consequences of adopting the error-statistical theory of evidence.

The first goal is important insofar as a necessary condition for any adequate theory of evidence is that it must yield coherent explications and criticisms of methodological intuitions. Only a theory that is able to meet this standard can serve as a basis for solving problems encountered in experimental practice. Of course, if other theories are able to do as well as the error-statistical theory at this task, then my argument will provide no reason to *prefer* the error-statistical theory, but only to show that one ought not eliminate it on this score. Because I do not attempt a comparison with other approaches here, I will have to content myself with the weaker claim that the error-statistical theory satisfies a necessary condition of adequacy with respect to this particular methodological problem. I leave it to others to demonstrate the ability of Bayesian or likelihood-based approaches, for example, to explicate the problem here considered, if that is indeed possible.

This brings me to the second goal. One point of the present essay is that adopting the error-statistical theory of evidence has consequences for experimental practice. As Deborah Mayo and Michael Kruse have recently shown with respect to the debate over stopping rules, differing principles of inference have different methodological consequences (Mayo and Kruse 2001).[5] One can thus judge theories of evidence by their consequences for scientific practice. Does adopting the theory at hand yield practices that help to achieve one's scientific aims? This question must be addressed if philosophers are ever to remedy their past failure to articulate theories of evidence that are relevant for scientists' concerns (this failure is discussed in Achinstein 2000; 2001, ch. 1).

3. THE SLT ANALYSIS

Recall that in the soft-lepton-tagging counting experiment, the aim was to identify $t\bar{t}$ events in which the W produced by one top quark decayed hadronically, producing jets, while the W from the other member of the $t\bar{t}$ pair decayed leptonically, producing a high-momentum lepton (e or μ), and a neutrino. A first selection of these "lepton + jets" events yielded 52 candidate events, but with an expected background of 46. To further eliminate background and discriminate any signal-over-background excess,

CDF pursued two strategies for identifying lepton + jets events containing b quarks, which would be produced alongside W bosons, in the initial decay of the $t\bar{t}$ pair. The SLT experiment sought to identify such events by picking out the relatively low-momentum ("soft") lepton produced in the b decay process.

The debate over the SLT analysis (previously discussed in Chapters 4 and 5) concerned one of the cuts used in the SLT search. Collaboration members questioned not only whether the choice for the minimum transverse momentum cut on the soft lepton met the standards that had been set for optimizing counting experiment cuts, but also the timing and rationale of the choice. Let us call the former *optimization* questions, and the latter *chronological* questions (as they concern the sequence of actions and events within the collaboration).

With respect to questions of optimization, critics of the SLT analysis charged that placing the minimum p_T cut at 2 GeV/c instead of 4 GeV/c opened the SLT search up to a signal-to-background ratio that was too low, as a result of the fact that background could be expected to dominate signal for events in which the soft lepton had transverse momentum in the 2- to 4-GeV/c range. Chronologically, these critics were suspicious of claims that placing the cut at 2 GeV/c was decided in advance of any knowledge regarding the difference the choice would make in the number of candidate events yielded by the SLT search. In Chapter 5, I discussed the extent to which such suspicions could be substantiated, and the kinds of facts deemed relevant to the debate.

My purpose here is to explore the reasons why such chronological matters would be relevant to the evaluation of evidence at all. In doing so, I wish to dispel two mistaken ways of approaching this question. The first mistake would be to insist that any consideration of "chronological" matters is an irrelevant artifact of the indoctrination into classical statistics (with its blind adherence to the predesignation rule) to which experimental scientists are subjected. The second mistake is to insist that all violations of the predesignation rule are epistemically vicious, precluding the possibility of learning anything from the experiment at hand. In order to show that both extremes are mistaken, let us consider more carefully the debate in CDF over the SLT results.

3.1. The Development of Soft-Lepton Tagging

To understand the disagreements over soft-lepton tagging, it will be helpful to review briefly its historical development. The SLT search was a descendant of an earlier search that selected top decay candidates by identifying a soft muon. CDF had employed that earlier form of soft-muon tagging in the 1992 *PRD* paper in which they reported the lower limit on the top mass to be 91 GeV/c^2 (Abe, Amidei, et al. 1992b). But the two searches differed in some important respects. Where the SLT algorithm used for analyzing the run Ia data was applied to a lepton + jets sample with three or more jets, the soft-muon search published in 1992 used a sample of lepton + two-jet events. The 1992 results also excluded information from the central muon upgrade and central muon extension chambers, since these parts of the detector had not yet been installed. The cuts for identifying muons were much looser in the earlier version of the algorithm, with background rejected primarily by kinematic requirements: It was required that the muon not be in one of the two highest-E_T jets, since these would almost always be from the decay of the W rather than the b if the top is not very heavy (Campagnari, personal communication). Both searches placed a lower limit on the transverse momentum of the soft lepton at 2 GeV/c.

As described previously (see Section 6.2 in Chapter 3) the SLT search was primarily the work of Claudio Campagnari and Avi Yagil. Campagnari began work on soft-lepton tagging in 1989 and had worked out enough of the basic strategy to tag soft muons in the 1988–9 data for the 1992 *PRD* mass limit paper just mentioned. Seeing that the strategy worked with that data, Campagnari decided that a refined version could be applied to the data expected from run Ia. Meanwhile, Avi Yagil had been working on identifying low-momentum photons and became interested in the top analysis. Campagnari and Yagil began to work together about a year prior to the beginning of Ia. By that time, the SVX b-tagging group had become rather large. Both physicists felt that the potential of the soft-lepton search was being neglected, leaving them an opportunity to develop their own approach rather than simply becoming two more piece-workers in the increasingly crowded SVX group (Campagnari, personal communication).

The kinematic cuts used in the soft-muon search in the 1992 search paper were not expected to be effective for the Ia data, since they had been useful only for a light top. With the increased data of Ia CDF expected to be sensitive to a much heavier top. So Campagnari took another look at the 1988–9 data to develop tighter muon cuts using the central muon detectors. Very early data from Ia was used to develop cuts for data coming from the newly installed central muon upgrade as well (the central muon extension was in place but not yet well understood, and hence would not be used until after the Evidence paper). Yagil extended his earlier work on soft photon identification to develop an algorithm for identifying soft electrons, relying primarily on the 1988–9 data. Information that had not been available during 1988–9 was incorporated,[6] again taking very early data from run Ia as a guide to developing the algorithm (Campagnari, personal communication).

The soft-lepton strategy is most effective for a relatively light top quark. If the mass of the top quark were only slightly greater than that of the W boson that is produced when the top decays, then there would be little energy available to give the accompanying b quark a "kick," so that the lepton that subsequently results from the decay of the *b* will not be expected to have a great deal of momentum. The more the mass of the top exceeds that of the *W*, the more energy will be "left over" from its decay, resulting in a more energetic *b*, which in turn will yield a more energetic lepton when it decays.

When the 1992 top search paper was written, CDF physicists thought it quite possible that the top would turn out to be only slightly heavier than the *W* ($M_W = 80 \text{ GeV}/c^2$). In the 1992 search paper, CDF noted that "for top-quark masses slightly above the *W* mass, the kinetic energy liberated in the top-quark decay is small, so that the P_T of the *b* quarks in the laboratory frame is small and the probability for reconstructing the *b* quark as a separate jet is not large" (Abe, Amidei, et al. 1992b, 3942). In other words, if the mass of the top were only slightly above that of the *W*, then one could expect that identifying *b*s from top quarks by soft-lepton tagging would be more effective than doing so by identifying *b* jets (which is what SVX *b*-tagging does). According to Claudio Campagnari, the SLT search would have yielded a very large top signal if the top had been lighter. Since the top quark turned out to be rather massive, the SLT results took a back seat to the SVX search, serving primarily as a kind of consistency check. Although the SLT search would not be likely to turn up a very significant signal for such a heavy top on its own, it was important that the findings in the SLT search turned out to be consistent with what one would expect given the SVX results (Campagnari, personal communication). However, the quantitative significance of the SLT channel in the Evidence paper (0.041) is at least comparable to that of the SVX channel (0.032). Critics in CDF charged that the significance level for the SLT results was misleading.

3.2. Issues in the Debate over Soft-Lepton Tagging

Next I will consider in detail some of the points on which critics and defenders of the SLT analysis disagreed. My concern here is not to try to settle the debate but to examine some of the issues on which the debate turned and discuss their relevance.

The background to the SLT search was broken down across the spectrum of the P_T of the tagged leptons, as was the expectation for signal events for various masses of the top quark (see Figure 4.2). Critics emphasize two facts about these distributions. First, while the expected signal in the 2- and 4-GeV/*c* bins is sizable for a light top (e.g., $M_{top} = 120$ or 140 GeV/c^2), it is quite small for a top with a mass as heavy as 160 or 180 GeV/c^2 (recall that the mass estimate reported in the Evidence paper is about 174 GeV/c^2). Second, the expected background in the 2- to 4- and 4- to 6-GeV/*c* bins is quite large compared to the other bins, and four of the seven tagged events have leptons with momenta in these two bins, three of them in the particularly high-background 2- to 4-GeV/*c* bin.

The first point is meant to indicate that, in light of the top mass estimate, the tags in the 2- to 4-GeV/c bin are unlikely to be actual top events because one does not expect many tags at this P_T level for such a heavy top. In Figure 4.2, the expected number of tagged top events in the SLT search for a 180-GeV/c^2 top is only 1.05 ± 0.18, compared to seven events tagged in the data. In the 2- to 4-GeV/c bin, about 0.2 tags are expected, compared to three found in the data.

However, defenders of the analysis replied, it was not known in advance that the top would be as heavy as it turned out to be. Hence, the search needed to be sensitive to a somewhat low-mass top quark as well. In any case, they pointed out, whatever Monte Carlo estimates may tell us to expect on the basis of a hypothesized top mass, the number of tags in the SLT search does significantly exceed what is expected from the *more reliable* background prediction.

This brings us to the critic's second point. Because the expected background is high in the 2- to 4- and 4- to 6-GeV/c bins, the four tags in those bins are more likely to be background than those in the other bins (note how low the expected background is in the bins occupied by the other three tags). Thus, the significance calculation that is based on simply summing the background across all the bins and then looking to see how far the observed number of tags departs from that total arrives at an artificially low significance estimate because it treats each event as making an equal contribution to the total. By this method, the same significance level would obtain if all seven events had been tagged in the very low–background region above 8 GeV/c. However, in that case, each tag individually would be much less probable assuming only background.

SLT's critics suggested two alternatives. One alternative would be to change the way that the significance of the SLT search was calculated. For example, one could give a partial breakdown according to P_T, calculating the significance of the events with $P_T < 6$ GeV/c and with $P_T \geq 6$ GeV/c separately and then combining the results. This would give a more conservative significance calculation (Yeh, personal communication). The other suggestion was to move the cut on P_T from 2 to 4 GeV/c. This would simply eliminate the largest source of background to the search (along with three of the candidate events). On either approach, the significance calculation for the SLT results (and for the counting experiments as a whole) would have yielded a higher value.

Because my aim here is not to arrive at a definitive resolution of the question, I will not undertake to present the entire argument, but will mention only two lines of defense of the SLT analysis's use of the 2-GeV/c cut before focusing on the part of the debate that has the most direct bearing on the methodological issues under consideration.

One reason given for the 2-GeV/c cut was that the systematic errors on the background are worse for the background in the high-P_T region, where the sources of background are not as well understood. By keeping the cut low, in a region where the systematic errors are smaller, the systematic error on the total background is proportionally smaller (Campagnari, personal communication); therefore, the overall result is presumably more reliable.

Claudio Campagnari further argued that, because the quantity $S/\sqrt{B}$, where S is the size of the expected signal and B is the size of the expected background, remains essentially constant as the placement of the cut varies, the choice between 2 and 4 GeV/c is a judgment call. But the critics of the SLT analysis pointed out that increasing the threshold to 4 GeV/c improved the ratio of signal to background, S/B, according to the SLT group's own calculations (Campagnari, Farhat, et al. 1993a). They concluded that the only justification for keeping the cut at 2 GeV/c would be if one were searching for a top quark with a mass in the neighborhood of 100 GeV/c^2.

What was the lower limit on the mass of the top quark prior to the Evidence paper analysis? Answering this question both illuminates the SLT debate and shows how some uses of data in violation of the predesignation rule were judged harmless by CDF, while they regarded others as threatening the epistemic success of the enterprise.

Here is an excerpt from the opening paragraph of the Evidence paper:

[D]irect searches at the Tevatron collider at Fermilab have placed a lower limit on [the top quark's] mass of 91 GeV/c^2 at the 95% confidence level [this refers to the limit set by CDF in Abe, Amidei, et al. (1992b)]. The limit was recently extended to 131 GeV/c^2 [this refers to the limit set by D-zero in Abachi, Abbott, et al. (1994)]. These searches assume the top quark decays predominantly to a W boson and a b quark. Limits independent of the decay mode come from measurements of the width of the W boson and require $M_{top} > 62$ GeV/c^2 at 95% C.L. Global fits to precision electroweak measurements yield a favored mass of $M_{top} = 177 \pm 11^{+18}_{-19}$ GeV/c^2. (Abe, Amidei, et al. 1994b, 2968)

Here CDF mentions several different judgments regarding the top mass, but they express no particular attitude regarding any of them. For the most part, this passage serves simply to inform the reader on the "state of play" in the world of top-quark searches.

Certainly, CDF granted the 91-GeV/c^2 limit some importance, since it was their own result. Naturally, they optimized their search for masses higher than this, so as not to repeat previous work. Although the 91-GeV/c^2 limit is dependent on the assumption that the top decays to a W and b, the search that is used in the Evidence paper also relies on this assumption. CDF would thus gain nothing by sticking to the decay-independent limit of 62 GeV/c^2 from the width of the W. The "favored mass" from electroweak measurements is not really a mass limit at all, and certainly the search used for the Evidence paper was sensitive to a top considerably lighter (or somewhat heavier) than 177 GeV/c^2. Subsequent to being mentioned in the opening paragraph, the 131-GeV/c^2 D-zero limit is ignored in the remainder of the Evidence paper. This is consistent with the commitment of both CDF and D-zero physicists to keeping the searches of the two collaborations as independent of one another as possible.

Toward the end of the discussion of the dilepton search that occupies section four of the Evidence paper, CDF mentions again their own previously published limit of 91 GeV/c^2 and notes that their counting experiments "concentrate on top masses in the range 120 GeV/c^2 and above where the event selection is reasonably efficient. This leaves a hole" in the 91- to 120-GeV/c^2 range (Abe, Amidei, et al. 1994b, 2982).

To fill this gap, the Evidence paper uses a dilepton analysis to improve the lower limit on the top mass. In August 1993, Paul Tipton had presented a mass limit of 113 GeV/c^2 at the Lepton-Photon conference (see Section 1 in Chapter 4). That limit was based on a search of the run Ia data in the dilepton channel alone. The Evidence paper improves on this result by using the 1988–9 data to reach a new limit of $M_{\text{top}} > 118$ GeV/c^2 at the 95% C.L. (Abe, Amidei, et al. 1994b, 2982).

This mass limit is obtained by means of a dilepton analysis, but it is not the same dilepton analysis that is used in the top search that serves as the basis for CDF's claim of evidence for the top quark. That search would not be sensitive to a top with mass less than 120 GeV/c^2 because it requires two jets. For a light top (i.e., one with a mass just slightly above the 91-GeV/c^2 limit), the jets from the b quarks produced by $t\bar{t}$ decay are not likely to have enough energy to be detected. In order to search in the mass range between 91 and 120 GeV/c^2, CDF removes the two-jet requirement, which does not let in any candidate events beyond the two found *with* the two-jet requirement in the Ia data, nor any events from the 1988–9 data. With the assumption that the two events selected by this looser dilepton search are both signal (a conservative assumption for the purposes of setting a mass limit), CDF derives an upper limit on the $t\bar{t}$ production cross section ($\sigma_{t\bar{t}}$). This estimate is weakly dependent on the actual top-quark mass because its calculation requires an estimate of the detection efficiency, which is somewhat higher for higher mass top quarks. By plotting the estimates for $\sigma_{t\bar{t}}$ on the same table as a plot of the strong theoretical dependence of $\sigma_{t\bar{t}}$ on the top mass, and then identifying the point at which the two curves intersect, CDF places a lower limit of $M_{\text{top}} > 118$ GeV/c^2 (at the 95% C. L.) on the top mass (see Figure 7.6) (Abe, Amidei, et al. 1994b, 2982).

The Evidence paper, then, contains two top-quark searches, using the same data. The first is a low-mass search in the dilepton channel, in which dilepton events are identified without the two-jet requirement. The lower limit on the mass for purposes of this search is 91 GeV/c^2, but the mass limit that emerges from the results of the low-mass search is 118 GeV/c^2. That limit in turn is used for the purposes of the second search, which includes the two lepton + jets searches and the dilepton search with the two-jet requirement. That the second set of counting experiments was considered to

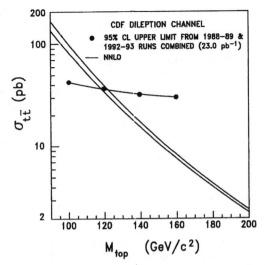

FIGURE 7.6. Results of the low-mass dilepton search. The upper limit on the $t\bar{t}$ production cross section based on the data (data points) overlaid with the theoretical lower bound and central value of the same (solid lines) (Abe, Amidei, et al. 1994b, 2982).

be relevant only in a mass range beginning at approximately 120 GeV/c^2 is confirmed by the efficiency calculations for the three counting experiments in question. Efficiencies depend on cross sections, and the cross section for top production depends on the mass of the top, so at every point that an efficiency is calculated for top events, one is presented with a range of values for different top masses. The format in which CDF presents these efficiency calculations in the Evidence paper for the two-jet dilepton search, the selection of $W + \geq 3$ jet events, the SVX search, and the SLT search is always the same: values are given for top masses of 120, 140, 160, and 180 GeV/c^2. In addition, numerous auxiliary calculations presented in the section of the Evidence paper describing the lepton + jets search (Section V) indicate that the range of interest is from 120 to 180 GeV/c^2, including the calculation of the expected P_T spectra of SLT tags from top events (Figure 4.2).

Apparently, then, a top quark with mass less than 120 GeV/c^2 had been decisively ruled out, which suggests that there was really no reason to keep the SLT low-P_T cut at 2 GeV/c. This cut had been put in place for the purposes of the 1992 top-quark search, which had been developed to look in the mass range of 80–100 GeV/c^2. With the lower limit raised well above this range, CDF no longer had a reason to retain a sensitivity to a 100 GeV/c^2 top, and hence had no reason for the 2-GeV/c P_T cut. Instead, the more pressing issue was to eliminate a significant source of background to the overall search.

One might view this issue differently, however. Even though it is true that in the Evidence paper the dilepton analysis raises the lower limit on the top mass to 118 GeV/c^2, this new limit is based in large part on the Ia data itself, the very data in which the top signal itself is being searched for. On the face of it, it would appear to be a potential violation of the predesignation rule to use this new limit, inferred from the small number of candidate events found with the looser dilepton search, as a reason to revise the cuts upward for a higher mass top. From this perspective, the searches from the two runs are to be treated more independently, and the SLT search has to begin on the Ia data where it left off on the 1988–9 data. Claudio Campagnari expressed just such an opinion:

The 1992 paper [Abe, Amidei, et al. 1992b] has a 2 GeV threshold on SLT. At the time we were sensitive only [to] top masses up to approximately 100 GeV. In this mass range the P_T of b-quarks is very soft, so we wanted to set the threshold as low as possible. The run Ia analysis picked up where the 1992 analysis ended, so we also needed to be sensitive to top quarks around 100 GeV. That's partly why we used 2 GeV once again. (Campagnari, personal communication)

Recall that the SLT analysis was not unchanged from the 1992 search. It included new channels, and overall the cuts were much tighter. For Campagnari, maintaining continuity in the search's sensitivity from one run to the next was an important consideration.

There is something peculiar here. The 91-GeV/c^2 lower limit on the top mass from earlier data suggested that the cuts for the search on the Ia data should be reasonably efficient for a top quark with a mass just slightly higher than that earlier limit. Yet, aside from the SLT search, the cuts used for the search in the Ia data were not very efficient for a top quark with a mass below 120 GeV/c^2. It is true that the dilepton search without the two-jet requirement fills in this gap in the sensitivity of the analysis to a lighter top quark. Since, however, the low-mass dilepton search was carried out on a data sample that included the Ia data used for the three main counting experiments, we may well wish to inquire about the relationship between the former search and the latter. We certainly ought to ask about this if our concern is over the predesignation requirement. Was the low-mass dilepton search done first, from which CDF inferred that they need not make their analysis efficient for a light top quark (and then chose their cuts accordingly), or did CDF set the cuts for the counting experiment, and then, noticing the gap in the experiment's sensitivity, decide that they ought to fill that gap?

What actually happened was more complicated than either of these scenarios. The cuts for selecting W + jets candidate events, the SVX tagging algorithm, and the dilepton search with the two-jet requirement were well-defined prior to the decision to employ a low-mass dilepton search. Henry Frisch described the low-mass search as an afterthought, regarded as necessary in light of the fact that the other searches were generally not sensitive for top quarks with mass below 120 GeV/c^2 (Frisch, personal communication).

The case of the SLT search is more complicated. Debate over the P_T cut continued at a time when the SLT authors had both the low-mass search results and the results of the initial SLT algorithm at hand. That this fact contributed to the problems that confronted the SLT tagging analysis was clear to Claudio Campagnari (Campagnari, personal communication). On the one hand, if the SLT analysts chose the 2-GeV/c cut, they could be charged with being influenced by the three tags in the 2-GeV/c bin, and hence of tuning on the signal. On the other hand, by choosing the 4-GeV/c cut (as appropriate for a heavier top quark), they would be subject to the charge of having let the results of the low-mass top search influence their choice of cuts. Either way they would be violating the predesignation rule.

Perhaps it is not clear why the latter type of scenario (first setting a mass limit, then claiming a positive result, based on a single set of data) would pose any difficulty, or undermine the reliability of any resulting evidence claim. Consider a more extreme hypothetical case, however. Suppose that, rather than using just the dilepton search to set a mass limit of 118 GeV/c^2, CDF had pulled out all the stops, combining all three counting experiment channels to set a mass limit of, say, 150 GeV/c^2 (assuming that they could have set such a limit using their data). Failing to find a significant excess of candidate events in this mass range, suppose the collaboration tightened its cuts, optimizing them for the range from 150 to 180 GeV/c^2. Supposing that with these much tighter cuts the background is greatly reduced while signal events are mostly retained, the significance calculation drops precipitously. Worse, imagine that CDF had used the very same events and the same mass analysis employed in the Evidence paper, and, estimating the mass of the top quark to be approximately 174 GeV/c^2, reoptimized their counting experiment with very tight cuts appropriate for a search in the 170- to 180-GeV/c^2 range. This would be a case of shooting the arrow first and painting the target afterward.

Although choosing to increase the P_T cut on the basis of the dilepton mass limit would certainly be much less extreme than either of these hypothetical cases, it was at least arguable that choosing the higher value on the basis of that limit would serve to bias the test of the top-quark null hypothesis by using the absence of candidate events in the low-mass range to justify cuts tailored to the high-mass range. Such a problem did not face the other searches, which were tailored to a higher mass top (in the range from 120 to 180 GeV/c^2) independently of that search.

Yet the debated adjustment to the SLT analysis differed from these hypothetical cases in another respect as well: We supposed in the imaginary scenarios that searching for candidates in a smaller region of the sample space yielded an exaggeration of the statistical strength of the excess by eliminating a great deal of background while keeping most of the candidate events. In the case of the SLT P_T cut, it was known already that, although the background expectation would decrease, the loss of candidates would be greater. There was no risk that the change of cuts for a higher mass top would artificially *reduce* the significance level of the results, because the collaboration already knew that it would make the significance calculation higher.

Clearly, there is more to this debate than simply finding out whether the predesignation rule was followed or not. As previous chapters have shown, the occasions when groups of CDF physicists strictly adhered to the rule were rather rare. Does this mean that CDF's results as a whole are to be regarded with suspicion? Should we declare their entire enterprise an evidential failure? Should we instead declare the predesignation requirement irrelevant? We should make neither declaration. To see why, we need a better understanding of the rationale for the predesignation rule.

4. WHAT EXPERIMENT DID WE JUST DO? BIAS, TUNING ON THE SIGNAL, AND THE REFERENCE CLASS

The officially quoted significance level for CDF's SLT search was 0.041. Let us suppose that it is true, based on CDF's assumptions, that if the null hypothesis were true, and one were to repeat infinitely many times an experiment using the same detector, using the same cuts, collecting the same amount of data, and so on, one would get as many as seven candidate events or more only 4.1% of the time.

However, if we know that the cuts used in this case were chosen in such a way as to exaggerate the apparent significance of the results, then we have statistically relevant information about the experimental procedure used to reach these results. Specifically, the procedure followed – including now the procedure for choosing the cuts – has different error characteristics than the procedure on which CDF based their significance estimate of 0.041. That estimate was based on the specification of a *reference class*, with respect to which the probability is calculated. The reference class used in calculating a significance level is a hypothetical population of repetitions of a certain experiment. The appropriateness of a particular reference class is therefore in part a matter of the testing procedure that has in fact been used. If experimenters know that they have tuned their cuts on the signal, then a reference class that would otherwise be appropriate would be the wrong reference class for calculating that probability. For example, if it were known that the SLT cuts had been chosen specifically in order to increase the value of the test statistic, and yet the statistical significance calculation were performed without taking this information into account, then the reference class chosen for purposes of that statistical assessment would not be correct.

Under such circumstances, the reference class used for calculating significance would not be appropriate because it would not be *homogeneous* with respect to the experimental outcome. The reference class would fail to be homogeneous because it could be further partitioned according to a factor statistically relevant to the outcome. If the cuts had been tuned on the signal, then the method of selecting cuts would be statistically relevant.

Wesley Salmon proposed a concept of homogeneity for use in his theory of statistical explanation that might seem promising here (1984). However, it is too stringent for use in cases of experimental inference. In Salmon's account, a reference class is homogeneous with respect to an explanandum partition just in case that class cannot be partitioned in any manner whatsoever relevant to the occurrence of any member of the explanandum partition. But in an experiment to test a hypothesis, the fact under investigation (whether or not top quarks are being produced in the generated collisions, for example) would constitute such a relevant (but presumably unknown) factor. Consequently, in precisely those cases where experimental inquiry is needed, such a requirement could not in principle be known to be satisfied. For a fully objective error-statistical theory of evidence, however, one feature of Salmon's concept should be retained: it does not suffice to satisfy the homogeneity requirement that one be unaware of any statistically relevant partition. The challenge is thus to articulate a notion of homogeneity that is objective, rather than epistemic, yet is suitable for purposes of statistical *inference.*

Such a notion remains to be fully specified. I can explicate the concept only in part here, and with less precision than I would like. A step toward a full explication is to stipulate the following *necessary* condition: A reference class A used in calculating the probability of an outcome E is homogenous with respect to E only if there is no factor B, under the control of the experimenter and present in that instance of the experiment that resulted in E, such that $p(E|A) \neq p(E|A\&B)$.

In calculating a significance level, one first supposes that the null hypothesis is true. One then asks, "Suppose I were to perform an infinite sequence of repetitions of this experiment, how often would I get such a result as this?" A great deal turns, however, on how *this experiment* is specified. Consider two possible sequences of experiments.

Sequence A consists of repetitions of the SLT top-quark search, each of which collects the same amount of data using an experimental setup identical to that used by CDF. In each member of A, the experimenters have a preference for a large excess of the number of candidate events over expected background, but they do not know which choice of soft-lepton momentum cut will yield a larger excess at the time that they make that choice. B is also a sequence of repetitions of the SLT top-quark search, each of which collects the same amount of data using the same experimental setup, and so on. But for each performance of the experiment in sequence B, the choice for the soft-lepton momentum cut was *caused* partly by the preference of the experimenters for a large excess of candidate events over the expected background (this is made possible perhaps by the experimenters' ability to examine the data before making that choice).

What is wrong with the experiments in sequence B? The type-B experiment is not an inherently bad experiment, but one for which calculating a significance level would be practically impossible.

For an experiment in A, a reliable model of the experiment yielding a probability distribution is available – that is the model used by CDF in their significance calculation (see Figure 7.7). Such probabilistic models of the experiment are a prerequisite for significance calculations. No such reliable model is available for an experiment in B, however. Such a model would have to take into account information about the intensity of the experimenter's motivation to increase the value of the test statistic, the magnitude of the desired enhancement, and so on. Were such models available (see Figure 7.8), tuning on the signal would not pose a problem. If you chose your cuts to maximize the value of the test statistic, you would simply need to remember to use the probability distribution for a type-B experiment rather than a type-A experiment.

The predesignation rule gets its force from the difficulty of generating such a distribution. It is not the act of peeking itself that is troublesome. What poses a difficulty to the experimenter is her inability to represent reliably the effect that peeking has on the probability of various experimental outcomes (cf. Mayo 1996, ch. 9). We simply cannot, practically speaking, generate a reliable probability distribution for the experiment in which the experimenter's zeal enters into the determination of the test statistic and the probability of getting an apparently positive outcome. Observing the predesignation rule helps to secure the conditions necessary for producing reliable probabilistic models of the experimental test, which are in turn necessary for generating significance calculations. When cuts are turned on the signal, significance calculations become unavailable, and so, it would seem, do severity assessments.

Note that the kind of failure brought on by tuning on the signal is not necessarily evidential failure. It *could* be the case that, although one has tuned one's cuts on the data (i.e., although the choice of test statistic is correlated with one's preference for a certain outcome), the resulting test is nevertheless severe, and the hypothesis in passing that test receives evidential support. But such evidential success is no guarantee of epistemic success. By tuning on the signal, the experimenter may put himself in a situation in which he cannot determine what the correct probability distribution *is* for the test being performed and, hence, cannot come to be justified in believing that the test carried out is severe. Although the nonpredesignated test is evidentially successful, the experimenter is not justified in believing this, so the experiment is an epistemic failure.

Probability distribution, number of candidate events (X$_{SLT}$)

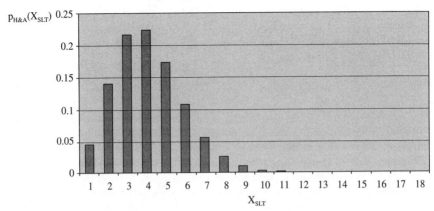

FIGURE 7.7. An approximation of the probability distribution used for calculating the SLT significance in the Evidence paper: A population of performances (sequence A) of the SLT experiment, in which experimenters' preferences *are not* causally related to choice of SLT p_T cut.

Probability distribution, number of candidate events (X$_{SLT}$)

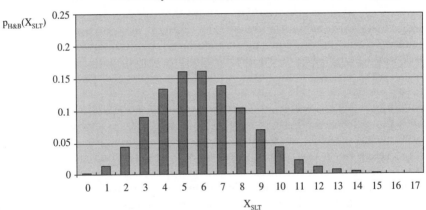

FIGURE 7.8. An imaginary probability distribution for the SLT experiment on the assumption that the SLT cuts were tuned on the signal: A population of performances (sequence B) of the SLT experiment, in which experimenters' preferences *are* causally related to choice of SLT p_T cut.

Violating the predesignation rule need not prevent epistemic success. One such scenario would be the case where, although predesignation was violated, the experimenter knows herself to have no strong preferences that might have influenced the choice of test statistic. Alternatively, she might recognize that she has such preferences, but have a justified belief that her preferences had no causal effect on her in the circumstances under which she chose the test statistic. She trusts herself, and is justified in doing so. Describing debates among CDF members over the propriety of looking at data before all of the decisions regarding the search algorithms had been made, Paul Tipton cast the issues in terms of the potentials for both self-knowledge and self-deceit:

[People working on the lepton + jets analysis] really did agree to not look at the data until distinct milestones, before major conferences...and I think that was good....There's some people that think that that's sort of preachy and silly because...it's as if we're virgins and we can't have sex before we're married – this attitude toward the data that we have to be pure. I really disagree with that. Some disagree with me because they think that you can look at the data and basically trust yourself to do the right thing, that if there's enough people watching, people won't be pathologically dishonest....I think that if you look at the history of science and all the wrong measurements made by very good people... the ability to fool yourself is pretty subtle. (Tipton 1995)

Knowing that one's preferences have not played a causal role in determining one's choices may be possible, even if difficult. Let us suppose that one attains such an enlightened state, so that one achieves epistemic success: although the data were known when the cuts were chosen, the test the hypothesis passes is severe, and one knows that this is so based on the knowledge that the probability model of one's testing procedure is accurate. (Less plausibly, an experimenter's preference for a certain outcome *is* statistically relevant to the choice of test statistics, but the experimenter is able to measure this effect and incorporate it into the probability distribution for the experiment.)

Under such conditions, the violation of predesignation may still present an obstacle to achieving *experimenters' success*. The experimenter may know that he has no preference with regard to the outcome, that he has such a preference but it has played no causal role in determining the test statistic, or that the statistical effect of his preferences is of a certain magnitude which, when incorporated into the model of the test procedure leaves him with a test that remains severe. In each of these cases, he may still face a difficulty in justifying to others that the test in question was indeed severe, and that the experiment was an evidential success with respect to the advertised hypothesis. For the audience interested in evaluating his experiment, knowledge of the relevant facts concerning his motivations will be elusive, and hence so will the knowledge of the severity of his test. As Peirce notes on a related point:

The drawing of objects at random is an act in which honesty is called for; and it is often hard enough to be sure that we have dealt honestly with ourselves in the matter, and still more hard to be satisfied of the honesty of another. (*CP* 2.727, *CE* 4, 428 [1883])

Distinctive methodological issues arise from the need to communicate with or persuade one's peers. Joseph Kadane and Teddy Seidenfeld make this point in their discussion of randomization in terms of Bayesian decision theory and statistics (Kadane and Seidenfeld 1999). In their decision-theoretic treatment, randomization (e.g., in the assignment of subjects to control and treatment groups in a clinical trial) reduces the cost of evaluation for the reader of the experimental report by rendering expensive information about the experimenter's utilities irrelevant (although methods other than randomization can also accomplish this aim). However, they find that randomization, for a Bayesian, is not called for when an experimenter is not engaged in attempting to persuade others of a result. In Kadane and Seidenfeld's terminology, one can give a rationale for randomization in "experiments to prove," but not in "experiments to learn." Although Kadane and Seidenfeld discuss randomization rather than predesignation, my view is similar in that I regard predesignation as a useful means to achieving a particular methodological aim the relevance of which arises because of the need to persuade another. However, there are also important differences. In my error-statistical analysis, the aim is not simply to lower the cost to the reader of evaluating the argument but to allow the reader to rule out a source of error that might otherwise remain troubling. Furthermore, because of the potential for fooling one's self, predesignation is typically valuable in experiments to learn, just as in experiments to prove. That the need to communicate with and persuade one's peers generates distinct methodological demands when viewed from such divergent evidence-theoretic perspectives as error statistics and Bayesian statistics is itself worth noting.

Are there steps that one can take to achieve experimenters' success in spite of having violated predesignation? I believe there are, and will explain why in the next section.

5. AFTER-TRIAL ATTEMPTS TO RESCUE THE EXPERIMENT

Sometimes an experimenter finds herself in a position where predesignation has been violated, and her scientific colleagues may reasonably be expected to raise questions about the role that some of her preferences may have played in determining certain test parameters. What can she do? A number of possibilities may be suggested:

1. Throw the data away and start over. The analysis has been tainted by the suspicion of bias, and experimenters' success is unobtainable.
2. Keep the data, but throw away the analysis. Try to come up with a new set of cuts in a way that will put any worries about bias to rest.
3. Keep the data and the analysis, but throw away the statistical significance assessment, as it simply creates a misleading impression of

precision. Simply present the number of candidate events and the background prediction, and leave the readers to draw their own conclusions about what it all means.

Each of these proposals may be problematic in certain contexts. The first remedy is the most radical, and the most costly. In cases where collecting data is difficult and expensive, it may not be practical to throw data away and start over. Not even the harshest critics of the analyses included in the Evidence paper wished to pursue option (1). It certainly is not necessary if there are less costly alternatives that will yield a reasonably reliable evaluation of the evidential import of the experimental result.

The second option could be carried out in various ways. The proposal to increase the SLT p_T cut amounted to a modest version of this strategy. While the severity of the test using the original SLT analysis might be impossible to assess reliably because of uncertainty about the means of selecting cuts, the effect of increasing the p_T cut was known. CDF would have to abandon or significantly weaken its evidence claim. (One could interpret the situation either as a matter of specifying a test that meets the standard of severity originally advertised, but which the top hypothesis then fails, or as a matter of acknowledging that the test passed by the top hypothesis is less severe than originally advertised.) Many in the collaboration believed, however, that this strategy would amount to "tuning out" the signal. G. P. Yeh presented after-trial arguments that the eliminated events were most likely not signal at all, but an upward fluctuation in the background. Not everyone found these arguments compelling, however. It was not certain that a desire to retain candidate events in the SLT sample was the reason for retaining the 2-GeV/c cut, as opposed to a desire to retain continuity with the earlier soft-muon search.

An objection to this strategy might be that it raises the question of how to decide what changes are enough to correct for the potential bias in the original cuts. If one simply tightens the cuts until the data no longer yield an excess over background worthy of notice, then this strategy nearly collapses into the "throw away the data and start over" strategy. However, given the history of the SLT analysis, in which the proposal to move the p_T cut to 4 GeV/c was toyed with before, during, and after the writing of the Evidence paper, this would not seem to be a valid objection to Yeh's proposal, which, he argued, was simply a matter of doing what the SLT's authors themselves had proposed as an improvement to the algorithm, which would at the same time protect CDF from the charge of pushing a stronger evidence claim than they could justify.

A more radical version of the second strategy would be to have someone unfamiliar with the data or the existing analysis come up with a new set of cuts without looking at any data. In a context such as CDF's, such an approach would encounter serious difficulties. It may not have been possible to find anyone in the collaboration sufficiently unfamiliar with the top search data. Even if such a person were found, he would have to be sufficiently knowledgeable about the physics of the SLT search to generate reasonable cuts. This approach also runs afoul of the rewards system of experimental high energy physics. The conceptual development of the SLT analysis was primarily the work of Avi Yagil and Claudio Campagnari. To turn the analysis over to other physicists, who would then receive recognition for the work, would have been regarded as unjust.

The third strategy would be the least radical. All data could be kept, and no changes need be made to the cuts. One simply reports the number of events observed, and the background estimate. However, the problem of *which* cuts to use in reporting these numbers remains. Furthermore, "back of the envelope" significance estimates can easily be generated from such information by any statistically literate reader, and these will mislead at least as much as any published significance levels. Finally, simply omitting the significance calculation leaves in place the question that significance assessments are intended to help address: Is this result evidence for a certain hypothesis? The reader of such an experimental report expects guidance from the authors in addressing this question.

The first and third strategies proposed here would not have been feasible for CDF. Although they certainly could have carried out some modest version of the second strategy such as that proposed by G. P. Yeh, which arguably would have helped to strengthen their justification for their evidence claim (by weakening the claim itself), they would still have been using a single statistical calculation to characterize their findings, and the resulting number would have been likely to be somewhat inaccurate.

There is another possibility. Faced with such a situation, in which experimenters have lost the ability to estimate accurately the probability that their testing procedure leads them into an error, one might abandon one's pretenses to precision, and instead seek to learn *how much* difference the involvement of fallible and interested humans in determining experimental procedure might make.

5.1. The Method of Counterfactual Significance Calculations

Error-statistical assessments of experimental results involve three elements: the model of the hypothesis, the model of the experiment, and the model of the data. The model of the hypothesis provides us with a probability distribution for some parameter describing the population being studied, on the assumption that the hypothesis being tested is true. The model of the experiment contains all the statistically relevant information about the experimental test itself and specifies the test's error probabilities for the hypothesis under test. The model of the data yields a test statistic. One important experimental strategy, emphasized by Mayo, involves holding the experimental model and the data model constant while varying the model of the hypothesis, in order to see what can be learned about various hypotheses from the testing results at hand.

In the strategy I wish to discuss, the model of the hypothesis and the data are held constant, while the model of the experiment is varied.[7] In a counting experiment context, this amounts to carrying out a significance calculation using a single data sample, but with a variety of cuts. I use the term "counterfactual" to refer to such a significance calculation in order to emphasize that what it yields is *not* in fact the significance level of the experimental result actually obtained. In the scenarios in which counterfactual significance calculations are appropriate, the introduction of statistically relevant factors the effect of which cannot be reliably accounted for in the model of the experiment puts the actual significance level beyond our reach.

Such "sensitivity analyses" are not novel, but I wish to explore the rationale for pursuing such an analysis.[8] The question that a counterfactual significance calculation allows one to address is this: How sensitive is my assessment of the severity with which this hypothesis passed a test to changes in the description of that test? This question becomes important when an experimenter is uncertain as to which of several distinct descriptions of the test is most accurate. The greater that uncertainty is, the more important this question becomes.

Sometimes, although one may be uncertain about which experiment one did (i.e., which reference class to use when calculating significance), one can nevertheless evaluate the experimental outcome counterfactually. On this approach, the experimenter evaluates a single set of data in the light of a number of different experiments that *might* have been done. The actual significance level of the result may remain forever unknown, but one can gain insight into just how sensitive the *apparent* significance level is to those aspects of experimental procedure about which one is uncertain.

In the method of counterfactual significance calculation, one hypothetically reconstructs the experiment without any link between the choice of cuts and the actual data at hand. Absent such a link, other cuts might have been chosen (within certain bounds – some choices would simply not be physically reasonable given the aims of the experimenter). In this reconstruction, the experimenter can address the following question: "Assume that we had not tuned our cuts on the signal; what cuts might we have chosen? What, then, would we now be saying about the significance level of our results?"[9]

The goal of this procedure is to determine whether one has evidence for a given hypothesis or not, and if so, to determine how strong that evidence is. In other words, the experimenter seeks to determine whether the experiment has been evidentially successful (a determinate fact independent of anyone executing the calculations in question) and, in the process, to achieve epistemic success by coming to know or be justified in believing that the experiment was an evidential success with respect to some particular hypothesis.[10] This goal is pursued by attempting – qualitatively – to evaluate the severity of the test that the hypothesis has passed. When an experimenter is uncertain about what the appropriate reference class is for her experimental results, she cannot specify an accurate significance level. However, on the error-statistical approach, significance levels are not an end in themselves, but a means to evaluating severity. Through counterfactual significance calculations, the experimenter may be able to gauge the severity of her test qualitatively although a quantitative determination is impossible.[11]

The method may be put to use to avoid both epistemic and experimenters' failure. If the experimenter is uncertain himself as to whether or not his preferences are statistically relevant to the outcome, a counterfactual significance calculation can be used to gauge how much such uncertainty threatens the accuracy of his assessment of the experiment's evidential success. Alternatively, if the experimenter, through self-reflection, knows either that he had no preferences that could have been statistically relevant, or that such preferences as he had were not causally active in determining his decisions regarding the testing procedure, but is unable to make an argument that would serve to convey this knowledge to a third party, he can (potentially) use counterfactual significance estimates to argue that his audience need not concern themselves over such matters: even if his preferences had been causally active in determining the outcomes of decisions regarding the test procedure, the experiment would be evidentially successful anyway.

5.2. Counterfactual Significance Estimates of CDF's Top-Quark Evidence

To illustrate this point, consider the significance calculations presented by CDF for different parts of their top search results, as well as other counterfactual calculations that they might have presented. In Table 7.1, I present CDF's calculated significance levels for each of the three parts of their top search, the dilepton (DIL), secondary vertex tagging (SVX), and soft-lepton tagging (SLT) searches, based on counting events (not tags, as in the official CDF top-quark results), along with my own calculations (KWS) for those same values and my calculations based on various changes that might have been made to the SLT analysis – specifically, changes in the soft-lepton p_T cut. (Although moving that cut to 6 GeV/c does not seem to have been seriously proposed, I include that choice for purposes of comparison.) My calculations are based on simple Poisson statistics, using CDF's data as published in the Evidence paper and ignoring systematic uncertainties. I do not reproduce their full statistical analysis, which involved subtleties beyond the simple application of Poisson statistics. My results are close to CDF's in the cases where they can be compared, but these numbers are meant only to

TABLE 7.1. *Counterfactual Significance Calculations Based on Counting Events*

Search Used (SLT Momentum Cut)	No. of Candidate Events	Expected Background	CDF's Significance Calculation[a]	KWS's Significance Calculation[b]
1. DIL	2	0.56	0.12	0.11
2. SVX	6	2.3	0.032	0.030
3. DIL+SVX	8	2.86	–	0.0091
4. SLT(2)	7	3.1	0.041	0.039
5. SLT(4)	4	1.7	–	0.093
6. SLT(6)	3	1.1	–	0.10
7. DIL+SVX+SLT(2)	12	5.7	0.016	0.014
8. DIL+SVX+SLT(4)	9	4.3	–	0.032
9. DIL+SVX+SLT(6)	9	3.7	–	0.014

[a] CDF's significance calculations come from Abe, Amidei, et al. (1994b).
[b] KWS's significance calculations are based on naïve use of Poisson statistics.

suggest the type of strategy involved and cannot be used to draw any reliable conclusions about CDF's results.

The apparent significance of the SLT search by itself is rather strongly dependent on the placement of the soft-lepton momentum cut. This can be seen in lines 4–6, where the significance calculation in the last column changes by well over a factor of two. Taken by itself, then, this suggests that if there are doubts as to whether the SLT cuts were tuned on the signal, then the calculated significance based on the 2-GeV/c cut may indeed be a poor indicator of the actual severity of the test.

Combining all three searches yields a different picture, however. The SVX search and the SLT search picked out some of the same events. Hence, the number of candidate events selected by at least one algorithm does not simply equal the sum of the numbers selected by each algorithm (line 7 does not equal the sum of lines 1, 2, and 4). None of the three events selected by the SLT search that fell into the momentum region between 2 and 4 GeV/c were tagged by the SVX algorithm, but three of the events in the higher momentum region were. Hence, in terms of number of events chosen by at least one search algorithm, the SLT contributes four events to the total (beyond those in the SVX + DIL sample) provided that the cut is kept at 2 GeV/c. However, although three of those events are lost by moving the cut to 4 GeV/c, there is also a significant decrease in the expected background. Hence a further increase in the cut to 6 GeV/c, which does not remove any events from the candidate sample, cuts out still more background, and the apparent significance is restored to what it was with the 2-GeV/c cut.

If they could be taken seriously, the preceding calculations would suggest that the results of the SLT present at best very weak evidence for the top quark.[12] The apparent significance of the SLT results depends strongly on

TABLE 7.2. *Counterfactual Significance Calculations Based on Counting Tags*

Search Used (SLT Momentum Cut)	No. of Tags	Expected Background	CDF's Significance Calculation[a]	KWS's Significance Calculation[b]
1. DIL+SVX+SLT(2)	15	5.96	2.6×10^{-3}	1.3×10^{-3}
2. DIL+SVX+SLT(4)	12	4.56	–	2.7×10^{-3}
3. DIL+SVX+SLT(6)	11	3.96	–	2.6×10^{-3}
4. SVX+SLT(2)	13	5.4	–	3.8×10^{-3}
5. SVX+SLT(4)	10	4.0	–	8.1×10^{-3}
6. SVX+SLT(6)	9	3.4	–	8.3×10^{-3}

[a] CDF's significance calculation comes from Abe, Amidei, et al. (1994b).
[b] KWS's use of Poisson statistics here is even more naïve than in Table 7.1.

where the SLT momentum cut is placed, which might have been reasonably chosen to take on a value that would have resulted in a much higher value for its apparent significance. For the evidence claim based on the results of all three searches, the picture is not so bleak. Although the placement of the soft-lepton momentum cut makes some difference in the apparent significance of the combined result, it is not so great as to undermine drastically whatever evidence claim might be made on the basis of these results.

Similar conclusions regarding the relative importance of the p_T cut are suggested when looking at results based on counting tags. (The "tags" count is really a count of candidate events in the dilepton channel, plus the number of tagged jets in the two lepton +jets searches.) In Table 7.2, CDF's reported, and my counterfactual, significance estimates are given for a count of tags. Here the discrepancy between CDF's calculation and mine, in the one case where they can be compared, is even greater. My calculation, again based on naïve Poisson statistics, does not take into account the SVX-SLT background correlations that are modeled in the Monte Carlo used for CDF's significance estimate. (If my earlier calculations for counting events were to be taken with a grain of salt, these ought to be accompanied by the entire shaker.) The significance estimate roughly doubles when the SLT p_T cut is moved from 2 to 4 GeV/c. Of course, one choice to which the significance estimate is *very* sensitive is the decision to count tags rather than events.

Note that such an evaluation could be carried out for other aspects of the counting experiment results. Some collaboration members raised questions about the choice of SVX b-tagging algorithm, for example. One could examine significance estimates for all three of the competing algorithms developed for the Ia data. Another dispute centered on the "pseudo e-μ" event: Did it belong in the SLT or dilepton sample? One would like to know how the significance estimate might change based on where that event landed.

In fact, to some extent CDF *did* carry out such calculations. The effect of the SLT p_T cut on the significance estimate for that channel, for example, was given in presentations to the collaboration by physicists working on the SLT analysis. These numbers were known in the collaboration, and even brought up later during interviews. However, they are not part of the way in which CDF presented their results in the Evidence paper. CDF did, in that paper, present background estimates and candidate event totals for all three of the SVX *b*-tagging algorithms, but they only gave a significance calculation for the jet-vertexing algorithm.[13]

So I am not proposing anything new in suggesting the use of counterfactual significance calculations to deal with potential biases from tuning on the signal. But it is worth looking more closely at how such methods were used. Although counterfactual statistical calculations were used within the collaboration to reassure themselves that their result was not overly dependent on particular decisions regarding cuts, CDF essentially took this to be a vindication of a single calculation used to summarize the import of the counting experiments – the 2.6×10^{-3} significance calculation highlighted in their abstract.

Some Fermilab physicists felt that the collaboration simply did not have a sufficiently precise understanding of the statistical characteristics of their testing procedure to support quoting a single significance level. Fermilab physicist Morris Binkley, in particular, objected that attempting to capture the evidential importance of the results in a single probability calculation was misguided, since "we had no idea... what the biases were in that number, and the number was fairly meaningless at the level we were quoting it." Consequently, the effort "to boil this down to one number" could be aptly described using H. L. Mencken's description of theology: "the attempt to put the unknowable in terms of the not-worth-knowing" (Binkley 1995).

Instead of such an approach, Binkley proposed that CDF present its results to the public using a variety of configurations of cuts, thus making systematic use of what I have been calling counterfactual significance calculations.

The use of counterfactual significance calculations is not a method that should always be pursued. When experimenters have adequately adhered to pretrial precautions that will permit the reliable construction of probability models for the testing procedure, and the accuracy of those models can be justified to other scientists interested in evaluating the outcome of the experiment, there may be no need to employ this more complicated, potentially less definitive means of assessing evidential strength. In addition, there will be contexts in which, given that such pretrial planning has failed, it really will be best to toss out the data and start again, or at least gather additional data to permit an independent check on the initial result.

The use of counterfactual significance calculations would seem to be called for in contexts with the following features: (1) the ability to calculate a significance level for one's result has been lost, or cannot be demonstrated to others owing to uncertainties regarding the reference class to be employed in that calculation; (2) data are expensive; and (3) there are strong reasons for reporting the evidential import of the data in hand.

Notice, however, that even though this description certainly applies to the context in which CDF found themselves, there is nothing about these contextual features that singles out particle physics in any way that is essentially related to its subject matter. It is indeed an empirical fact that high energy particle physics is done primarily in large collaborations in which there are ample opportunities for controversies and uncertainties to arise regarding how test parameters have been chosen, and that there is consequently an ever-present risk of finding that "what experiment did we do?" might become a pertinent, and nontrivial question. It is also true that data at a particle accelerator are expensive. Furthermore, high energy physics is a "high-profile" field in which new discoveries receive a great deal of attention. But these same features could quite easily be found to apply in other disciplines as well. Furthermore, the particular goal of this method, which is to determine whether the hypothesis in question has passed a severe test, whether the apparent effect being detected is an artifact, is decidedly *not* domain-specific.

It should also be noted that for collaborations there may be better pretrial precautions than predesignation for avoiding the problem of tuning on the signal. Even within the collaboration, scientists might find it difficult to verify that their colleagues did indeed adhere to the predesignation rule. In some cases, collaborations may be able to enforce a strict adherence to predesignation by means of a "blind box" approach, in which the scrutiny of data falling into a "signal region" is delayed until after calculation of backgrounds. CDF has used the blind box approach in a recent search for supersymmetric particles (Affolder, Akimoto, et al. 2002), following other collaborations that have adopted the technique (Alavi-Harati, Albuquerque, et al. 1999; Vaitaitis, Drucker, et al. 1999; Formaggio, Zimmerman, et al. 2000). However, this approach rules out the use of potential signal data for the discovery of possible flaws in the algorithm, something that the developers of the SVX algorithms in run Ia considered important in light of the novelty of the silicon vertex detector.

Under such conditions, securing both an efficient approach to discovery and the kind of verifiable reliability required for experimenters' success will require either the use of more elaborate mechanisms for handling data ("social technology," as it might be called – see Pitt 1999, esp. 9–12), or a more systematic use of posttrial techniques for evaluating evidence such as counterfactual significance calculations. Allan Franklin gives an example of the former approach (Franklin 1986, 170), in which experimenters arrange for a random number unknown to the experimenters to be added to each measurement, and then subtract that number only after events are selected. Franklin's example comes from the search by W. M. Fairbank and colleagues for fractional charges (LaRue, Phillips, et al. 1981). This practice has been adopted by other groups of experimental physicists as well (Glanz 2000).

6. OBJECTIVITY, IN SPITE OF IT ALL

The predesignation rule may call to mind the "novelty" requirement that has long been a focus of philosophical debate over theory confirmation. In fact, both the novelty requirement and the predesignation rule suggest dimensions of evidential assessment that appear initially troubling to anyone who holds that evidential relationships are objective. When considering predesignation or novelty, questions about who knew what, and when they knew it, become relevant. In this section, by subjecting novelty and predesignation considerations to an analysis from the perspective of the error-statistical theory of evidence, I wish to clarify just how the epistemic and preferential states of experimenters can be relevant to evaluating the *objective* evidential import of experimental outcomes.

6.1. Novelty

Numerous philosophers of science have argued for a principle of confirmation according to which hypotheses are confirmed by facts they predict or explain that are in some sense "novel." (See Musgrave 1974 for a discussion of the earlier history of this view.) I will not survey here all of the variations on the novelty requirement. In some versions, novelty is a necessary condition for confirmation; in others, novel facts simply provide better confirmation of hypotheses than nonnovel facts. Accounts of novelty also vary. Early versions of the view state the requirement simply in terms of when a hypothesis is known, for example in William Whewell's requirement that a hypothesis "ought to foretel phenomena which have not yet been observed" (Whewell 1989 [1858], 151). More recent work has tended to distinguish this purely temporal view from a "heuristic" or "use" notion of novelty that Whewell seems to have anticipated with his "consilience" requirement: "the evidence in favour of our induction is of a much higher and more forcible character when it enables us to explain and determine cases of a *kind different* from those which were contemplated in the formation of our hypothesis" (ibid., 153). The heuristic view in contemporary form is exemplified in the account given by Elie Zahar, according to which a fact is considered novel with respect to a hypothesis "if it did not belong to the problem-situation which governed the construction of the hypothesis" (Zahar 1973, 103). I will focus on a related "use" notion of novelty proposed by John Worrall as an improvement on Zahar's heuristic account (Worrall 1985; 1989) because it seems to lend itself to a modest reformulation that might appear to capture the underlying motivation for the predesignation requirement. I will argue, however, that it does not. Instead, both novelty and predesignation requirements are motivated by error-statistical considerations. Neither novelty nor predesignation, however, is *necessary* to satisfy the requirements for error-statistical evidence. Instead, they are *means*, which may in some contexts be especially effective, for satisfying (or justifying the claim to have satisfied) evidential requirements.

We might, in the spirit of Worrall, express the use novelty requirement (UN) in the following way:

UN: (i) *h* entails or is a good fit with *e*,
(ii) *e* is not used in the construction of *h*. (Mayo 1991, 524; 1996, 259)

However, this is too crude to apply to the problem of tuning on the signal. There was no question of the top-quark hypothesis itself having been constructed in a way that used the experimental result generated by CDF from the Ia data. Fortunately, the hierarchy of experimental models described in the previous chapter allows us to do better. Recall that for the purposes of bringing data to bear on a hypothesis, one typically has to employ a *model* of the hypothesis. Following the analysis in Section 5.1 of Chapter 6, we can represent CDF's null hypothesis in the Evidence paper (there is no top quark) by means of the following discrete (Poisson) probability distribution:

$$H: f(s, \lambda) = \frac{\lambda^s e^{-\lambda}}{s!}, \quad \text{for } s = 0, 1, 2, \ldots, \text{with } \lambda = \lambda_0.$$

Because the alternative hypothesis that asserts the existence of the top quark does not determine a unique probability distribution, we represent that possibility by means of a family of Poisson distributions:

$$J: \left\{ f(s, \lambda) = \frac{\lambda^s e^{-\lambda}}{s!}, \text{for } s = 0, 1, 2, \ldots : \lambda > \lambda_0 \right\},$$

where different values of λ yield models of distinct possibilities, corresponding to different values for the rest mass of the top quark.

At the other end of the hierarchy of models, we can refer to the model of the data, D: $[\Xi, S(\Xi)]$, where Ξ is an array of all the particular measurements made using the detector during run Ia, and $S(\Xi)$ is an operation on Ξ that yields the value of the test statistic – in this case the number of candidate events s.

Concerns about tuning on the signal seemed to turn on the uses to which experimenters had put the data in deciding on the definition of the test statistic. Of course, the choice of test statistic had implications for the statistical models to be used in testing the primary hypotheses. In the spirit of the use novelty requirement, then, we might articulate the ban on tuning on the signal in the following way:

UN': In order for a test of a hypothesis Θ based on a comparison of outcome s ($= \mathbf{S}(\Xi)$) to statistical model H to be successful, it is necessary that:
   (i) H accurately represents the probabilities of various outcomes on the assumption that hypothesis Θ is true, and
   (ii) Ξ is not used in the construction of *H*.

Applied to the top search, the first condition amounts to the requirement that the background prediction be an accurate description of what would be found in the data if there were no top quark. The second requirement stipulates that the data to which the top search algorithms are to be applied must not be used to determine what that prediction is. Since background depends on the choice of cuts, one effect of UN' is to prohibit the use of data in setting cuts.

Worrall notes that in light of the use novelty requirement, "the intuitions behind the notion of a genuine test cannot be captured in purely logical terms but must involve consideration of how the theory concerned was constructed" (Worrall 1985, 149). Since CDF and D-zero physicists considered the way in which cuts were chosen relevant to their assessment of the evidential strength of the outcome *e*, we might conclude that Worrall's claim is vindicated. But does use novelty as reformulated in UN' account properly for *why* these physicists regarded (or ought to have regarded) such information as relevant?

Deborah Mayo (Mayo 1991; 1996, ch. 8) has already argued that in those cases where scientists have judged successful predictions of "novel" facts to constitute impressive evidence for a theory, it is because those successful novel predictions constituted severe tests of the theory in question. Use novelty, however, is neither sufficient nor necessary for severely testing a hypothesis. A test can be severe and provide strong evidence for a hypothesis and yet violate the use novelty requirement (confidence interval construction methods, such as those used to arrive at estimates of the top quark's mass, provide examples of such severe tests in violation of use novelty – see Mayo 1996, 272–4). Furthermore, a test may satisfy the use novelty requirement and yet fail to be severe. (Trivially, one need only specify a test that a hypothesis is likely to pass, even if it is false, but that involves the collection of new data, such as a test that passes some version of string theory if there is an earthquake somewhere in the next year.)

I wish to argue for two further points: (1) The notion of severity explains the need for CDF physicists to avoid tuning their cuts on the signal better than the requirement of use novelty because, unlike novelty, severity is directly linked to a notion about which CDF members expressed concern in connection with that issue. (2) If understood in terms of severity, arguments given in the context of the debate over the SLT analysis are rendered consistent that, if understood in terms of use novelty, would be inconsistent.

CDF members who were concerned about cuts used in the Evidence paper being tuned on the signal explicitly questioned whether the resulting significance calculation was misleading. The significance level is meant to represent the probability that the testing procedure rejects the null hypothesis (that there is no top quark) when it is true. These physicists worried that tuning the cuts on the data would result in a misleadingly small value for this number. Significance is directly related to the severity of the test that the top-quark hypothesis passed based on the data CDF gathered. The severity of that test for passing the top-quark hypothesis with the result obtained can be expressed as one minus the probability of getting a result that accords as well with the hypothesis that there is a top quark, when in fact there is not (i.e., when the null hypothesis is true).

In short, to worry about the accuracy of one's significance estimate *is* to worry about how well one understands the severity of the test being used. Viewing the problem in terms of severity, therefore, accords well with the physicists' own explicit concerns raised in their discussions of tuning on the signal.

Nevertheless, one might object, it remains the case that to the extent that CDF physicists worried about specific physicists having used particular features of the data as reasons for selecting particular cuts, they were concerned with a kind of use novelty as expressed in UN'. This brings me to my second point.

Not only do we know that concerns about significance (and hence severity) drove these debates over the use of data in choosing cuts, but we also know that CDF members did not regard the use of data in choosing cuts as intrinsically troubling, but *only* objected when cuts were chosen in a manner that raised doubts about the reliability of the resulting significance estimates.

Consider the debate over the SLT analysis. Critics such as G. P. Yeh expressed concern over the problem of misleading significance calculations. Furthermore, there can be no doubt that Yeh's concern was *not* about use novelty because his own proposed remedies included options that would have violated use novelty! Yeh reported not having paid attention to the details of the SLT analysis until the first draft of the Evidence paper was distributed in December 1993. Only then did he ask for a soft lepton p_T plot for both the data and the expected background and signal. After seeing these plots, with the three candidate events in the high-background, low-signal region of 2–4 GeV/c, Yeh proposed that the SLT cuts be changed. In effect, Yeh proposed that facts about the data *should* be used as grounds for altering the cuts. By doing so, he argued, the estimate of the statistical importance of those data would be less likely to be misleading. In other words, CDF would not face the difficulty of being unable to justify their claims about the severity of the test passed by the top hypothesis.

If his concerns over the soft lepton p_T cut were motivated by use-novelty considerations, then we would have to regard Yeh as trapped in a rather obvious inconsistency. This apparent inconsistency is easily resolved, however, if we see the debate as concerning severity rather than use novelty. Without any means of addressing worries that CDF's desire to find evidence for the top quark was causally connected to the choice of cuts in the SLT analysis, and hence to the number of candidate events that they found in that counting experiment, CDF might not be able to justify either to themselves or to their audience that the model they used to calculate the error probabilities for their testing procedure was accurate. Using the 2-GeV/c value for the p_T cut therefore raised a concern that CDF was ignoring a part of their testing procedure that potentially made that procedure less severe than advertised. Raising the cut to 4 GeV/c, although it would violate use novelty, would raise the value of the significance estimate to a value that (allegedly) would be a more realistic reflection of the actual error probabilities of the test procedure CDF employed.

6.2. Predesignation, "Subjective Circumstances," and the Historical Assessment of an Evidence Claim

C. S. Peirce, who advocated the predesignation requirement once observed:

[I]n demonstrative reasoning the conclusion follows from the existence of the objective facts laid down in the premises; while in probable reasoning these facts in themselves do not even render the conclusion probable, but account has to be taken of various subjective circumstances – of the manner in which the premises have been obtained, of there being no countervailing considerations, etc.; in short, good faith and honesty are essential to good logic in probable reasoning. (*CP* 2.696; *CE* 4, 410 [1883])

I am hesitant to call such circumstances "subjective" myself, as it seems to grant too much to those who insist that they cannot be relevant to an objective assessment of evidence.[14] What is more important, however, is to understand in what way such matters are relevant, and why.

On this matter Peirce expressed himself very clearly. Writing of an experimenter faced with the task of selecting a sample S from a population of Ms, he noted:

The volition of the reasoner (using what machinery it may) has to choose S so that it shall be an M; but he ought to restrain himself from all further preference, and not allow his will to act in any way that might tend to settle what particular M is taken, but should leave that to the operation of chance. . . . Now in choosing the instance S, the general intention (including the whole plan of action) should be to select an M, but beyond that there should be no preference; and the act of choice should be such that if it were repeated many enough times with the same intention, the result would be that among the totality of selections the different sorts of *M*'s would occur with the same relative frequencies as in experiences in which volition does not intermeddle at all. (*CP* 2.696; *CE* 4, 411 [1883])

Peirce identifies as the desideratum in making such a selection, not that intentions should not figure at all, but rather that they should not correlate with any particular features of the population from which one is sampling.[15] This is in stark contrast to pronouncements from advocates of some schools of statistical inference who insist that considerations such as rules for stopping data collection, the processes of generating hypotheses, and the pre-designation of a test statistic are never relevant because they are "purely external" to the evaluation of evidence (Edwards 1992, 30), and the experimenter's intention "is locked up inside his head" (Savage 1962, 76).[16]

The motivation for such proclamations is the likelihood principle. According to both personalist Bayesians such as L. J. Savage and likelihood theorists such as A. W. F. Edwards, the evidential impact of some experimental outcome is strictly and exclusively a function of the likelihood ratio. Hence, the likelihood ratio expresses everything about the experimental outcome that is relevant to assessing the probability of the hypothesis. As Savage puts it, for a hypothesis about a parameter λ, and an experimental outcome x,

According to Bayes's theorem,[17] $Pr(x/\lambda)$, considered as a function of λ, constitutes the entire evidence of the experiment, that is, it tells all that the experiment has to tell. More fully and more precisely, if y is the datum of some other experiment, and if it happens that $Pr(x/\lambda)$ and $Pr(y/\lambda)$ are proportional functions of λ (that is, constant multiples of each other), then each of the two data x and y have exactly the same thing to say about the values of λ. (Savage 1962, 17; see Berger and Wolpert 1984 for an excellent general discussion and defense of the likelihood principle)

Whether in fact the likelihood principle renders stopping rules, predesignation, and hypothesis generation techniques evidentially irrelevant I leave for its adherents to debate. For my part, I wish to address the charge that a theory of statistical inference according to which such matters *are* evidentially relevant is somehow less than fully objective.

It is not, after all, only Bayesians and likelihood theorists who have called into question the relevance of Peirce's "subjective circumstances" for an objective theory of evidence. Consider the response, for example, of John Maynard Keynes. In his *Treatise On Probability*, Keynes takes up an example of Peirce's, which Peirce offers to illustrate his argument for the relevance of predesignation. Consulting a biographical dictionary, Peirce takes the first five poets in alphabetical order and notes the ages at which they perished. Upon inspection, the ages of these five poets at death reveal certain shared features, such as "the difference of the two digits composing the number, divided by three, leaves a remainder of one." Peirce argues that similar factual detritus can be found common among any "lot of objects" that we happen to choose, so that simply looking for any shared characteristic, and finding one, is no grounds at all for a statistical inference.[18] Keynes argues that, to the contrary, the inference in Peirce's example is weak, not because predesignation is absent but because the inference is "based on a small number of instances" and is "out of relation to the rest of our knowledge" (Keynes 1921, 305).

Keynes concludes, in a categorical rejection of Peirce's advocacy of predesignation:

> The peculiar virtue of prediction or predesignation is altogether imaginary. The number of instances examined, and the analogy between them are the essential points, and the question as to whether a particular hypothesis happens to be propounded before or after their examination is quite irrelevant. (Keynes 1921, 305)

In a similar vein, Laura Snyder, defending the claim that scientists' use of evidence commits them to an "impersonal" concept of evidence, goes on to argue that this use of impersonal evidence rules out the "historical thesis" that "on any plausible theory of evidence, the time at which *e* is known relative to the invention of *h* is relevant to whether *e* is evidence for *h*" (Snyder 1994, 102). The error-statistical concept of evidence as I have described it is meant to be an impersonal concept in Snyder's sense (see Chapter 6). On the error-statistical account, I have argued, when a test result *e* is evidence for a hypothesis *h*, it provides a "reason to believe *h*, whether or not any particular person does, in fact, believe *h* for the reason that *e*, or whether any person should, given his other beliefs and knowledge, believe *h* for the reason that *e*" (Snyder 1994, 103). This is Snyder's criterion for an impersonal concept of evidence.

However, if Snyder's argument is correct, then the error-statistical concept of evidence cannot both be impersonal in this sense and also entail the relevance of Peirce's "subjective circumstances." For if such matters are relevant on the error-statistical account, then it makes questions of when *e* (the data) is known relative to the invention of *h* (the model of the hypothesis), potentially relevant to whether *e* is evidence for *h*, and hence puts error statistics in agreement with the historical thesis. Error-statistical evidence might be impersonal or historical, but not both.

My task, then, is to show how "subjective circumstances" can be relevant to an impersonal concept of evidence, in a way that responds to the complaints of the Bayesians and likelihoodists, and answers Snyder's argument.

The Historical Thesis and Error-Statistical Evidence. As a first step, consider Peter Achinstein's discussion of the historical thesis.[19] Achinstein argues that in some cases where historical facts seem relevant to the evaluation of an evidence claim, the relevant facts concern what "selection procedure" was used in obtaining the putative evidence in question, where a selection procedure is a rule "determining how to test, or obtain evidence for, an hypothesis" (Achinstein 1995, 157). This concept of a selection procedure can shed light on the present case, as well, but we will have to take a broad view of what can be included in a selection procedure.

On the error-statistical view, choosing an appropriate selection procedure amounts to defining a severe test. Suppose we regard the cuts used in CDF's counting experiments as part of the selection procedure they used in testing the top hypothesis. Then CDF's deliberations over which cuts to use was a matter of determining an appropriate selection procedure so that, should the top hypothesis pass the test, it would pass an appropriately severe test.

This, however, does not account for all of the historical factors relevant to the case. As CDF physicist Morris Binkley put it, in order for a significance calculation to be credible "you have to know that you have no biases in the way you selected the data, and the way you selected the algorithm to analyze the data" (Binkley 1995).

In order to use the notion of a selection procedure to account for the relevance of historical facts regarding the selection of cuts, we need to expand the notion of a selection procedure to incorporate, not only how the data were selected for testing purposes but also how the procedure for selecting the data was itself selected. Thus, in some circumstances, describing the selection procedure simply in terms of the cuts used in the various counting experiments omits relevant information (as I argued in Section 4 of this chapter). A representation that omits information relevant to the assessment of evidence claims based on data selected by a certain procedure might be thought of as an "under-description" of that selection procedure. In the present case, the historical thesis is borne out as a result of the need to appeal to the selection procedure, for it was precisely facts about the knowledge states of members of the collaboration that, for some, raised doubts about how well they understood the error probabilities of their testing procedure.

Drawing a distinction here will allow us both to see how an impersonal concept of evidence can be compatible with the historical thesis and to reformulate Snyder's thesis in a way that does not run afoul of our present considerations.

At the heart of Snyder's argument is the claim that the epistemic states of individuals are not relevant to the evaluation of evidence in its impersonal sense. But here "relevance" is ambiguous. Person P's epistemic state S might be relevant to whether e is evidence for h in the sense that, whether e is P's evidence for h depends on whether or not P, *in her role as an epistemic agent*, is in state S. Any theory of evidence that makes epistemic states relevant in this sense (let us call it *epistemic relevance*) would clearly be in conflict with the impersonal concept of evidence that Snyder advocates.

But epistemic states might be evidentially relevant in another way. Person P's epistemic state S might be relevant to whether e is evidence for h in the sense that, insofar as *e* is the outcome of a testing procedure carried out (in part) by *P* (*in her role as an experimental agent*), whether *e* is evidence for *h* depends on the effect of *P*'s epistemic state *S* on the probabilities of various possible test outcomes.[20] If epistemic states are relevant in this sense (*causal relevance*), the evidence concept may nevertheless be impersonal insofar as, given the information about *P*'s epistemic state *qua* experimenter, the evidential relationship is not relativized to any person or any epistemic state.

That is to say, if *e* is evidence for *h* given the error probabilities of the testing procedure as a whole, where such probabilities may be affected by the epistemic states of experimental agents, then *e* constitutes a reason for anyone to believe *h*, regardless of the epistemic state of the person in question. Of course, this just *is* to say that *e* is evidence for *h* in the impersonal sense as Snyder defines it.

We can extend this notion of causal relevance to other facts about the experimental agent that might, due to their causal connections with the experimental outcome, make a difference to the error probabilities of the testing procedure, such as his preference for a particular outcome. (In tuning-on-the-signal cases, it is the experimenter's epistemic and preferential states acting in concert that create the problem.) Furthermore, if there are states of social entities that are nonreducible to states of individuals and have properties that alter the error probabilities of the testing procedure, these also can be causally relevant. (Whether or not such social states are reducible, it may be easiest to evaluate their effects on a testing procedure without carrying out such a reduction. For example, a person suspicious of the procedure by which the SVX *b*-tagging algorithm was chosen will most likely be motivated by facts about the structure and procedures of the group working on the problem and the selection mechanisms acting within that group.)

Returning to Snyder's thesis, we might reformulate it as such: No theory of evidence that makes the time at which *e* is known relative to the invention of *h epistemically* (rather than causally) relevant to whether *e* is evidence for *h* can capture the impersonal concept of evidence. Hence no such theory can be an adequate theory of evidence.

We now are in a position to respond to Keynes as well. Peter Achinstein quotes Keynes's categorical claim of the irrelevance of predesignation approvingly in his discussion of the historical thesis, but Achinstein's argument itself seems only to support the more modest and quite unexceptionable claim that predesignation is not *always* relevant. Or, to put it another way, a prediction sometimes yields worse evidence than an explanation of a known fact, and vice versa. Claiming that predesignation is never relevant goes much further, and is in any case mistaken, for reasons that Peirce saw quite clearly.

Keynes diagnoses Peirce's example of the five dead poets in terms of two weaknesses: First, the inference is based on a small sample. Second, the inference is "out of relation to the rest of our knowledge." That is, we find the conclusion absurd in light of "the knowledge we actually possess that the ages of poets are not in fact connected by any such law" (Keynes 1921, 305). Both replies fail to address Peirce's argument, however. First, Peirce claims that *any* sample, of any size, will yield some common characteristic, however recondite, but most of the inferences formed on the basis of such observations would be false. Finding *some shared characteristic or another* in a sample is guaranteed, independently of that characteristic being common to the entire population.

Keynes's second point of attack claims that our background knowledge gives us reason to reject the dead poets inference. Everyone already knows that, for example, it is not a universal truth regarding the ages at which poets die that "the sum of the prime factors of each age, including one as a prime factor, is divisible by three." Keynes claims that, given our background knowledge, we must regard the inference in question as entirely lacking in direct evidence (or for a Bayesian, having a very low prior probability). If we simply did not know anything at all about the ages at which poets die apart from the information given in Peirce's example, then Peirce's induction "would be as valid as any other which is based on a very weak analogy and a very small number of instances and is unsupported by evidence" (Keynes 1921, 305).

The analysis of this chapter can help to explain why Keynes is mistaken on this point. According to the present view, evidential relationships supervene on the error characteristics of testing procedures, and the testing procedure in Peirce's example does not have the same error characteristics as one in which the characteristic to be observed is specified in advance. In fact, the procedure in Peirce's example is a highly unreliable one, which, as Peirce notes in his argument, will almost always yield a false conclusion.

Nevertheless, some might feel, with Keynes, that part of the force of the dead poets example lies in the nature of the characteristics mentioned by Peirce, and that invoking our background knowledge about that very characteristic explains the weakness of this induction. Peirce recognizes that the "recondite" nature of this characteristic makes a difference. However, that difference is not to be explained by appealing to some prior knowledge state (or degree of belief) with respect to the hypothesis inferred. Instead, following Peirce, we can understand this difference, too, as a matter of testing procedure.

In most circumstances in which a researcher selects a sample without predesignating the characteristic with respect to which she will scrutinize them, she nevertheless will not be examining the sample for any and all characteristics that they might share. Rather, her search space will encompass just those characteristics of interest to her. The obscure shared characteristics of the terminal ages of poets mentioned by Peirce are practically certain not to be examples of such characteristics. This limitation on the number of characteristics about which one would care to draw a conclusion lowers the probability of reaching a false conclusion (and the more familiar or "striking" the characteristic, the larger this effect is), and that probability can be further lowered, Peirce notes, by increasing the sample size when carrying out such an exploratory investigation of the properties of a sample.

Thus, Peirce concludes that

> if a large number of samples of a class are found to have some very striking character in common, or if a large number of characters of one object are found to be possessed by a very familiar object, we need not hesitate to infer, in the first case, that the same characters belong to the whole class, or, in the second case, that the two objects are practically identical; remembering only that the inference is less to be relied upon than it would be had a deliberate predesignation been made. (*CP* 2.740; *CE* 4, 439 [1883]; see also *CP* 6.409; *CE* 3, 314 [1878])

We can add that if one wishes to calculate an error probability for one's testing procedure, it will be vital to take into account the fact that the characteristic in question was not predesignated, and that a number of other characteristics, if they had been found present, might have been projected instead of that actually chosen. In any case, it would be highly unreliable to cook up just any recondite characteristic that can be mapped onto the sample and project *that*: "This is no doubt the precise significance of the rule sometimes laid down, that a hypothesis ought to be *simple*, – simple here being taken in the sense of familiar" (*CP* 2.740; *CE* 4, 439 [1883], emphasis in original).

Thus, contrary to Keynes, the "virtue" of predesignation is ultimately not particularly "peculiar" (and certainly not "imaginary"). Predesignation is simply one technique among many for ruling out error and ensuring the reliability of one's test procedure. The experimenter is part of the machinery of the experimental procedure. The experimenter is able to reflect on her own potential contributions to the possibility of error (something that a phototube cannot do!). She can then use that knowledge, just like knowledge of the other parts of the experimental apparatus, to engineer the experiment's causal linkages, including those in which she herself is involved, to ensure its overall reliability.

The Likelihood Principle, Causally Relevant Epistemic and Intentional States, and Stopping Rules. The distinction between the epistemic and causal relevance of epistemic states of experimenters may also help to clarify the debate over the meaning of the likelihood principle. As indicated earlier, some have taken the likelihood principle to entail the irrelevance of the intentions and epistemic states of experimenters, since ultimately only the probability of the actual outcome of the experiment on the hypothesis being tested – relative to its probability on all competing hypotheses – matters. Contrary to the argument quoted by Savage, however, the experimenter's intention is not "locked up inside his head." Deborah Mayo has already argued persuasively that this "argument from intentions" is fallacious (Mayo 1996, 346–50). My present aim is to examine this dispute in light of the distinction between the epistemic and causal relevance of an experimental agent's intentional and epistemic states. Further, I wish to point out how casting this issue in terms of the relevance of intentions misdiagnoses the point of disagreement between error statisticians and advocates of the likelihood principle.

When intentions are causally effective, they make an objective difference to the probabilities of various experimental outcomes. For the error statistician, such causal relevance is relatively straightforward, as I have argued: It shows up in the probabilities of various possible test outcomes (the experimental or sampling distribution, as it is often called). Thus, acknowledging the relevance of the intentions and epistemic states of experimenters does not, within the error-statistical framework, invite subjectivism.

The present point can be seen in the case of stopping rules. Some personalist Bayesians see the consideration of stopping rules as an example of error statistics allowing irrelevant information into the assessment of evidence.[21] Nevertheless, in their discussion of stopping rules, Howson and Urbach do allow that from the Bayesian perspective "sometimes events attending the sampling process . . . have significance as evidence" (Howson and Urbach 1993, 366). They give the following as an example of "an inductively relevant stopping rule" (ibid., 214): In an experiment to estimate the mean height of a group of cooks occupied at preparing lunch, the stopping rule of the experimenter is that the measurement will be made when lunch preparation is complete. However, it is known that tall chefs cook more quickly than short chefs. Howson and Urbach note that "[i]n this instance, exactly when the trial was concluded would depend on the unknown parameter and so would be some guide to the stature of the cooks. Ignoring the stopping rule in such a case would be overlooking relevant information" (Howson and Urbach 1993; see Berger and Wolpert 1984, 88–9 for a similar example). However, they insist that, unlike the position of "classical" frequentist statisticians, they do not hold "that the scientist's *intention* to stop the trial at a particular point is of any inductive significance" (Howson and Urbach 1993, 366).

However, to the extent that this argument shows that, for Bayesians, stopping rules are sometimes evidentially relevant, it also undermines the Bayesian's attempt to exploit the stopping rule controversy as an argument against error statistics. What could Howson and Urbach mean by saying that here the stopping rule is relevant? On one interpretation of the example, it is not the stopping rule at all that is relevant, but instead the time of data collection, relative to the beginning of lunch preparation. On this interpretation, the stopping rule is identified with the intentional state of the experimenter *taken in isolation*. The relevant datum is the *outcome* of the stopping rule, which is held apart from the rule itself. This interpretation would vindicate Howson and Urbach's contention that, in their example, the intention of the experimenter is not by itself evidentially significant.

However, if that is what is meant by "stopping rule," then neither are stopping rules *in this sense* typically relevant for error statisticians. For them, the experimenter's intention (or epistemic state, where predesignation is at issue) *taken in isolation* is not relevant. Rather, the *effects* of intentions are of interest to experimenters.

If, on the other hand, Howson and Urbach mean seriously to defend the claim that in their example the stopping rule is relevant, then we have to take them to mean that it is relevant insofar as it is causally connected with the outcome obtained. But then they no longer can claim that the manner in which the stopping rule is relevant is fundamentally different for them than it is for the error statistician. Certainly they are not entitled to say that a stopping rule is an objective public datum when they refer to it, and a ghostly private intention when an error statistician refers to it. For both Bayesians and error statisticians, stopping rules ought to be regarded in terms of their causal propensities. The primary aim for an error statistician is to ensure that the error probabilities derived from the experimental model do not mislead one as to the severity of the test (i.e., that those probabilities serve as a good guide to the severity of the testing procedure actually employed, and hence are a good partial indicator of the strength of one's evidence).

Hence, the fact that CDF did not explicitly articulate a stopping rule at the beginning of run Ia, before collecting any data, should not by itself generate concern about their statistical assessment of their evidence. To the extent that CDF could be regarded as following any stopping rule during run Ia, that rule might be something like: "Collect data until the lab administration shuts down the collider (which the lab administration does for maintenance, or for purposes of running fixed-target experiments, according to schedules that can stretch or shrink under political pressures within the lab, among other things). Then release analyses of the data."[22] Such a stopping rule does not obviously correlate the termination of data collection with any particular kind of test outcome (except when the desire to reach a certain result becomes part of the political negotiations over ending a run, as when a collaboration at CERN recently lobbied their lab administration to extend a collider run because they felt they were close to having enough data for a Higgs boson discovery). Hence, when calculating error probabilities, it was safe for CDF to assume that the same amount of data would be collected in each hypothetical repetition of their experiment. That is to say, calculations based on such an assumption would not be misleading. In the case where an experimenter *does* intend to stop collecting data when he obtains a result that is, say, five standard deviations from the null expectation, but is taking data "blind," so that he does not know when that result has been achieved, the intention in question would not be causally effective (not being conjoined with an epistemic state of the required sort) and could be safely ignored by the error statistician no less than a Bayesian.

That intentions matter only insofar as they affect the probabilities of various outcomes is recognized by the objective Bayesian Roger Rosenkrantz. Discussing an example offered by Savage intended to illustrate the irrelevance of stopping rules for a Bayesian, Rosenkrantz notes that "a more careful Bayesian analysis shows that, where those intentions [regarding when to stop collecting data] materially reduce or alter the originally contemplated sample space, both the likelihood function and the import of the data are affected" (Rosenkrantz 1977, 199).[23] For Bayesians, too, intentions make a difference to the probabilities of various outcomes. But, whereas the error statistician continues to care about the experimental (sampling) distribution after the experiment is performed, the Bayesian or likelihoodist loses interest, focusing exclusively on the probability of the single result obtained.

Thus, the real point of disagreement between error statisticians and adherents to the likelihood principle has nothing to do with the relevance of intentions per se. The difference lies entirely in the interest that the error statistician has in the various ways that the experiment *could have* gone wrong. This makes the whole experimental distribution relevant, whereas the likelihood principle does entail that only the probability of the actual outcome matters. For the error statistician, the stopping rule needs to be considered as to its implications for various possible outcomes, and the same goes for considerations of predesignation. An adherent to the likelihood principle may therefore object to the consideration of counterfactual outcomes, but not to the relevance of epistemic or intentional states of the experimenter.[24] In short, the question at issue is not "does the experimenter's intention regarding when to stop collecting data make a difference?" but "what kind of a difference matters?"

6.3. Evidential Considerations Made Invisible – and Bringing Them to Light

Many CDF members worried that their cuts might have been tuned on the signal, although most whom I interviewed were convinced in the end that the problem was not serious. However, the fact that they worried about it at all is interesting, for it means that within CDF the question of whether they did have good evidence for the top quark could not be settled merely on the basis of the cuts that they used and the number of events that passed those cuts. Such information did not indicate whether the cuts had been tuned on the signal (although, as discussed in Chapter 5, one can reanalyze data to look for possible biases). Faced with the possibility that their own group structures and interests may have biased their test, some CDF members sought to rule out that source of error by reasoning about those structures and the motivations of themselves and their colleagues.

In fact, the very same features of CDF that prompted suspicions in some of tuning on the signal also helped to justify, for others, the claim that it had not happened. For example, questions arose over the choice of SVX *b*-tagging algorithm during run Ia in part because there were three groups in a heated competition to be "the tagger of choice." However, Mel Shochet, a cospokesperson for CDF at the time of Evidence paper, saw the competitive nature of CDF as a reason to conclude that tuning on the signal was very unlikely in CDF. While the competitive environment, in which people could freely develop their own approaches to analyzing the data, may have tempted people to try to come up with cuts that would give the "right" answer, Shochet argued that getting away with this would be difficult. "There were an awful lot of people there who were ready to hit the roof if it looked as if anybody was tuning cuts on [signal] data" (Shochet 1995).

I do not wish to exaggerate the importance of these considerations in the thinking of CDF physicists. For the most part, CDF scrutinized their top-quark results by carrying out the kind of detailed checks discussed in earlier chapters: running their algorithms on Monte Carlo and control sample data, looking in fine detail at the physical features of the candidate events, and so on. In the end, however, they found themselves in a situation in which small statistics would serve as the basis for a result of considerable significance for the physics community. Hence arose a significant concern that the desire to find the top would cause physicists to select their cuts in a biased way, either consciously or unconsciously, especially as it was widely felt that some study of signal events (e.g., in developing the SVX *b*-tagging algorithm for run Ia) was necessary to understand how these complex algorithms were working. These factors combined to make the context in which cuts were chosen relevant to the evaluation of the evidence.

This would not be a consideration if CDF physicists were perfectly rational and disinterested reasoning machines. If this were the case, then it is unlikely that these decisions regarding the placement of particular cuts would correlate with any particular outcome. It is because they are not such disinterested reasoning machines, and know that they are not, that experimentalists take steps to correct for the biases that may result. Most CDF physicists, in discussing their work on the top search, stressed that they took a very "conservative" approach to analyzing the top-quark data. They saw this as a kind of preventive action against possible biases introduced by tuning on the signal. In a situation where choices could be retroactively justified as "reasonable," and yet still give a biased result, CDF physicists report that they attempted to make choices so as to correct for such possible biases. Their solutions included some drastic measures, such as the run Ib "data moratorium" described in Chapter 5.

It is easy to miss the relevance of such considerations if one studies science only through published research papers. As real as concerns over tuning on the signal were for members of CDF, they do not mention it anywhere in the otherwise extaordinarily detailed Evidence paper. Yet CDF physicists certainly discussed it, amongst themselves, and with me during interviews. Scientific research papers, however, are governed by conventions that may well exclude discussions of evidentially relevant factors.

As though to correct for this, the very publication of a claim to have found evidence for the top quark was justified in part by most CDF physicists I interviewed by the fact that, as far as they were concerned, no tuning on the signal had taken place. Under the conventions of the research report, one cannot explicitly include one's worries about tuning on the signal and the reasons for believing it did not happen. Instead, one tries to satisfy oneself prior to publication that this worry does not count against the evidence claim in question. Because scientists cannot take it explicitly into account, they must be sure that they can safely ignore it. To the extent that researchers conscientiously satisfy this prepublication requirement, such a norm can safeguard the reliability of the mechanism of publication by which scientific results are disseminated. However, it has the negative effect (particularly for historians, sociologists, and philosophers of science) of obscuring an important piece of reasoning underlying the published claim. Furthermore, the exclusion of such considerations from the content of scientific papers effectively places them outside of the realm of science's public discourse, thus posing an obstacle to both critical discussion and informative dissemination of techniques for preventing or ruling out experimenter-induced bias. Such open discussion seems particularly important in an era in which experiments are increasingly performed in socially complex settings.

Admittedly, reasoning about the effects of individual or group propensities toward particular types of behavior might not fit easily into the canons of reasoning that govern the publication of results in physics. The process that is at the center of the problem is a human process, that of choosing a selection procedure. A detailed formal characterization of the problem might be difficult, insofar as it would require the characterization of both social mechanisms whereby subgroups of physicists in the collaboration made decisions regarding the analysis, and of psychological mechanisms employed by individual physicists during the decision-making process. However, as the case of CDF shows, such a detailed mathematical treatment may not be necessary. Unable to make such a formal assessment, CDF physicists sought to assure themselves that those social and psychological mechanisms were such that they could safely be left out of a formal assessment of the experiment's results. This they did, however, not by ignoring these mechanisms, but by reflecting on how they worked.

I do not claim that the inclusion of such reflections in published research reports is a prerequisite to enhancing the reliability of experimental knowledge. Careful pretrial planning can make such considerations irrelevant, and I have already suggested how counterfactual error-statistical calculations can assist the assessment of severity even where tuning on the signal has occurred. However, when the mental states of individual experimenters and the mechanisms by which they act in collaboration are evidentially relevant, our understanding of particular scientific results and scientific knowledge in general will benefit from a clear view of such factors.

7. SUMMARY

In these last two chapters, my aim has been to progress toward an episte-mology of experimental scientific knowledge. I have attempted to articulate a view of experimental evidence that takes seriously the causal linkages be-tween experimenters and the results they produce, while also focusing on those aspects of such connections that are relevant to experimenters' epis-temic aims. I do not claim that determining what they collectively have reason to believe based on the results of experiment (i.e., achieving ex-perimenters' success) is the only epistemic aim that a scientific community ought to take seriously. Researchers will typically have an interest in cer-tain kinds of knowledge that they consider more worthwhile than others, whether such preferences manifest themselves as a preference for a par-ticular form of data or as a desire to give certain kinds of explanations or achieve theoretical unification. A full understanding of the epistemology of science must encompass the wide range of intellectual cravings that scien-tists indulge in pursuing their research. However, for a science to give up interest in evidence would be like a chef turning wanton over the edibility of her creations. A theory of evidence may not exhaust, but must be central to, the epistemology of science.

My contribution to this project has been to attempt a conceptual clarifi-cation of the error-statistical program in philosophy of science, an approach with roots in the works of C. S. Peirce, Egon Pearson, and Karl Popper, but most recently articulated and advocated by Deborah Mayo. I have attempted this clarification by treating the error-statistical theory, more explicitly than it has been treated previously, as a theory of *evidence*. The chief advantage of casting the theory in these terms is to clarify its relation to *belief*, drawing on a number of Peter Achinstein's insights into scientific evidence.

To summarize: My claim is that the error-statistical theory of evidence satisfies the following four theses:

1. If *e* is evidence for *h*, then *e* provides a reason to believe *h*.
2. Whether *e* is evidence for *h* is an objective matter in that whether *e* is actually evidence for *h* is epistemically independent of whether anyone believes *e*, *h*, or that *e* is evidence for *h*.
3. Nevertheless, whether *e* is evidence for *h* is a fact that may partially supervene on the intentional and epistemic states of experimenters investigating *h*.
4. Evaluating evidence resulting from an experimental procedure re-quires consideration of the probabilities of various outcomes of that

procedure, based on the assumption of various hypotheses under consideration.

From 4 it follows that

5. The construction of probability models of experimental procedures aids in the evaluation of evidence.

On the assumption that experimenters do wish to evaluate evidence resulting from their experimental procedures, it follows that

6. It is wise to follow methodological rules that permit the construction of reliable probability models.

The following observation regarding our epistemic limitations has been confirmed in the present study:

7. When information about the intentional and epistemic states of experimenters is necessary for the calculation of probabilities of various outcomes, the construction of adequate probability models, or the communication of reasons for claiming that a model is adequate, becomes difficult or impossible.

(In this context, an adequate probability model is one that will not yield a misleading impression regarding the strength of experimental evidence, or more typically that will not mislead one into believing the evidence to be much stronger than it is.) From 6 and 7 it follows that

8. It is wise to follow methodological rules that help make the intentional and epistemic states of experimenters statistically irrelevant to experimental outcomes.

In other words, in the pursuit of experimenters' success, researchers ought to follow those methodological rules that facilitate the construction of probability models that they can justify as adequately reflecting the error probabilities of their testing procedures. No particular methodological rule will be absolutely required in every testing context.[25] Neither will every violation of a methodological rule, even when that rule would be called for by the testing context, result inevitably in some form of epistemological failure.

The present account does emphasize, however, the importance of determining the various ways in which a testing procedure may lead to error, and stresses how the probability models of experimental tests employed in standard statistical assessments can help rule out such errors, leading to the establishment of evidence claims, and hence to experimenters' success. Such models must themselves be understood as products of the experimenter's engagement and interaction with the world, an engagement which, if it is to lead to experimenters' success, must in a sense leave no traces of itself. That is to say that such success is promoted by rendering the problematic workings of individual or collective experimenters' psyches statistically *irrelevant*. Other kinds of relevance matter for philosophers and historians of the sciences, however. In particular, by focusing on the abstract dimension of probability models, one must not lose sight of the fact that, in order to isolate a single factual thread, experimenters necessarily tangle themselves up in the web that they are trying to unweave – but carefully and deliberately, so that in the end they can distinguish their own contributions from the rest of the pattern.

Epilogue

Writing about living science involves risk. When a theory has been decisively disproved, historians can treat its discussion as an autopsy – the cause of death may be in dispute, but the status of the subject is settled – though fallibly. The unrefuted claim, however, threatens to confound the author who takes it as his subject – especially when it has only been made fairly recently and is subject to ongoing scrutiny.

The CDF and D-zero collaborations continued their efforts as I wrote this book, making the tale of the top a continuing story. I merely relate the first chapter. Both collaborations continued to refine their analyses and reported their results as they finalized their data samples from run Ib. They published numerous studies of the top quark subsequent to the events I have described. Furthermore, as I write, CDF and D-zero are in the midst of run II, collecting more data with significantly upgraded detectors for both collaborations. Here I wish to summarize the status of the top quark based on these subsequent developments.

Both collaborations have found additional confirmation of the top quark's existence and have measured its main properties to greater precision. Two such properties have been the principal concern of these studies: the mass and the pair production cross section. These quantities have been inferred by both collaborations on the basis of data from individual decay channels as well as from all available data. Such further study has yielded results resoundingly in concert with the existence of the top quark as reported in their earlier papers.

(All of the results discussed here are based on the complete run Ia + Ib data set. Depending on the analysis and the trigger paths involved, the data sets range in size from about 109 to 125 pb^{-1}.)

One important result subsequent to the 1995 Observation papers was the appearance of a significant excess over background in the all-jets final state, which CDF reported in a 1997 publication. Using SVX b-tagging, CDF found 222 tagged jets, with an expected background of about 165 tags. An

293

alternative method requiring two SVX tags in a single event counted 157 candidate events, with about 123 expected from background. Previously, the large background in this decay mode had prevented its use in gathering evidence for the top. A mass estimate based on these candidate events yielded $m_{top} = 186 \pm 10 \pm 12$ GeV/c^2 (Abe, Amidei, et al. 1997).[1]

Three *PRL* papers published by CDF in 1998 explored different routes to understanding the top quark. One focused just on the dilepton decay channel. Using dilepton cuts similar to those in the Evidence and Observation papers, they found 9 candidate events in the run I data, with 2.4 ± 0.5 expected from background. This resulted in a mass estimate of $m_{top} = 161 \pm 17 \pm 10$ GeV/c^2 and a production cross section estimate of $\sigma_{t\bar{t}} = 8.2 ^{+4.4}_{-3.4}$ pb (Abe, Amidei, et al. 1998a). Another paper derived a mass estimate from lepton + jets decays: $m_{top} = 175.9 \pm 4.8 \pm 4.9$ GeV/c^2 (Abe, Amidei, et al. 1998b). Both lepton + jets and all jets final states were used in an estimate of the production cross section, which yielded $\sigma_{t\bar{t}} = 7.6 ^{+1.8}_{-1.5}$ pb (Abe, Amidei, et al. 1998c). CDF improved the dilepton analysis of 1998a by using more of the information in each event. The new mass estimate from dilepton events alone was $m_{top} = 167.4 \pm 10.3 \pm 4.8$ GeV/c^2. When combined with the previous estimate from lepton + jets and all jets decays, this gave an overall mass estimate of $m_{top} = 176.0 \pm 6.5$ GeV/c^2 (Abe, Amidei, et al. 1999a).

In a study of the kinematics of top events published in *PRD*, CDF compared their data to the predictions of various Monte Carlo models. CDF had been using these all along, but they sometimes worried over the trustworthiness of these models. This comparison, which focused especially on the HERWIG Monte Carlo, found that their models "reproduced the data well" (Abe, Amidei, et al. 1999b). One could see such agreement as conferring credit either upon the top quark or upon the Monte Carlo, depending on which model one regards as more in need of support.

Two *PRD* articles published in 2001 further refine CDF's estimates of mass and production cross section. A mass estimate combining a revised lepton + jets analysis with dilepton and all jet decay channels gave $m_{top} = 176.1 \pm 6.6$ GeV/c^2 (Affolder, Akimoto, et al. 2001a). At the center of a revised estimate of cross section were separate treatments of SVX and SLT b-tagged lepton + jets events. This yielded results from SVX events: $\sigma_{t\bar{t}} = 5.1 \pm 1.5$ pb, and from SLT events: $\sigma_{t\bar{t}} = 9.2 \pm 4.3$ pb. When combined with results from the dilepton and all jet channels, this gives $\sigma_{t\bar{t}} = 6.5 ^{+1.7}_{-1.4}$ pb (Affolder, Akimoto, et al. 2001b). (One historically interesting aspect of this paper is that it uses the older jet-probability algorithm for tagging secondary vertices as a cross check on the main SVX analysis using SECVTX.)

D-zero's publications in the wake of their Observation paper exhibit a similar pattern of piecemeal probing of the top's characteristics, with an emphasis on the production cross section and the mass. A pair of papers from 1997 reported on a mass estimate from the study of lepton + jets decay events and estimated the production cross section based on both dilepton and lepton + jets events. These gave a mass estimate of $m_{top} = 173.3 \pm 5.6 \pm 6.2$ GeV/c^2 and a cross section estimate of $\sigma_{t\bar{t}} = 5.5 \pm 1.8$ pb (Abachi, Abbott, et al. 1997; Abbott, Abolins, et al. 1997).

A subsequent improvement in the mass estimate derived from an analysis of dilepton events. Looking at six dilepton candidates, D-zero estimated $m_{top} = 168.4 \pm 12.3 \pm 3.6$ GeV/c^2 (Abbott, Abolins, et al. 1998a; 1999a), which they then combined with a revision of their previous estimate from lepton + jets events to arrive at $m_{top} = 172.1 \pm 5.2 \pm 4.9$ GeV/c^2 (Abbott, Abolins, et al. 1998b).

A 1999 revision zeroed in on the production cross section by extending D-zero's analysis to all jet final states. This analysis, which employed the technique of neural networks in the development of a more efficient identification of top events, gave an estimate in the all jets channel of $\sigma_{t\bar{t}} = 7.1 \pm 2.8 \pm 1.5$ pb. D-zero combined this with their estimate based on lepton + jets events for an overall estiamte of $\sigma_{t\bar{t}} = 5.9 \pm 1.2 \pm 1.1$ pb (Abbott, Abolins, et al. 1999b; 1999c). A subsequent paper by D-zero expands their lepton + jets and dilepton analysis, combines that with their all jets estimate, and gives a "final" run I estimate of $\sigma_{t\bar{t}} = 5.69 \pm 1.21 \pm 1.04$ pb (Abazov, Abbott, et al. 2002).

All of this may make it seem as though top-quark analysis has become dull and routine at CDF and D-zero, as physicists seek simply to move the number of significant figures further up the decimal ranks. However, the outward appearance of a routine accumulation of more precise measurements of the same parameter is entirely deceptive. Physicists are watching the studies of the top quark being carried out at the Tevatron with great interest because the top quark itself may hold the key to future theoretical developments. The plausibility of various candidates for the next phase of fundamental physical theory (supersymmetry and technicolor, for example) rests on just such details of the top quark (see Bhat, Prosper, and Snyder 1998 for an overview).

In a paper recently presented in Poland, CDF physicist Krys Sliwa discussed some reasons for thinking surprises may be awaiting physicists at the Tevatron as run II proceeds. Noting that in the small statistics related to top neither collaboration found "significant disagreements" with standard model predictions, Sliwa described "a few hints that the simplest hypothesis that the top candidate events are just the $t\bar{t}$ events and SM background may not be entirely correct" (Sliwa 2002, 3866).

These "hints" include the fact that indirect estimates of the mass of the top quark based on standard model assumptions yield slightly lower values (in the neighborhood of 150–167 GeV/c^2) than the measurements just cited from CDF and D-zero top data samples. Also, there are more double-tagged W + jet events in CDF's sample than the standard model would lead one to expect. Also mentioned by Sliwa are two CDF and one D-zero dilepton events that "yield poor fits to the $t\bar{t}$ hypothesis" and have "unexpectedly large" amounts of missing transverse energy as well as large transverse energy leptons. Furthermore, when individual top events are plotted by mass, both collaborations find more than expected at higher masses. With the larger data sets and slightly higher energies of run II, Sliwa noted that such hints "should be monitored carefully, as they may be offering us glimpses of new physics" (ibid., 3866).

Sliwa struck a hopeful note with regard to just those things in the data that threaten our current understanding. Even when experimenters find that they have achieved experimenters' success, they look more closely to see the interesting flaw in their achievement – the discrepancy that will mean, not failure necessarily, but the possibility of some new success to strive for. Just as the prelude to discovery should not be seen in terms of a monotonic preparation for the discovery that occurs, so the aftermath of discovery should not be seen as the straightforward unfolding of the consequences of the knowledge thus gained. CDF and D-zero went to great lengths to establish that they now know some things about the world that were previously unknown. Just as important is the probing of this new realm of information to find out just what it is that we do not yet know.

Notes

Chapter 1

1. See Crane (1980) for citation statistics on this and a number of other important papers from this period. More recent statistics on citations are kept at the SLAC–SPIRES (Stanford Public Information Retrieval System) website: http://www.slac.stanford.edu/library/topcites.
2. For example, "Indeed, such an additional [third] generation [of quarks and leptons] had already been predicted in 1973 by Makoto Kobayashi and Toshihide Maskawa" (Brown, Riordan, et al. 1997, 13).
3. Until the recent experimental findings of the DONUT (Direct Observation of Nu-Tau) Collaboration, announced in late 2000 (Kodama, Ushida, et al. 2001), the evidence for the elusive tau neutrino (ν_τ) was indirect, based on data on τ decays (see, e.g., Feldman 1981).
4. For an informative and detailed treatment of the developments described here, see Franklin (1983).
5. An excellent resource for information on the Sakata model's development, and on Japanese elementary particle physics in the middle of the twentieth century in general, is Brown, Kawabe, et al. (1991), which includes interviews, recollections, and translations into English of numerous important documents. On the Sakata model in particular, see especially Hayakawa (1991), Low (1991), Maki (1991), and Nakagawa (1991).
6. This result is credited to Gamba, Marshak, and Okubo, who presented it in a paper given at a conference in Kiev (Gamba, Marshak, and Okubo 1959).
7. The reason for the prohibition of such decays is that, as will be seen, the proton and neutron in the urbaryon scheme each have $N_{\chi_0} = 1$, while mesons have $N_{\chi_0} = 0$. However, other observable baryon states have $N_{\chi_0} = 0$, and the decay of such a baryon into a nucleon and meson would be a process with $\Delta N_{\chi_0} = 1$.
8. Bjorken and Glashow could not have cited the paper by Amati, Bacry, et al., which was received at *Physics Letters* just three days after their paper, although it was published earlier by about two weeks.
9. See Ne'eman (1974) and Ogawa (1985) for earlier discussions, and Hirokawa and Ogawa (1989) for a more detailed account of Sakata's intellectual development.

297

10. "The electron is as *inexhaustible* as the atom, nature is infinite, but it infinitely *exists*. And it is this sole categorical, this sole unconditional recognition of nature's *existence* outside the mind and perception of man that distinguishes dialectical materialism from relativist agnosticism and idealism" (Lenin 1927, 269, emphasis in original). In a paper originally published in 1965 and presented by Fujimoto at the 1970 Lenin Symposium, Sakata wrote, "One may quote the following two points as remarkable features of the physics of the present century. The first is the *recognition of the strata-structure of Nature*, in particular the discovery of a series of new strata of the microscopic world. . . . The second is the recognition of a limit of validity of the physical laws, in particular the discovery that the Newtonian mechanics is not the eternal truth of perfection. As a result, it established the following point of view for Nature: *there exist in Nature an infinite number of strata with different qualities* amongst each other. . . . Each of those strata is governed by its respective and proper laws of physics, and all of the strata are always in the middle of creation and annihilation, and they compose Nature as one and whole unified existence through their correlation and mutual dependence among themselves. *This point of view is called the dialectic-philosophical view of Nature, and it was already put forward by Engels at the end of the nineteenth century.* One may say as a conclusion that the atomic physics of the twentieth century re-discovered the dialectic-philosophical view of Nature" (Sakata 1971b, 200, emphasis in original).

11. For more on Sakata's philosophical views and their development, see Ogawa (1985) and Taketani (1991). Morris Low (1991) discusses the role of the three-stage theory as well as Bohr's correspondence principle in the development of the Sakata model.

12. In fact, they wrote that the Nagoya model belonged to "the next level hidden behind the urbaryonic level." However, as Maki pointed out years later, this was simply a mistake, as the urbaryon concept was not even introduced until the 1964 paper of Maki and Ohnuki discussed earlier (Maki 1989, 89).

13. Against the charge that Sakata was blinded by his philosophical "dogmatism" (see Section 8 of this chapter), it deserves notice that he and his collaborators considered a "core type model" in which one of the constituents of the lepton-B^+ system is assumed to be "the point-like core of the other" among the possible models for the relationship between leptons and B-matter, although this ran strictly contrary to Sakata's rejection of pointlike particles based on his dialectical materialism (Maki, Nakagawa, et al. 1960, 1179).

14. The system of units used in particle physics is based on units of eV (electron-volts), the amount of energy gained by an electron accelerated across an electromagnetic potential of 1 volt. One gigaelectron-volt (GeV) is the same as one billion electron-volts; other commonly used denominations are KeV (one thousand electron-volts), MeV (one million electron-volts), and TeV (one trillion electron-volts). Momentum and mass are then represented in terms of these units as well. Momentum is given in units of eV/c, while mass is expressed in units of eV/c^2, where c is a constant representing the speed of light in a vacuum. For convenience, particle physicists set $c = 1$. As a consequence of this convention, c is sometimes not written explicitly and energy, momentum, and mass are all given in units of eV. Context is nearly always sufficient to indicate the intended quantity.

15. As written, ADONE looks like an acronym. This is somewhat misleading. ADONE's predecessor at Frascati was the early electron-positron storage ring, the Anello di Accumulazione (AdA or ADA), which had been proposed in 1960 by Frascati's Bruno Touschek. It began operation in March 1961. When Frascati supplanted AdA with the first really large electron-positron storage ring, they named it ADONE, which is an Italian way of saying "Big AdA" (see Pellegrini and Sessler 1995).

16. In fact, it took considerable follow-up analysis and further experiments to reach agreement on the precise nature of the J/Ψ. For subsequent experiments that were of importance, see Goldhaber, Pierre, et al. (1976); Peruzzi, Piccolo, et al. (1976); Feldman, Peruzzi, et al. (1977); and Rapidis, Gobbi, et al. (1977). For an extensive historical discussion, see Pickering (1984).

17. A post–November revolution paper that did discuss the X-particle was published in 1976 by T. K. Gaisser and F. Halzen. They discuss events with long-lived particle tracks recorded in emulsion chambers in the context of the charm proposal. They also cite Hayashi et al.'s paper on the X-particle (Hayashi, Kawai, et al. 1972a), describing their interpretation as an instance of "production of a charmed hadron" (Gaisser and Halzen 1976, 3155). In 1979, Kotaro Sawayanagi, of Waseda University, reported in *PRD* on a search for cosmic ray events similar to the X-particle, noting that "It is now believed that some of the cosmic-ray X particles are charm particles, though not all of them" (Sawayanagi 1979, 1037).

   In another interesting exception to the trend, Emmanuel Paschos, referring to "models which introduced charm into weak interactions," places three footnotes. The first cites only the 1964 paper by Bjorken and Glashow. The second cites the GIM paper of 1970. The third footnote cites Amati, Bacry, et al. (1964); Hara (1964); Maki and Ohnuki (1964); Tarjanne and Teplitz (1963); and – surprisingly – Katayama, Matumoto, et al. (1962), one of the two 1962 papers to extend the Nagoya model to accommodate two neutrinos. This is the only case I know of in which an author of a paper in the *Physical Review* cites one of the original "extended" Nagoya model papers in connection with charm (Paschos 1977).

18. Andrew Pickering mentions the Nagoya model only once in passing in his *Constructing Quarks* (Pickering 1984, 86). He does not mention the X-particle at all. Cahn and Goldhaber's very useful book tracing the experimental supports for the standard model (with reprints) mentions Niu's event briefly (Cahn and Goldhaber 1989, 266), but not the Nagoya model. Two good "popular" treatments, by Michael Riordan and by Robert Crease and Charles Mann, do not mention the X-particle. Although both make passing reference to the early Sakata model, neither discusses the Nagoya model (Crease and Mann 1986; Riordan 1987).

   The physicist A. I. Sanda has gone further than most in recognizing the importance of Niu et al.'s work in some recent comments. Sanda has noted that in addition to the 1971 X-particle discovery, Niu et al. found another 20 events of the same type, which they were unable to publish in a journal, but presented at the Fourteenth International Cosmic Ray Conference at Munich in 1975. Sanda writes, "Niu and his group should be recognized as co-discoverers, if not *the* discoverers, of the charm particle" (Sanda 1998, 370).

19. My discussion draws upon Gottfried and Weisskopf (1984, vol. 1, 165–6). See also Ellis, Gaillard, and Nanopoulos (1976) and Cahn and Goldhaber (1989, 339–41).

20. Technically, the "left-handed" and "right-handed" components refer to the two solutions to the Dirac equations for a massive particle of spin $\frac{1}{2}$:

$$i\left[\frac{\partial}{\partial t} + \sigma \bullet \nabla\right] \chi^{(+)} = m\chi^{(-)},$$

$$i\left[\frac{\partial}{\partial t} - \sigma \bullet \nabla\right] \chi^{(-)} = m\chi^{(+)}.$$

See Gottfried and Weisskopf (1984, vol. 2, 198–204, 469–74) for a helpful discussion.

21. A quantity is a "scalar" if its value does not depend on the orientation of the coordinate system and is also invariant under reflection. As applied to particles, the term indicates that it is a particle with zero spin and positive parity. The Higgs field was hypothesized by Weinberg and Abdus Salam (Weinberg 1967; Salam 1968) – using an idea first suggested by Peter Higgs (1964a; 1964b; 1966) – to impart mass to the "intermediate vector bosons," while avoiding anomalous massless particles known as "Goldstone bosons." The Higgs field is invoked in the standard model to explain the masses of the elementary particles generally. The physical particle corresponding to the Higgs field is the "Higgs boson," which is yet to be detected experimentally.

22. Kobayashi's retrospective comments in 1997 suggest a more assertive attitude: "We felt excitement in the fact that we could conclude the inevitable existence of new particles from quite logical reasoning" (Kobayashi 1997, 139).

23. These figures are based on a search of both *PRD* and *PRL*.

24. One caveat regarding citation statistics deserves mention. Relying only on the number of occurrences of citations of a work in the "references" lists in other publications may lead one to underestimate that work's importance. By 1980, it is possible to find papers referring to the "Kobayashi-Maskawa angle" and the "Kobayashi-Maskawa matrix" without citing the KM paper at all. That evolutionary biologists use terms such as "Darwinian" without citing any of Darwin's publications probably does not indicate that the influence of the great naturalist is *declining*. In time, citations become superfluous.

25. "Since muons exist in nature for no apparent reason," Tsai's paper began (calling to mind I. I. Rabi's comment upon the discovery of the muon: "Who ordered that?"), "it is possible that other heavy leptons may also exist in nature" (Tsai 1971, 2821).

26. Notation conventions were not at this time uniform. Where some used p, n, and λ to denote the first three quarks proposed by Gell-Mann and Zweig, others employed the u, d, and s that came to be standard. Meanwhile, it took some time for physicists to settle on c for the charmed quark, as opposed to p', r, or ζ, all of which were also common.

27. As in Harari's scheme, Pakvasa and Sugawara's proposal represented the newly discovered charmed resonances at 3.1, 3.7, and 4.1 GeV as mixed states carrying

not only charm, but also the quantum numbers of the fifth and sixth quarks (Pakvasa and Sugawara 1976, 306).

28. Somewhat different proposals attempting to account for *CP* violation using additional Higgs fields were published previously by T. D. Lee (Lee 1973) and at around the same time as Weinberg by P. Sikivie (Sikivie 1976). Indeed, the proposal was to an extent anticipated by Kobayashi and Maskawa themselves, who had suggested additional scalar doublet fields as an alternative *CP*-violation scheme to the six-quark proposal.

29. Accounts conflict over the brand of the champagne. According to Yoh, it was Moët (Yoh 1998b, 35). Lederman reports that it was Mumm's (Lederman 1997, 106).

30. Neutral weak currents had been discovered by physicists at the Gargamelle bubble chamber at CERN in Geneva, Switzerland, in 1973 (Hasert, Kabe, et al. 1973), followed, after some difficulties, by confirmation from Experiment IA at Fermilab (Aubert, Benvenuti, et al. 1974; Benvenuti, Cline, et al. 1974). For a striking account of these events see Galison (1987, ch. 4).

31. For a discussion of experiments on parity violation and their implications for electroweak theory see Pickering (1984, 294–302). Prescott's own recollections of these experiments have been published as well (Prescott 1997).

32. Distinguishing the Nagoya approach from the "dogmatic" dialectical materialism of "Soviet philosophers," Taketani wrote, "I am almost a pragmatist. Dialectical materialism I found most useful, both in politics and in physics" (Taketani 1991, 93).

33. Holton himself alludes to dialectical materialist natural philosophy as a thema when he writes, "There are publications insisting, for example, that nature exists in an infinite number of strata with different qualities, each stratum being governed by its own laws of physics and each always in the midst of creation and annihilation" (Holton 1988, 15).

34. Sakata himself acknowledged the *usefulness* of the "bootstrap" or "*S*-matrix" approach pioneered by Berkeley theorist Geoffrey Chew but claimed that his own viewpoint was superior (see Sakata 1971c, 208).

35. Worse yet (for Ne'eman's assessment), the Maki, Nakagawa, and Sakata paper of 1962, whose suggestion of a fourth fundamental baryon was so uniformly ignored in the West, also contained a prescient discussion of neutrino mixing, a phenomenon prominently confirmed experimentally by the "Super-Kamiokande" experiment (Fukuda, Hayakawa, et al. 1998). In a significant change of fortunes for that paper, the matrix used to describe neutrino mixing is now referred to as the "MNS" (Maki-Nakagawa-Sakata) matrix, thus taking its place alongside the CKM (Cabibbo-Kobayashi-Maskawa) matrix used to describe weak decays.

36. "[I]t is difficult not to conclude that publishing a paper is roughly equivalent to throwing it away" (Hull 1988, 360).

37. According to the SLAC-SPIRES database of journal articles in high energy physics, the KM paper was the third most frequently cited article in the literature for the period from 1974 to the end of 2001, with over 3,500 citations in that literature to date. For the SLAC-SPIRES citation statistics see http://www.slac.stanford.edu/library/topcites.

Chapter 2

1. The Particle Data Group, in the 2002 edition of the *Review of Particle Physics*, estimates the masses of the strange, charm, and bottom quarks to be 80 to 155 MeV, 1.0 to 1.4 GeV, and 4.0 to 4.5 GeV, respectively (Hagiwara, Hikasa, et al. 2002, 26).

2. Although Fermilab achieved 500 GeV protons in 1976, energy costs constrained them to operate at 400 GeV (Wilson and Kolb 1997, 363, n.21).

3. Since placing a bar over the symbol for a particle denotes its antiparticle, the suffix "-bar" is used as an equivalent means of specifying an antiparticle.

4. Although the proton-antiproton scheme had not been definitively ruled out at the time, a 1977 article in *Physics Today* suggests that Wilson did not favor the idea. In discussing the prospects for colliding-beam experiments at Fermilab, he mentioned two intersecting ring schemes for proton-proton collisions, but not proton-antiproton collisions (Wilson 1977).

5. Henceforth, citations of the minutes of the weekly CDF meetings will be denoted by *CDF*, followed by the date of the meeting.

6. The CDF detector still exists, and the collaboration has continued to make changes to it, some of them quite substantial. Many of the things I say here about the CDF detector as it was in 1987 are still true, but some are not. I will not attempt to keep track of changes made after the events described in this history, however. The detector described here is already a thing of the past. Hence, I speak of it in the past tense, but keep in mind that neither the CDF detector nor the CDF collaboration is a relic of history. Both are very much alive at the time that I write this.

7. In colliding-beam physics, the conventional spatial coordinates are defined as follows: the z axis coincides with the beam axis, the x axis points horizontally away from the center of the ring, and the y axis points vertically, orthogonal to the x and z axes. But x, y, and z are not the coordinates that are most commonly used. More convenient coordinates are the *polar angle* (θ), measured relative to the proton beam axis, and the *azimuthal angle* (ϕ), measured relative to the plane of the accelerator. Finally, there is *pseudorapidity*, η, defined as $\eta = -\ln(\tan(\theta/2))$, with θ ordinarily measured assuming that the event's *primary vertex*, or original collision point, is located at $z = 0$ (see Figure 2.5).

8. Galison (Galison 1997b) interprets this difference as a fundamental epistemological discontinuity between the image and logic traditions. In his view, the difference is so radical that the two groups constituted distinct linguistic communities. Thus, hybridization of the two traditions could only come about through the development of a "pidgin" language enabling communication and cooperation. I argue against Galison's "epistemological discontinuity" claim, and his consequent appeal to a physics "pidgin" (Staley 1999a).

9. CDF kept a library of papers prepared for the collaboration called "CDF Notes." These were numbered sequentially, and would eventually number in the thousands. The CDF note on the top by Fox and Romans (CDF-70) bears no date, but based on its place in the sequence of CDF notes, it was distributed some time in 1980.

Chapter 3

1. One barn is 10^{-24}cm^2. Thus the integrated luminosity, expressed in units of inverse barns, can be thought of as a measure of the density of collisions accumulated over time within a cross section of the beam. Typically, as the integrated luminosity increases, it is expressed by going to smaller units of area. It is as if the U.S. Census Bureau reported population in terms of people per square yard rather than the total number of people. As population increases, the area available per person diminishes. One might begin reporting data in terms of people per square yard, then people per square foot, and then per square inch. The advantage of this system of units is that integrated luminosity can be calculated more or less directly from the constantly monitored instantaneous luminosity. As CDF collected ever more data, they began reporting their accumulated data in terms of inverse picobarns (10^{-12} barns) instead of inverse nanobarns (10^{-9} barns).

2. In a Lego display as seen in Figure 3.1, the cylindrical geometry of the calorimeters surrounding the event is projected onto a two-dimensional grid, viewed in perspective. Towers arising from squares in the grid represent energy deposits in calorimeter segments. Towers are divided into two colors representing hadronic calorimeter energy and electromagnetic calorimeter energy.

3. At the time of the 1981 Design Report, 12 Pisa physicists belonged to CDF (Franco Bedeschi as a fellow at Fermilab). Of these, 8 had been coauthors on the NA1 collaboration's paper describing the silicon microstrip detector that appeared in *Nuclear Instruments and Methods in Physics* (Amendolia, Batignani, et al. 1980).

4. The Pisa silicon detector group, which included both CDF and ALEPH members, noted in a 1984 article on "High Resolution Silicon Detectors for Colliding Beam Physics" that "High density preamplifiers with low power consumption are under study and first prototypes have been produced. Output signals are transmitted to ADCs [analog-to-digital converters] through especially [sic] developed twisted pair cables made from 50 μm diameter coated copper wire.... The number of readout lines must be reduced by local multiplexing" (Amendolia, Batignani, et al. 1984, 84).

5. This was pointed out to me by John Yoh.

6. However, as in that case, it would be a mistake to conclude, as Galison there does, that this is a matter of two groups, previously operating in different linguistic frameworks, achieving communication by forging a "pidgin" that allows them to use the same terms while attaching different meanings to them. See Staley (1999a).

7. This is an inclusive estimate of data from all trigger paths. For individual results, usually not all of the data set is used. Thus, for example, the top quark mass limits from the 1988–9 run are based on a smaller data set of 4.1 pb^{-1}.

8. Statistical significance will be discussed more carefully in Chapter 6.

9. See Edwards (1992) for a discussion and defense of likelihood methods in scientific inference.

10. Although I will defend significance testing on philosophical grounds later in this work, this should not be taken as an indictment of the likelihood analyses

described here. Particular uses of likelihoods may be perfectly compatible with the philosophical theory of evidence that I advocate.

11. When I interviewed him in the fall of 1995, Kuni Kondo was continuing to work on developing a likelihood method for reconstructing events, in the belief that such a method will prove useful in the long run. "My method is alive," he said (Kondo 1995).

12. I am thankful to Krys Sliwa for explaining these aspects of the analysis to me in correspondence.

13. Henceforth, references to the minutes of the top group meetings will be indicated by *Top*, followed by the date of the meeting.

14. This argument assumed the accuracy of the standard CDF top analysis, which, at the time that I corresponded with him (1996), Sliwa regarded with some skepticism. Sliwa eventually did have his top identification method "blessed" by the collaboration.

15. This last claim is especially contentious, as it assumes that the only justification for the Tevatron upgrade was the search for the top quark. In spite of my own focus on that endeavor in this study, there were many other physics problems being pursued by CDF and by Fermilab's collider program in general, not to mention the fixed-target experiments that also benefited from an improved Tevatron. In any case, the data came from the Tevatron regardless of who claimed credit for the discovery.

16. In an uncharacteristically ambiguous piece of nomenclature, CDF used the acronym SVX to denote both the silicon vertex detector itself and the method of tagging events with secondary vertices that makes use of data from that detector. Context typically provides sufficient information to determine which sense of the term is intended.

17. Since the b quark always hadronizes before decaying, it would be more accurate to describe these decay processes as starting with a b-carrying meson or baryon, rather than with a b quark *simpliciter*. However, to avoid rendering the description of these searches more opaque than necessary, I will ignore this complication.

18. In Chapter 5, I will examine in more detail the concerns that some CDF members had over the choice of SVX b-tagger.

Chapter 4

1. Authorship identifications within CDF should be taken with a grain of salt, particularly with respect to the Evidence paper. Published CDF papers listed *all* collaboration members as authors, regardless of their contribution to the work presented. The analyses used in the top search emerged within working groups of varying sizes. CDF had no formal mechanism for deciding who would actually write what, and the only basis for making any attribution of this sort is agreement among collaboration members regarding who was primarily responsible for writing various papers or parts of papers.

2. This title had its origins in an earlier draft circulated among the "core group" of authors. Alluding to the movie "Dr. Strangelove," it bore the title: "Evidence for Top Quark Production in $p\bar{p}$ Collisions at $\sqrt{s} = 1.8$ TeV, or, How I Learned to Stop Worrying and Love the Bomb."

3. The scope of this pronouncement of Sliwa's is meant to apply not only to the Evidence paper, but to CDF's Observation paper as well (see Chapter 5). D-zero's Observation paper would likewise fail this test. The historical basis for this claim is difficult to verify, as many bubble-chamber discoveries did not include an estimate of statistical significance in terms of numbers of standard deviations.

4. Historically, the use of "Evidence" in a title as a more moderate term than "Observation" is not a novelty. Some historically significant examples of particle physics papers with such titles are Rochester and Butler (1947); Steinberger, Panofsky, et al. (1950); Maglic, Alvarez, et al. (1961); Pevsner, Kraemer, et al. (1961); Hanson, Abrams, et al. (1975); and Bagnaia, Banner, et al. (1983).

5. I mean "textual" in a broad sense, including the tables and figures that convey much of the information presented in scientific publications.

6. Wray presents this in criticism of the "invisible hand" explanation of the success of science offered by David Hull (1988). In an invisible hand explanation, "a particular outcome is described as an unintended consequence of the intentional behavior of a number of individuals. The individuals have one end in mind, and act accordingly; but their concerted efforts give rise to a consequence that was no part of their intentions" (Wray 2000, 164). Although Wray makes a valuable point about the intentionality underlying the establishment of scientific institutions, Hull is correct in that the contributions of a particular scientist or group of scientists to the epistemic aims of science need not be explained by appealing to the primary intentions of the scientist or scientists in question. As such, Hull's invisible hand explanations could be maintained while also accepting Wray's claim. Even in Wray's account of Hull's argument, the "structure of science" is "given" (Wray 2000, 166). Accordingly, Wray's point could be reformulated: An invisible hand explanation may explain why particular scientists manage to contribute to the epistemic aims of science, in spite of their nonveritistic intentions, given the structure of scientific institutions. However, one cannot invoke an invisible hand explanation of why scientific institutions have that kind of structure.

7. Recall that the godparent system had been put in place by Roy Schwitters in 1982 as a system to review the work being done by various groups on parts of the detector (see Section 6 in Chapter 2). Even then, a godparent committee's principal responsibilities included verifying that the group over which it watched was meeting the design expectations for the detector and reporting problems as they arose.

Chapter 5

1. According to CDF member Mark Timko, systematic error should have been reported to be ± 15 GeV$/c^2$. The discrepancy was the result of a resolution factor on corrections to jet energy measurements that had not been factored into the published result (Timko 1998).

2. Such empirical testing of evidence claims is to be expected if, as Peter Achinstein has argued, evidence claims are themselves empirical claims rather than logical truths to be known a priori (Achinstein 1995).

3. See Chapter 6 for a more thorough discussion of this point.
4. To make the results easier to read at an intuitive level, I have shown the negative logarithm of the significance calculations, rather than the significance estimates themselves. For ease of comprehension, I will speak of "increases in significance" where it is to be understood that the actual value of the significance estimate has decreased, meaning that the probability of obtaining such a result, on the assumption of a fair coin, has in fact decreased.
5. The calculations used in compiling Figure 5.3 are rough, back-of-the-envelope calculations using data presented informally at several CDF meetings. Systematic uncertainties have been ignored. Hence, these are *not* a substitute for a full and careful study of the sort used for official CDF publications. It was also impossible for me to take into account changes, mostly minor, in the algorithms employed by the counting experiments during 1995. The data from July 1998, which I have included for comparison, are based on algorithms that are significantly different from those used in reporting earlier results (Tollefson 1998). Already by the time of the Observation paper, CDF had switched from method 1 to method 2 for estimating the SVX background, and the July 1998 dilepton results incorporate searches including τ leptons, which had previously been excluded from the search. Thus, I do not presume here to offer a substantive analysis of CDF's data justifying conclusions about physics. My aims instead are to (1) engage in a historical project of explaining the judgments of CDF members regarding their own results and (2) set the stage for a philosophical exploration of issues regarding biased tests.
6. Also of interest here is the use of mass reconstruction as a cross-check on the counting experiments. In July 1995, with about 100 pb^{-1} worth of data, CDF looked at plots of the mass estimates derived from the candidates in the SVX and SLT samples. Single- and double-tagged SVX candidates formed well-defined peaks in the neighborhood of 170–180 GeV/c^2. No such peak was exhibited by events tagged by the SLT algorithm alone, for which the distribution was somewhat flat over the 140- to 180-GeV/c^2 region.

Chapter 6

1. Nor *should* they have done so, according to Peter Achinstein, who argues that most theories of evidence offered by philosophers have rested on a pair of faulty assumptions: (1) that evidence is a weak concept and (2) that the evidential relationship is a priori. Any theory resting on *these* assumptions will prove inadequate for the purposes of working scientists (Achinstein 2000; 2001, ch. 1).
2. I recall this demonstration having been given by Lowell Herr, from whom I learned physics in high school.
3. The example under discussion by Suppes is a test of a linear response learning theory. Obviously, different ceteris paribus conditions will have relevance for different kinds of experiments. Bad odors are fairly routine in high energy physics labs and are not regarded as threatening to the reliability of experimental results.
4. This term derives from the neologism introduced by J. H. Woodger (1962) to describe horse(*equus*)–rider(*homo*) pairings: *equomos*.

5. I do not claim that no philosophers have held the view criticized by Gooding, but I do maintain that the text in which he characterizes the "received view" fails to identify any philosophers who would agree with the claim that the relationship between theory and experiment is a logical relationship between propositions.
6. I illustrate this point in the next chapter with the problem of "tuning on the signal."
7. For other scientific uses of inaccurate models see Wimsatt (1987).
8. To the best of my knowledge, this is the only place in their analysis where CDF used anything approaching a Bayesian statistical technique. Because the Gaussian distribution referred to here is a distribution over population parameters, it is not a proper frequentist probability distribution. However, this Bayesian technique served an error-statistical aim: the enhancement of the severity of the test to which they were subjecting the top-quark hypothesis. The effect of the Gaussian smearing of the background parameter was to "spread" the Poisson-distributed Monte Carlo results, raising the probabilities on the tail of the null distribution, where CDF's results lay. In effect, CDF was hedging against their own uncertainties regarding their background estimate by making the significance calculation more "conservative."
9. As Peter Galison relates, "model sampling" has at times been used to describe Monte Carlo techniques, particularly by statisticians disputing the novelty of early work by Stanley Ulam, John von Neumann, and Nicholas Metropolis (Ulam and von Neumann 1947; Metropolis and Ulam 1949), who pioneered the technique that Metropolis dubbed "Monte Carlo." Some statisticians insisted that it was simply a variant of the "model sampling" that statisticians had been using for quite some time (since the seventeenth century, according to one mathematician) (Galison 1997b, 771–4). Regardless of the priority dispute, the term "model sampling" is an apt description.
10. Such use of the data to decide, after the fact, whether to report a confidence limit or to claim a significant result and report a confidence interval on the resulting measurement leads to confidence intervals that fail to include the true value of the measured parameter more frequently than indicated by the confidences officially reported on the basis of such findings. This has been established by Gary Feldman and Robert Cousins, who also propose a method for avoiding the problem by following a single interval-construction method that is applicable for both purposes while retaining the accuracy of quoted confidence levels (Feldman and Cousins 1998; this reference was brought to my attention by Martin Krieger).
11. These issues are treated also by Mayo (1996, esp. 192–8), but I here rely on the more detailed treatment in her previous work (1985).
12. Here and in what follows, I have altered the notation for consistency with my usage.
13. Mayo gives a similar example involving fish populations (1985).
14. Some qualification may be needed here. Achinstein proposes to understand the objectivity of potential evidence as entailing that if *e* is potential evidence for *h*, then *e* is a good reason to believe *h* in an "abstract sense" that makes no reference to any person or epistemic situation (Achinstein 2001, 25, 28). I prefer to say that such evidence provides an objectively good reason to believe *h* in the sense that any person in an *ideal* epistemic situation would be rational

to believe *h* for the reason that *e*. The latter concept is compatible with the error-statistical account in a way that the former is not. I argue for this position in Staley 2003a.

15. I have Peter Achinstein himself to thank for bringing this issue to my attention.
16. That individual experimental results, as well as many general hypotheses, have this character of comprising strictly incompatible possibilities was lucidly argued by Pierre Duhem ([1914] 1954, 152–3; see also Duhem [1892] 1996, 90–1).
17. When the hypotheses in question are unrelated, a "degenerate case" comparative assessment can be made only insofar as the pairs (*e*, *h*) and (*e′*, *h′*) fall into different categories with regard to the classificatory concept. That is, one might then say that *e* is stronger evidence for *h* than *e′* is for *h′*, when *e′* is not evidence for *h′* at all, while *e* is evidence for *h*. However, nothing seems to be gained by such mistreatment of our language. To say "Herbie Hancock is a better musician than my cat" sounds funny precisely because it invites thinking of my cat as a musician, which he certainly is not in any ordinary sense.
18. The unavoidable reliance on models of the very systems used to carry out the experiment in the first place (such as the models of the interactions of protons and antiprotons used to simulate the processes of *b* production and decay in the HERWIG Monte Carlo mentioned in Section 5.4) illustrates a point made by David Resnik: Hacking's attempt to ground realism in the experimental practice of intervention independently of the representational activity of scientific investigators fails to take into account the role of theory in experimental practice (Resnik 1994). Here experimenters' use of protons and antiprotons to investigate the possible existence of top quarks required them to represent theoretically the interactions between protons and antiprotons. They could not otherwise have made sense of their results with any confidence.

Chapter 7

1. In what follows, when giving citations to works by Peirce, I will give citations in the standard manner for both the *Collected Papers* (*CP*, followed by volume and paragraph number) and the *Chronological Edition* of Peirce's writings (*CE*, followed by volume and page number), for example: *CP* 2.696; *CE* 4, 411.
2. It may happen that the experimenters themselves do not know that their experiment is evidentially successful with respect to a particular hypothesis, but someone else does know this. We might then make the relativization to a particular epistemic agent explicit and say that the experiment is epistemically successful with respect to that hypothesis *for* such and such a person.
3. As will be seen, tuning on the signal can be a cause of experimenters' failure alone, or of experimenters' and epistemic (but not evidential) failure, or it can, by causing evidential failure for some hypothesis of interest, induce all three types of failure at once. Allan Franklin's valuable discussion of bias arising from tuning cuts to produce a desired outcome (Franklin 1998; 2002) includes a number of illuminating examples, including a number of examples in which physicists pursue the strategy of "varying selection criteria," which I explore in detail later.

4. Here I have assumed a Poisson distribution and calculated significance in the same manner as described in the previous chapter.

5. Insofar as the predesignation of stopping rules is simply another aspect of the predesignation of statistical tests in general, the difficulties discussed in Mayo and Kruse (2001) in explicating methodological intuitions regarding stopping rules from the Bayesian and likelihood-theory perspectives already provide reasons to wonder whether those theories can give a coherent account of the issues discussed here. Daniel Steel has argued that for some Bayesian confirmation measures stopping rules are evidentially relevant, but only on the condition that the likelihood principle is abandoned (Steel 2003).

6. The central preradiator (see Section 4 in chapter 3) had since been installed, and the central tracking chamber had been modified to yield information about the change of energy of particles over distance (dE/dx), a useful aid to identifying particles.

7. The model of the data in its entirety cannot be held exactly constant under the proposed variation in the experimental model. When the experimental model changes, this changes the definition of the test statistic, which is also part of the model of the data. However, the data that determine the value of the test statistic can be held constant. In the present case, holding the data model roughly constant while varying the experimental model amounts to supposing that all measurements made on particles produced in collisions were to remain the same, but supposing that different choices were made regarding the cuts imposed on those measurements.

8. See A. Franklin (1998; 2002) for more examples.

9. As discussed in Chapter 5, some of these calculations were in fact carried out within CDF and shown at collaboration meetings. They did not at that time become part of CDF's official presentation of the results of their top search, although such a step had been advocated by at least one member of CDF, Fermilab physicist Morris Binkley (Binkley 1995).

10. The method may also be useful in cases where an experimenter already knows that an experiment is evidentially successful but wishes to convince a skeptical reader.

11. A similar approach can be employed where questions arise about whether a predetermined "stopping rule" has been followed: From the data collected, sample smaller subsets of data and calculate apparent significance levels based on those subsets. In this way, the sensitivity of the significance level to the precise stopping point in gathering data can be evaluated.

12. I do not intend to suggest by this that CDF ever suggested that the SLT results by themselves constitute evidence for the top, only that whatever severity the SLT test of the top hypothesis might appear to have would have to be itself regarded as quite uncertain, based on these calculations. It would be a mistake to regard the CDF evidence claim in the Evidence paper as resting only on the outcome of the counting experiments combined. Although only those results are incorporated into the significance estimate, the claim is supported as well by appeals to the kinematics of the events and the fact that plotting mass estimates from reconstructing the chosen events as top quarks yields a reasonably well-defined peak.

13. My naïve calculations indicate the following apparent significance levels: jet vertexing – 0.030 (reported as 0.032 in the Evidence paper); jet probability – 0.201; d-ϕ – 0.036.

14. Taking "subjective" in its usual sense would also violate the spirit of Peirce's theory of probable inference. Accordingly, I would propose interpreting his phrase "subjective circumstances" to refer simply to facts regarding the *subject* (i.e., person) who gathered the information that forms the basis of the inference. Of course, such facts may be as objective as you like.

15. Although Peirce clearly recommends the use of randomization ("leave that to the operation of chance"), he equally clearly states that the desideratum is simply an unbiased sampling distribution, with randomization merely a tool toward achieving that aim.

16. Here Savage is describing an argument due to G. A. Barnard, and admits that he has never been "comfortable with that argument" (Savage 1962, 76). It is not clear, however, whether Savage is expressing hesitancy regarding the claim that the experimenter's intention is irrelevant because it is an isolated mental state ("locked up inside his head") or with the claim that the experimenter's intention is irrelevant because it is epistemically inaccessible for anyone besides the experimenter ("cannot be known to those who have to judge the experiment"). Possibly he is skeptical about both claims. In any case, Savage is perfectly comfortable with the *conclusion* of Barnard's argument: "The likelihood principle . . . affirms that the experimenter's intention to persist does not change the import of his experience" (Savage 1962, 18).

17. Savage uses a very general "informal" statement of Bayes's theorem: $\Pr(\lambda|x) \propto \Pr(x|\lambda) \cdot \Pr(\lambda)$, where $\Pr(\lambda)$ is the prior probability that some unknown parameter has value λ, $\Pr(x|\lambda)$ is the probability of observing datum x given that the parameter in question has value λ, and $\Pr(\lambda|x)$ is the posterior probability that the parameter has value λ, given that datum x has been observed.

18. The argument for this claim differs in Peirce's two discussions of the example, as does the scope of the claim itself. In "The Order of Nature," published as part of his "Illustrations of the Logic of Science" series in *Popular Science Monthly* in 1878, Peirce states the point categorically: "any plurality or lot of objects whatever have some character in common (no matter how insignificant) which is peculiar to them and not shared by anything else" (*CP* 6.402; *CE* 3, 310 [1878]). Peirce's argument relies on allowing "negative characteristics" to count as characters. By the time of Peirce's "Theory of Probable Inference" of 1883 – to which Keynes is replying – Peirce presents his argument without the nominalistic reference to "negative characteristics": "all the qualities of objects may be conceived to result from variations of a number of continuous variables." So, just as "in geometry a curve can be drawn through any given series of points without passing through any one of another given series of points" independently of the number of dimensions involved, Peirce concludes that "any lot of objects possesses some character in common, not possessed by any other." Peirce then qualifies this, by admitting that this would not be strictly true "if the universe of quality is limited," but notes that this makes little difference for purposes of un-predesignated inference since we nevertheless find ourselves unable to set any limits on the frequency with which we draw erroneous conclusions.

Under such circumstances "we could not reason that if the *M*'s [members of the population under investigation] did not generally possess the character *P* [the character being projected], it would not be likely that the *S*'s [members of the sample] should all possess this character" (*CP* 2.737; *CE* 4, 434–5 [1883]). A clearer invocation of the severity requirement can hardly be sought.

19. Achinstein reformulates the historical thesis as "Whether some claim *e*, if true, is evidence for an hypothesis *h*, or how strong that evidence is, depends on certain historical facts about *e*, *h*, or their relationship" (Achinstein 1995, 156). Nothing in this argument turns on whether the historical thesis is interpreted in Snyder's version or in Achinstein's formulation.

20. Note that the test procedure here may include the process of hypothesis construction (as well as the construction of a model of the hypothesis), so that this also takes into account the relevance of an investigator's epistemic states where there is suspicion that hypotheses are being concocted ad hoc to fit the data in an unreliable way (also known as "gellerization"; see Mayo 1996, 201–4).

21. For a thorough discussion of this issue from the error-statistical standpoint, see Mayo (1996, ch. 10).

22. The stopping rule for run Ib was not so straightforward, and in fact was a subject of debate within CDF, as discussed in Chapter 5.

23. Peculiarly, this comes after Rosenkrantz has written without qualification that the likelihood principle "implies, among other things, the irrelevance of the experimenter's intentions when to stop sampling, of his hopes of confirming (or of disconfirming) the theory, of outcomes that were not but might have been observed, and of whether the theory was formulated in advance or suggested by the observations themselves" (Rosenkrantz 1977, 121). Rosenkrantz's own more nuanced discussion later in his book indicates that his principles commit him only to the third item on this list – the irrelevance of outcomes not observed.

24. I do not take such objections to be well founded. I am in fact puzzled when such complaints issue from Bayesians who adopt the "counterfactual" solution to the old evidence problem. On that account, if I already know *e* prior to assessing its degree of support for a hypothesis *h*, then I should, in applying Bayes's theorem, consider not the probability that I actually assign to *e*, which would be (very close to) one, but rather the probability that I *would* assign to *e* if I did not already know it to be true, based on my *other* background knowledge. It is unclear why it should be kosher to incorporate information about beliefs I do not have, but would have under different circumstances (not an easy task under the best circumstances), but objectionable to consider the probabilities of outcomes that might have, but did not, occur in a given experiment (information that, with a little careful planning, can be straightforwardly ascertained).

25. Thus the present view accords with John Stuart Mill's general views about rules of scientific method, as implicit in both his essay *On Liberty* and in his *System of Logic* (Staley 1999b).

Epilogue

1. When two errors are given in the following, the first will be the statistical error and the second the systematic error.

Methodological Appendix

Any time that a philosopher sets out to discuss a historical episode, sensible readers will be suspicious. Has the episode at hand been chosen simply because it fits the philosopher's existing commitments? Is the story being slanted to make those views look more plausible than they would if the story were told differently? To address these suspicions, let me comment briefly on the philosophical methodology of the present study. One of the themes of this study is the importance of selection procedures and the potential for error that arises from using biased selection procedure. It is therefore only fair to ask about the selection procedures used in the present study.

How was this episode chosen? In a sense, I did not choose this episode for study; rather, it chose me. More precisely, the origins of this project lie in a striking case of serendipity when I was searching for a dissertation topic as a graduate student. I wished to carry out a study of some experiment or other at Fermi National Accelerator Laboratory. I chose that laboratory because I had worked there during summers as an undergraduate, and I still knew some people there. I was ready to study any experiment that sounded sufficiently complicated to provide interesting philosophical problems. My expectation was that *any* experiment being done at Fermilab would qualify.

I called the physicist who had been in charge of summer programs during my undergraduate days, Drasko Jovanovic. I did not know that he was a member of CDF. I asked him what was happening at Fermilab. He replied that there was "exciting" news, but he could not tell me what it was. "Call back in a week," he said. That was April 19, 1994. On April 26, the headline "Top Quark, Last Piece in Puzzle of Matter, Appears to Be in Place" was on the front page of the *New York Times*. CDF had announced finding evidence for the existence of the top quark, and I had an experiment to study that would prove to pose many interesting challenges. Approximately one year later, both CDF and D-zero released announcements claiming the "observation" of the top quark. But by that time I had already begun collecting information.

How were informants chosen? Much of this study is based on interviews that I conducted with physicists involved in the search for the top quark. Most of these physicists are members of the CDF collaboration, although I also interviewed several members of the neighboring D-zero collaboration. However, at the time of the interviews, CDF and D-zero had about 450 member physicists each. Clearly I could not interview everyone. Nor would it have made sense to do so, since I was interested chiefly in the search for the top quark, and not everyone in either collaboration was directly involved in that search. Drasko Jovanovic had told me to talk to Henry Frisch, a CDF member intimately familiar with the top search, who also had an interest in preserving the historical record of the collaboration. Henry gave me a list of people he thought I should talk to. Contacting people at first by e-mail or phone, I then started accumulating contacts by asking each person I contacted for the names of other persons who were knowledgeable about or involved in the top search. I also sought the names of those who had been in the collaboration from very early in its history or who had played important roles in building the CDF detector. I wished to speak also with some members of the collaboration who were not directly involved in the top search. Eventually, I had enough contacts to visit Fermilab and talk to people. On the first such visit, in June 1995, I simply took notes. On two subsequent visits, in October 1995 and July 1998, I recorded my interviews.

Because it was a CDF member whom I contacted initially and because CDF was the first group to announce evidence for the top quark, my investigation at first focused exclusively on CDF. However, several CDF members suggested that I should also talk to members of D-zero. I was already immersed in the details of CDF's history and realized that I had to make a decision as to how much attention I was going to give to the D-zero collaboration. I realized that I could not in any reasonable time frame write about both collaborations at the level of detail that I was hoping to achieve. Furthermore, CDF's initial evidence for the top quark was laid out in a lengthy, detailed article in *Physical Review D*, the writing of which had involved the collaboration in a lengthy internal debate over the significance of their findings. When both collaborations later announced the "observation" of the top quark, they released much shorter articles, in *Physical Review Letters*, which did not create nearly as much controversy. As a philosopher interested in the concept of evidence, I decided that the earlier, more hotly debated evidence claim issued by CDF would prove to be a richer topic for my studies. Consequently, this work centers on CDF and their top search, particularly the genesis and subsequent fate of their first evidence claim.

To what extent are the scientific controversies described colored by the author's own opinions? The study presented here should not be taken to constitute any kind of *evaluation* of any of the work done by physicists involved in the search for the top quark. Nevertheless, you will find that I discuss several controversies that did arise within CDF over their findings, and that I go into some detail regarding some of them. This I do with some trepidation, as the controversies themselves were in some instances the locus of strong feelings, and physicists, like others, quite reasonably prefer to put bitter disputes with their peers behind them whenever possible. However, my aim here is in part to understand just such disputes as arise amongst experimenters engaged in a collaborative effort to find the answer to pressing empirical questions, particularly with regard to the role that different factors, both epistemic and nonepistemic, might play in such disagreements. Such an investigation cannot be pursued without discussing the development of discord. My presentation of the facts about how physicists argue amongst themselves cannot make them "look bad" unless one were to start out with a dangerously unrealistic expectation of what scientific conduct looks like. My aim is to present considerations that were regarded by the physicists in question themselves as relevant for the evaluation of evidence, within the context in which such evaluation was carried out. I also seek to understand whether such considerations are relevant, and if so, why they are, from the perspective of a philosophical theory of evidence.

Consequently, I have no particular stake in identifying one opinion or another amongst those held by CDF physicists as being correct, except insofar as I found on several occasions particular physicists making methodological comments with which my own views are in agreement (not that I take such agreement by itself to be good grounds for holding the opinions in question). Agreement on such methodological points, however, should not be taken as a sign of agreement in debates over particular claims regarding CDF's results. As an interviewer, I attempted to project a general attitude of neutrality, tempered by a sympathetic interest in the arguments of those I interviewed. I often found that interviewees were keen to persuade me of their own opinions. Although I cannot say that I kept myself free of all opinions regarding any of the controversies here discussed, I can say that I have a quite low degree of confidence in whatever opinions I do have on these matters of scientific fact. The physicists who worked on the search for the top quark put years of their lives into understanding their work. My aim is not to second-guess them.

References

[Abbreviations: *PRL* = *Physical Review Letters*, *PRD* = *Physical Review D*, *PTP* = *Progress of Theoretical Physics*, *PL* = *Physics Letters*]

Abachi, S., B. Abbott, et al. [D-zero] (1994). "Search for the Top Quark in $\bar{p}p$ Collisions at $\sqrt{s}$ = 1.8 TeV." *PRL* **72** (14): 2138–42.
    (1995a). "Search for High Mass Top Quark Production in $\bar{p}p$ Collisions at $\sqrt{s}$ = 1.8 TeV." *PRL* **74** (13): 2422–6.
    (1995b). "Observation of the Top Quark." *PRL* **74** (14): 2632–7.
    (1997). "Direct Measurement of the Top Quark Mass." *PRL* **79**: 1197.
Abazov, V. M., B. Abbott, et al. [D-zero] (2002). "$t\bar{t}$ Production Cross-Section in $\bar{p}p$ Collisions at $\sqrt{s}$ = 1.8 TeV." Preprint posted at: arXiv:hep-ex/0205019v2, 19Sep2002.
Abbott, B., M. Abolins, et al. [D-zero] (1997). "Measurement of the Top Quark Pair Production Cross Section in $\bar{p}p$ Collisions." *PRL* **79**: 1203.
    (1998a). "Measurement of the Top Quark Mass Using Dilepton Events." *PRL* **80**: 2063.
    (1998b). "Direct Measurement of the Top Quark Mass by the D0 Collaboration." *PRD* **58**: 052001.
    (1999a). "Measurement of the Top Quark Mass in the Dilepton Channel." *PRD* **60**: 052001.
    (1999b). "Measurement of the Top Quark Pair Production Cross Section in $\bar{p}p$ Collisions Using Multijet Final States." *PRD* **60**: 012001.
    (1999c). "Measurement of the Top Quark Pair Production Cross Section in the All-jets Decay Channel." *PRL* **83**: 1908.
Abe, F., D. Amidei, et al. [CDF] (1988). "The CDF Detector: An Overview." *Nuclear Instruments and Methods in Physics Research* **A271**: 387–403.
    (1990a). "Search for the Top Quark in the Reaction $\bar{p}p$ → electron + jets at $\sqrt{s}$ = 1.8 TeV." *PRL* **64**: 142–6.
    (1990b). "Search for New Heavy Quarks in Electron-Muon Events at the Fermilab Tevatron Collider." *PRL* **64**: 147–51.
    (1991). "Top-Quark Search in the Electron + Jets Channel in Proton-Antiproton Collisions at $\sqrt{s}$ = 1.8 TeV." *PRD* **43**: 664–86.

317

(1992a). "Lower Limit on the Top-Quark Mass from Events with Two Leptons in $\bar{p}p$ Collisions at $\sqrt{s} = 1.8$ TeV." *PRL* **68**: 447–51.

(1992b). "Limit on the Top-Quark Mass from Proton-Antiproton Collisions at $\sqrt{s} = 1.8$ TeV." *PRD* **45**: 3921–48.

(1994a). "Evidence for Top Quark Production in $\bar{p}p$ Collisions at $\sqrt{s} = 1.8$ TeV." *PRL* **73**: 225–31.

(1994b). "Evidence for Top Quark Production in $\bar{p}p$ Collisions at $\sqrt{s} = 1.8$ TeV." *PRD* **50**: 2966–3026.

(1995a). "Observation of Top Quark Production in $\bar{p}p$ Collisions with the Collider Detector at Fermilab." *PRL* **74** (14): 2626–31.

(1995b). "Search for Second Generation Leptoquarks in $\bar{p}p$ Collisions at $\sqrt{s} = 1.8$ TeV." *PRL* **75**: 1012–16.

(1995c). "Measurement of the W Boson Mass." *PRL* **75**: 11–16.

(1997). "First Observation of the All Hadronic Decay of $t\bar{t}$ Pairs." *PRL* **79**: 1992.

(1998a). "Measurement of the Top Quark Mass and $t\bar{t}$ Production Cross Section from Dilepton Events at the Collider Detector at Fermilab." *PRL* **80**: 2779.

(1998b). "Measurement of the Top Quark Mass." *PRL* **80**: 2767.

(1998c). "Measurement of the $t\bar{t}$ Production Cross Section in $\bar{p}p$ Collisions at $\sqrt{s} = 1.8$ TeV." *PRL* **80**: 2773.

(1999a). "Measurement of the Top Quark Mass with the Collider Detector." *PRL* **82**: 271.

(1999b). "Kinematics of $t\bar{t}$ Events at CDF." *PRD* **59**: 092001.

Abrams, G. S., D. Briggs, et al. (1974). "Discovery of a Second Narrow Resonance in e^+e^- Annihilation." *PRL* **33**: 1453–5.

Achinstein, P. (1968). *Concepts of Science: A Philosophical Analysis.* Baltimore, The Johns Hopkins University Press.

(1983). "Concepts of Evidence." In *The Concept of Evidence*, ed. P. Achinstein. Oxford, Oxford University Press, 145–74.

(1995). "Are Empirical Evidence Claims A Priori?" *British Journal for Philosophy of Science* **46**: 447–73.

(2000). "Why Philosophical Theories of Evidence Are (and Ought to Be) Ignored by Scientists." *Philosophy of Science* **67** (Supplement, PSA 1998): S180–92.

(2001). *The Book of Evidence.* New York, Oxford University Press.

Adeva, B., D. P. Barber, et al. (1983). "Search for Top Quark and a Test of Models without Top Quark up to 38.54 GeV at PETRA." *PRL* **50**: 799–802.

Affolder, T., H. Akimoto, et al. [CDF] (2001a). "Measurement of the Top Quark Mass with the Collider Detector at Fermilab." *PRD* **64**: 032003.

(2001b). "Measurement of the $t\bar{t}$ Production Cross Section in $\bar{p}p$ Collisions at $\sqrt{s} = 1.8$ TeV." *PRD* **64**: 032002.

(2002). "Search for Gluinos and Scalar Quarks in $\bar{p}p$ Collisions at $\sqrt{s} = 1.8$ TeV Using the Missing Energy plus Multijets Signature." *PRL* **88**: 4108.

Alavi-Harati, A., I. F. Albuquerque, et al. [KTeV] (1999). "Observation of Direct CP Violation in $K_{S,L} \to \pi\pi$ Decays." *PRL* **83**: 22.

Alberti, L. B. ([1435] 1966). *On Painting.* Tr. J. R. Spencer. New Haven, Connecticut, Yale University Press.

Alibini, E., S. R. Amendolia, et al. (1982). "Electronic Measurement of the Lifetime of $D^{+/-}$ Mesons." *PL* **110B**: 339–43.

Amati, D., H. Bacry, et al. (1964). "$SU(4)$ and Strong Interactions." *PL* **11**: 190–2.

Amendolia, S. R., G. Batignani, et al. (1980). "A Multi-electrode Silicon Detector for High Energy Experiments." *Nuclear Instruments and Methods in Physics Research* **176**: 457–60.

(1984). "High Resolution Silicon Detectors for Colliding Beam Physics." *Nuclear Instruments and Methods in Physics Research* **226**: 82–4.

Amidei, D. (1995). Oral History Interview by K. Staley. Tape Recording. November 21, 1995. By telephone from University of Michigan.

(1998). Oral History Interview by K. Staley. Tape Recording. July 14, 1998. Fermilab.

Amidei, D., P. Azzi, et al. (1994). "The Silicon Vertex Detector of the Collider Detector at Fermilab." *Nuclear Instruments and Methods in Physics Research.* **A350**: 73–130.

Anjos, J. C., J. A. Appel, et al. (1989). "A Study of the Semileptonic Decay Mode $D^0 \rightarrow K^+ e^+ \nu_e$." *PRL* **62**: 1587–90.

Ankenbrandt, C., C. Atac, et al. (1978). "Preliminary Design of a Magnetic Detector Facility for Colliding Beams at Fermilab" (CDF-11). Batavia, Illinois, Fermilab.

Arnison, G., A. Astbury, et al. [UA1] (1983). "Experimental Observation of Isolated Large Transverse Energy Electrons with Associated Missing Energy at $\sqrt{s} = 540$ GeV." *PL* **122B**: 103–16.

(1984). "Associated Production of an Isolated Large-Transverse-Momentum Lepton (Electron or Muon) and Two Jets at the CERN $\bar{p}p$ Collider." *PL* **147B** (6): 493–507.

Atac, M., M. Breidenbach, et al. (1977). "Report of the Detector Group." In *Summer Study on Colliding Beam Physics at Fermilab*, ed. J. K. Walker. Batavia, Illinois, Fermilab, 1–166.

Aubert, B., A. Benvenuti, et al. (1974). "Further Observation of Muonless Neutrino-induced Inelastic Interactions." *PRL* **32**: 1454–7.

Aubert, J. J., U. Becker, et al. (1974). "Experimental Observation of a Heavy Particle J." *PRL* **33**: 1404–6.

Augustin, J.-E., A. M. Boyarski, et al. (1974). "Discovery of a Narrow Resonance in $e^+ e^-$ Annihilation." *PRL* **33**: 1406–8.

Bacci, C., R. Balbibi Celio, et al. (1974). "Preliminary Result of Frascati (ADONE) on the Nature of a New 3.1 GeV Particle Produced in $e^+ e^-$ Annihilation." *PRL* **33**: 1408–10.

Bagnaia, P., M. Banner, et al. [UA2] (1983). "Evidence for $Z^0 \rightarrow e^+ e^-$ at the CERN $\bar{p}p$ Collider." *PL* **129B**: 130.

Banner, M., R. Battiston, et al. [UA2] (1983). "Observation of Single Isolated Electrons of High Transverse Momentum in Events with Missing Transverse Energy at the CERN $\bar{p}p$ Collider." *PL* **122B**: 476–85.

Barger, V., and S. Pakvasa. (1979). "Weak Isospin of the *b*-Quark and Possible Non-existence of a *t*-Quark." *PL* **81B**: 195–9.

Barnes, V. E. (1987). "Status and Prospects of the Fermilab Collider Detector, TeVatron Collider, and Antiproton Source." In *Collider Physics: Current Status and Future Prospects*, eds. J. E. Brau and R. S. Panvini. Singapore, World Scientific, 141–62.

Barnes, V. E., P. L. Connolly, et al. (1964). "Observation of a Hyperon with Strangeness Minus Three." *PRL* **12**: 204–6.

Barnett, B. (1996). Oral History Interview by K. Staley. Tape Recording. March 8, 1996. Johns Hopkins University.

Barnett, R. M. (1975). "Model with Three Charmed Quarks." *PRL* **34**: 41–3.

(1976). "Evidence for New Quarks and New Currents." *PRL* **36**: 1163–6.

Bedeschi, F. (1998). Oral History Interview by K. Staley. Tape Recording. July 16, 1998. Fermilab.

Bellettini, G. (1995). Oral History Interview by K. Staley. Tape Recording. October 18, 1995. Fermilab.

(1998). Oral History Interview by K. Staley. Tape Recording. July 16, 1998. Fermilab.

Benvenuti, A., D. Cline, et al. (1974). "Observation of Muonless Neutrino-Induced Inelastic Interactions." *PRL* **32**: 800–3.

Berger, J. O., and R. L. Wolpert (1984). *The Likelihood Principle.* Hayward, California, Institute of Mathematical Statistics.

Bhat, P., H. B. Prosper, and S. S. Snyder (1998). "Top Quark Physics at the Tevatron." *International Journal of Modern Physics A* **13**: 5113–218.

Bigi, I. I., and A. I. Sanda (2000). *CP Violation.* New York, Cambridge University Press.

Binkley, M. (1995). Oral History Interview by K. Staley. Tape Recording. October 19, 1995. Fermilab.

Birnbaum, A. (1977). "The Neyman-Pearson Theory as Decision Theory, and as Inference Theory; With a Criticism of the Lindley-Savage Argument for Bayesian Theory." *Synthese* **36**: 19–49.

Bjorken, J. D., and S. L. Glashow (1964). "Elementary Particles and $SU(4)$." *PL* **11**: 255–7.

Bown, W. (1992). "Transatlantic Row Keeps Top Quark Hidden." *New Scientist,* June 27, 1992, 10.

Brandelik, R., W. Braunschweig, et al. [TASSO] (1979). "Energy Scan for Narrow States in e^+e^- Annihilation at C.M. Energies between 29.90 and 31.46 GeV." *PL* **88B**: 199–202.

Broad, W. J. (1994). "Top Quark, Last Piece in Puzzle of Matter, Appears to Be in Place." *New York Times,* April 26, 1994.

Brown, L., and H. Rechenberg (1996). *The Origin of the Concept of Nuclear Forces.* Philadelphia, Institute of Physics Publishing.

Brown, L., M. Riordan, et al. (1997). "The Rise of the Standard Model: 1964–1979." In *The Rise of the Standard Model: Particle Physics in the 1960s and 1970s,* eds. L. Hoddeson, L. Brown, M. Riordan, and M. Dresden. New York, Cambridge University Press, 3–35.

Brown, L. M., R. Kawabe, et al., eds. (1991). *Elementary Particle Theory in Japan, 1930–1960, PTP Supplement* **105**.

Budker, G. I. (1967). "An Effective Method of Damping Particle Oscillations in Proton and Antiproton Storage Rings." *Soviet Journal of Atomic Energy* **22**: 438–40. Rpt. in *The Development of Colliders,* eds. C. Pellegrini and A. M. Sessler. New York, American Institute of Physics, 1995, 252–40.

Cahn, R. N., and F. J. Gilman (1978). "Polarized-electron–Nucleon Scattering in Gauge Theories of Weak and Electromagnetic Interactions." *PRD* **17**: 1313–22.

Cahn, R. N., and G. Goldhaber (1989). *The Experimental Foundations of Particle Physics.* New York, Cambridge University Press.

Campagnari, C., B. Farhat, et al. (1993a). "Search for Top in the Lepton + Jets Channel with a Lepton Tag, a First Look" (CDF-1961). Batavia, Illinois, Fermilab, January, 1993, CDF/ANAL/TOP/1961.

———— (1993b). "Muon Identification in the Lepton + Jets + Muon Tag Analysis" (CDF-2098). Batavia, Illinois, Fermilab, September 13, 1993, CDF/ANAL/TOP/CDFR/2098.

———— (1993c). "Update on the Search for Top in the Lepton + Jets Channel with a Lepton Tag" (CDF-2150). Batavia, Illinois, Fermilab, July 11, 1993, CDF/ANAL/TOP/CDFR/2150.

———— (1993d). "Update on the Search for Top in the Lepton + Jets Channel with a Lepton Tag, Volume 2" (CDF-2174). Batavia, Illinois, Fermilab, July 21, 1993, CDF/ANAL/TOP/CDFR/2174.

Campbell, D. T., and D. W. Fiske (1959). "Convergent and Discriminant Validation by the Multitrait-Multimethod Matrix." *Psychological Bulletin* **56**: 81–105.

Carnap, R. (1962). *The Logical Foundations of Probability*, 2nd ed. Chicago, University of Chicago Press.

Cartwright, N. (1983). *How the Laws of Physics Lie*. New York, Oxford University Press.

CDF (1981). *Design Report for the Fermilab Collider Detector Facility*. Batavia, Illinois, Fermilab.

———— (1987). "Status Report on CDF." *Fermilab Report*, May/June 1987, 15–21.

———— (1993a). "Search for the Top Quark Using Secondary Vertexing Tags in $W +$ Multijet Events in $\bar{p}p$ Collisions at $\sqrt{s} = 1.8$ TeV" (CDF-2267). Batavia, Illinois, Fermilab, October 4, 1993, CDF/PHYS/TOP/CDFR/2267.

———— (1993b). "A Search for the Top Quark in Events with Two High P_T Leptons in $\bar{p}p$ Collisions at $\sqrt{s} = 1.8$ TeV" (un-numbered CDF note). Batavia, Illinois, Fermilab, October 4, 1993.

———— (1993c). "Search for $t\bar{t}$ Production at CDF" (un-numbered CDF note). Batavia, Illinois, Fermilab, September 30, 1993.

Christenson, J. H., J. W. Cronin, et al. (1964). "Evidence for the 2π Decay of the K_2^0 Meson." *PRL* **13** (4): 138–40.

Cobal, M., H. Grassmann, and S. Leone (1994). "On Exploiting the Single-Lepton Event Structure for the Top Search." *Il Nuovo Cimento* **107A**: 75–83.

Contreras, M. (1995). Oral History Interview by K. Staley. Tape Recording. October 17, 1995. University of Chicago.

Crane, D. (1980). "An Exploratory Study of Kuhnian Paradigms in Theoretical High Energy Physics." *Social Studies of Science* **10**: 23–54.

Crease, R. P., and C. C. Mann (1986). *The Second Creation: Makers of the Revolution in Twentieth-century Physics*. New York, Macmillan.

Cronin, J. (1997). "The Discovery of *CP* Violation." In *The Rise of the Standard Model: Particle Physics in the 60s and 70s*, eds. L. Hoddeson, L. Brown, M. Riordan, and M. Dresden. New York, Cambridge University Press, 114–36.

Culp, S. (1995). "Objectivity in Experimental Inquiry: Breaking Data-Technique Circles." *Philosophy of Science* **62**: 430–50.

Dalitz, R. H., and G. R. Goldstein (1992a). "The Analysis of Top-Antitop Production and Dilepton Decay Events and the Top Quark Mass." *PL* **B287**: 225–30.

———— (1992b). "The Decay and Polarization Properties of the Top Quark." *PRD* **45**: 1531–43.

322 References

Danby, G., J.-M. Gaillard, et al. (1962). "Observation of High-Energy Neutrino Reactions and the Existence of Two Kinds of Neutrinos." *PRL* **9**: 36–44.

DeRújula, A., H. Georgi, and S. L. Glashow (1975a). "Changing the Charmed Current." *PRL* **35**: 69–72.

(1975b). "Vector Model of the Weak Interactions." *PRD* **12**: 3589–605.

Duhem, P. ([1892] 1996). "Some Reflections on the Subject of Experimental Physics." In *Essays in the History and Philosophy of Science*, eds. R. Ariew and P. Barker. Indianapolis, Hackett, 75–111.

([1914] 1954). *The Aim and Structure of Physical Theory*. Translated by Philip P. Wiener. Princeton, New Jersey, Princeton University Press.

Eaton, M. W. (1987). "Status of CDF." In *6th Topical Workshop on Proton-Antiproton Collider Physics, Aachen, Germany, 30 June–4 July 1986*, eds. K. Eggert, H. Faissner, and E. Radermacher. Singapore, World Scientific, 440–50.

Economist (1994). "Quarks Don't Bark." *The Economist*, March 26, 1994, 103.

Edwards, A. W. F. (1992). *Likelihood*, expanded edition. Baltimore, The Johns Hopkins University Press.

Ellis, J., M. K. Gaillard, and D. V. Nanopoulos (1976). "Left-Handed Currents and *CP* Violation." *Nuclear Physics* **B109**: 213–43.

Ellis, J., M. K. Gaillard, et al.(1977). "The Phenomenology of the Next Left-handed Quarks." *Nuclear Physics* **B131**: 285–307.

Engels, F. (1940). *Dialectics of Nature*, tr. Clemens Dutt. New York, International Publishers.

Feldman, G. J. (1981). "The Lepton Spectrum." In *Particles and Fields – 1981: Testing the Standard Model*, eds. C. A. Heusch and W. T. Kirk. New York, American Institute of Physics, 280–303.

Feldman, G. J., and R. D. Cousins. (1998). "Unified Approach to the Classical Statistical Analysis of Small Signals." *PRD* **57**: 3873–89.

Feldman, G. J., I. Peruzzi, et al. (1977). "Observation of the Decay $D^{*+} \rightarrow D^0\pi^+$." *PRL* **38**: 1313.

Fernow, R. C. (1986). *Introduction to Experimental Particle Physics*. New York, Cambridge University Press.

Fisher, R. A. (1955). "Statistical Methods and Scientific Induction." *Journal of the Royal Statistical Society (B)* **17**: 69–78.

Flam, F. (1992). "Researchers Quell Quark Rumour: The Top Is Still at Large." *Science*, July 24, 1992, 475–6.

Flaugher, B. (1998). Oral History Interview by K. Staley. Tape Recording. July 14, 1998. Fermilab.

Formaggio, J. A., E. D. Zimmerman, et al. [NuTeV] (2000). "Search for a 33.9 MeV/c^2 Neutral Particle in Pion Decay." *PRL* **84**: 4043–6.

Fox, G. C., and L. J. Romans (1980). "Longer Note on Top Quarks" (CDF-70). Batavia, Illinois, Fermilab.

Franklin, A. (1983). "The Discovery and Acceptance of *CP* Violation." *Historical Studies in the Physical Sciences* **13**: 207–38.

(1986). *The Neglect of Experiment*. Cambridge, Cambridge University Press.

(1997). "Instrumental Loyalty and the Recycling of Expertise." *Philosophy of Science* **64** (Proceedings of PSA'96): S42–S52.

(1998). "Selectivity and the Production of Experimental Results." *Archive for History of Exact Sciences* **53**: 399–485.

(2002). *Selectivity and Discord: Two Problems of Experiment.* Pittsburgh, University of Pittsburgh Press.

Franklin, M. (1998). Oral History Interview by K. Staley. Tape Recording. July 14, 1998. Fermilab.

Frisch, H. (1995). Oral History Interview by K. Staley. Tape Recording. October 17, 1995. University of Chicago.

Fritzsch, H., M. Gell-Mann, and P. Minkowski (1975). "Vectorlike Weak Currents and New Elementary Fermions." *PL* **59B**: 256–60.

Fukuda, Y., T. Hayakawa, et al. (1998). "Evidence for Oscillation of Atmospheric Neutrinos." *PRL* **81**: 1562–7.

Gaillard, M. K., B. W. Lee, and J. Rosner (1975). "Search for Charm." *Reviews of Modern Physics* **47**: 277–310.

Gaisser, T. K., and F. Halzen (1976). "Long-lived Tracks in Emulsions: New Hadrons or Background?" *PRD* **14**: 3153–66.

Galison, P. (1987). *How Experiments End.* Chicago, University of Chicago Press.

(1997a). "Pure and Hybrid Detectors: Mark I and the Psi." In *The Rise of the Standard Model: Particle Physics in the 1960s and 1970s*, eds. L. Hoddeson, L. Brown, M. Riordan, and M. Dresden. New York, Cambridge University Press, 308–37.

(1997b). *Image and Logic.* Chicago, University of Chicago Press.

Gamba, A., R. E. Marshak, and S. Okubo (1959). "On a Symmetry in Weak Interactions." *Proceedings of the National Academy of Sciences* **45**: 881–5.

Gell-Mann, M. (1961). "The Eightfold Way: A Theory of Strong Interaction Symmetry." Cal Tech Synchrotron Laboratory report CTSL-20. Rpt. in *The Eightfold Way*, eds. M. Gell-Mann and Y. Ne'eman. New York, W. A. Benjamin, 1964.

(1964). "A Schematic Model of Baryons and Mesons." *PL* **8**: 214.

Gell-Mann, M., and A. Pais (1955). "Behavior of Neutral Particles under Charge Conjugation." *Physical Review* **97**: 1387–9.

Georgi, H., and S. L. Glashow (1972). "Unified Weak and Electromagnetic Interactions without Neutral Currents." *PRL* **28**: 1494–7.

Gerdes, D. (1995). Oral History Interview by K. Staley. Tape Recording. October 6, 1995. Fermilab.

Giere, R. (1969). "Bayesian Statistics and Biased Procedures." *Synthese* **20**: 371–87.

Glanz, J. (2000). "New Tactic in Physics: Hiding the Answer." *New York Times*, August 8, 2000.

Glashow, S. (1961). "Partial-Symmetries of Weak Interactions." *Nuclear Physics* **22**: 579–88.

Glashow, S., J. Iliopoulos, and L. Maiani (1970). "Weak Interactions with Lepton-Hadron Symmetry." *PRD* **2**: 1285–92.

Goldhaber, G., F. M. Pierre, et al. (1976). "Observation in e^+e^- Annihilation of a Narrow State at 1865 MeV/c^2 Decaying to $K\pi$ and $K\pi\pi\pi$." *PRL* **37**: 255–9.

Goldstein, G. (1992). Letter to the Editor. *Science*, November 6, 1992.

Goldstein, G. R., K. Sliwa, and R. Dalitz (1992). "On Observing Top Quark Production at the Tevatron" (CDF-1750). Batavia, Illinois, Fermilab, CDF/HEAVYFLAVOUR/ANA/CDFR/1750.

(1993). "Observing Top-quark Production at the Fermilab Tevatron." *PRD* **47**: 967–72.

Gooding, D. (1990). *Experimentation and the Making of Meaning.* Cambridge, Cambridge University Press.

Goodman, N. (1968). *Languages of Art.* Indianapolis, Bobbs-Merrill.

Gottfried, K., and V. Weisskopf (1984). *Concepts of Particle Physics.* New York, Oxford University Press.

Grannis, P., B. Klima, et al. (1998). Oral History Interview by K. Staley. Tape Recording. July 10, 1998. Fermilab.

Groves, T. H. (1979). "Physics Advisory Committee Summer Meeting." *Fermilab Report,* July 1979, 1–9.

Gürsey, F., and P. Sikivie (1976). "E_7 as a Universal Gauge Group." *PRL* **36**: 775–8.

Hacking, I. (1983). *Representing and Intervening.* New York, Cambridge University Press.

Hagiwara, K., K. Hikasa, et al. (2002). *Review of Particle Physics. PRD* **66**: 010001.

Han, M.-Y., and Y. Nambu (1965). "Three-triplet Model with Double $SU(3)$ Symmetry." *Physical Review* **139** (4B): 1006–10.

Hanson, G., G. S. Abrams, et al. (1975). "Evidence for Jet Structure in Hadron Production by e^+e^- Annihilation." *PRL* **35**: 1609–12.

Hara, Y. (1964). "Unitary Triplets and the Eightfold Way." *Physical Review* **134**: B701–4.

Harari, H. (1975). "A New Quark Model for Hadrons." *PL* **57B**: 265–9.

Hasert, F. J., S. Kabe, et al. (1973). "Observation of Neutrino-like Interactions without Muon or Electron in the Gargamelle Neutrino Experiment." *PL* **45B**: 138–40.

Hayakawa, S. (1991). "Sakata Model and Activities in Nagoya." In *Elementary Particle Theory in Japan, 1930–1960,* eds. L. M. Brown, R. Kawabe, M. Konuma, and Z. Maki. *PTP Supplement* **105**: 120–2.

Hayashi, T., T. Karino, et al. (1973). "On the $\Delta S = 0$ and $\Delta N = 1$ Non-Leptonic Interaction in the New Nagoya Model." *PTP* **49**: 293–303.

Hayashi, T., E. Kawai, et al. (1972a). "A Posssible Interpretation of the New Event in the Cosmic Ray Experiment." *PTP* **47**: 280–7.

(1972b). "A Possible Interpretation of the New Event in the Cosmic Ray Experiment. II." *PTP* **47**: 1998–2014.

Hayashi, T., M. Kobayashi, et al. (1971). "Apparent Violation of the $\Delta S = \Delta Q$ Rule in High Energy Neutrino Processes." *PTP* **46**: 1944–5.

Hayashi, T., Y. Koide, and S. Ogawa (1968). "On the Weak Interaction and the Structure of Sakaton." *PTP* **39**: 1372–4.

Hayashi, T., M. Matsuda, and S. Ogawa (1973). "Effect of the Form Factor on the New Particle's Decay." *PTP* **49**: 353–4.

Hayashi, T., M. Nakagawa, et al. (1973). "On the Selection Rule of Weak Processes and Decays of New Particles." *PTP* **49**: 350–2.

Herb, S. W., D. C. Hom, et al. (1977). "Observation of a Dimuon Resonance at 9.5 GeV in 400 GeV Proton-Nucleus Collisions." *PRL* **39**: 252–5.

Higgs, P. W. (1964a). "Broken Symmetries, Massless Particles and Gauge Fields." *PL* **12**: 132–3.

(1964b). "Broken Symmetries and the Masses of Gauge Bosons." *PRL* **13**: 508–9.

(1966). "Spontaneous Symmetry Breaking without Massless Bosons." *Physical Review* **145**: 1156–63.

Hirokawa, S., and S. Ogawa (1989). "Shoichi Sakata – His Physics and Methodology." *Historia Scientiarum* **36**: 67–81.

Hoddeson, L. (1987). "The First Large-Scale Application of Superconductivity: The Fermilab Energy Doubler, 1972–1983." *Historical Studies in the Physical Sciences* **18**: 25–54.

Holmes, F. (1991). "Argument and Narrative in Scientific Writing." In *The Literary Structure of Scientific Argument*, ed. P. Dear. Philadelphia, University of Pennsylvania Press, 164–81.

Holton, G. (1988). *Thematic Origins of Scientific Thought: Kepler to Einstein*, revised ed. Cambridge, Massachusetts, Harvard University Press.

Hom, D. C., L. M. Lederman, et al. (1976). "Observation of High-Mass Dilepton Pairs in Hadron Collisions at 400 GeV." *PRL* **37**: 1374–7.

Howson, C., and P. Urbach (1993). *Scientific Reasoning: The Bayesian Approach*, 2nd ed. Chicago, Open Court.

Hughes, R. I. G. (1997). "Models and Representation." *Philosophy of Science* **64** (Proceedings of PSA '96): S325–36.

Hull, D. (1988). *Science as a Process: An Evolutionary Account of the Social and Conceptual Development of Science.* Chicago, University of Chicago Press.

Huth, J. (1995). Oral History Interview by K. Staley. Tape Recording. October 17, 1995. Hinsdale, Illinois.

Ikeda, M., S. Ogawa, and Y. Ohnuki (1959). "A Possible Symmetry in Sakata's Model for Bosons-Baryons System." *PTP* **22**: 715–24.

Incandela, J. (1995). Oral History Interview by K. Staley. Tape Recording. October 16, 1995. Fermilab.

(1998). Oral History Interview by K. Staley. Tape Recording. July 10, 1998. Fermilab.

Jacob, M. (1977). "Physics at Collider Energies." In *Colliding Beam Physics at Fermilab*, ed. J. K. Walker. Batavia, Illinois, Fermilab, 227–60.

Jensen, H. (1995). Oral History Interview by K. Staley. Tape Recording. October 18, 1995. Fermilab.

Jovanovic, D. (1995). Oral History Interview by K. Staley. Tape Recording. October 20, 1995. Fermilab.

Kadane, J. B., and T. Seidenfeld (1999). "Randomization in a Bayesian Perspective." In *Rethinking the Foundations of Statistics*, eds. J. B. Kadane, M. J. Schervish, and T. Seidenfeld. New York, Cambridge University Press, 293–313.

Kane, G. (1993). *Modern Elementary Particle Physics*, updated edition. Reading, Massachusetts, Addison-Wesley.

Katayama, Y., K. Matumoto, et al. (1962). "Possible Unified Models of Elementary Particles with Two Neutrinos." *PTP* **28**: 675–89.

Keynes, J. M. (1921). *A Treatise on Probability.* London, MacMillan and Co.

Kitcher, P. (1995). "The Cognitive Functions of Scientific Rhetoric." In *Science, Reason, and Rhetoric*, eds. H. Krips, J. E. McGuire, and T. Melia. Pittsburgh, University of Pittsburgh Press, 47–66.

Klima, B. (1995). Oral History Interview by K. Staley. Tape Recording. October 12, 1995. Fermilab.

Kobayashi, M. (1997). "Flavor Mixing and *CP* Violation." In *The Rise of the Standard Model: Particle Physics in the 1960s and 1970s*, eds. L. Hoddeson, L. Brown,

M. Riordan, and M. Dreseden. New York, Cambridge University Press, 137–42.

(1998). "CP Violation in a Six-quark Model." In *Twenty Beautiful Years of Bottom Physics: Proceedings of the b20 Symposium*, eds. R. A. Burnstein, D. M. Kaplan, and H. A. Rubin. Woodbury, New York, American Institute of Physics, 15–25.

Kobayashi, M., and T. Maskawa (1973). "*CP*-Violation in the Renormalizable Theory of Weak Interaction." *PTP* **49**: 652–7.

Kobayashi, M., M. Nakagawa, and H. Nitto (1972). "Quartet Models Based on Fundamental Particles with Fractional Charge." *PTP* **47**: 982–95.

Kodama, K., N. Ushida, et al. [DONUT] (2001). "Observation of Tau Neutrino Interactions." *PL* **504B**: 218–24.

Kondo, J., Z. Maki, and T. Maskawa (1972). "A Note on the Leptonic Decays of Charmed Mesons." *PTP* **47**: 1060–2.

Kondo, K. (1988). "Dynamical Likelihood Method for Reconstruction of Events with Missing Momentum. I. Method and Toy Models." *Journal of the Physical Society of Japan* **57**: 4126–40.

(1991). "Dynamical Likelihood Method for Reconstruction of Events with Missing Momentum. II. Mass Spectra for $2 \rightarrow 2$ Processes." *Journal of the Physical Society of Japan* **60**: 836–44.

(1995). Oral History Interview by K. Staley. Tape Recording. October 10, 1995. Fermilab.

Kondo, K., T. Chikamatsu, and S. Kim (1993). "Dynamical Likelihood Method for Reconstruction of Events with Missing Momentum. III. Analysis of a CDF High P_T $e\mu$ Event as $t\bar{t}$ Production." *Journal of the Physical Society of Japan* **62**: 1177–82.

Krige, J. (2001). "Distrust and Discovery: The Case of the Heavy Bosons at CERN." *Isis* **92**: 517–40.

Kuhn, T. (1970). *The Structure of Scientific Revolutions.* Chicago, University of Chicago.

Laenen, E., J. Smith, and W. L. van Neerven (1994). "Top Quark Production Cross Section." *PL* **B321**: 254–8.

Lande, K., E. T. Booth, et al. (1956). "Observation of Long-Lived Neutral V Particles." *Physical Review* **103**: 1901–4.

LaRue, G., J. Phillips, et al. (1981). "Observation of Fractional Charge of $(1/3)e$ on Matter." *PRL* **46**: 967–70.

Latour, B. (1987). *Science in Action: How to Follow Scientists and Engineers through Society.* Cambridge, Massachusetts, Harvard University Press.

Lederman, L. (1997). "The Discovery of the Upsilon, Bottom Quark, and *B* Mesons." In *The Rise of the Standard Model: Particle Physics in the 1960s and 1970s*, eds. L. Hoddeson, L. Brown, M. Riordan, and M. Dresden. New York, Cambridge University Press, 101–13.

Lee, T. D. (1973). "A Theory of Spontaneous *T* Violation." *PRD* **8**: 1226–39.

Leipuner, L. B., W. Chinowski, et al. (1963). "Anomalous Regeneration of K_1^0 Mesons from K_2^0 Mesons." *Physical Review* **132**: 2285–91.

Lenin, V. I. (1927). *Materialism and Empirio-Criticism: Critical Comments on a Reactionary Philosophy.* New York, International Publishers.

Lindley, D. V. (1957). "A Statistical Paradox." *Biometrika* **44**: 187–92.

Liss, T. (1995). Oral History Interview by K. Staley. Tape Recording. October 12, 1995. Fermilab.

(1998). Oral History Interview by K. Staley. Tape Recording. July 13, 1998. Fermilab.

Litke, A. M., and A. S. Schwarz (1995). "The Silicon Microstrip Detector." *Scientific American*, 75–81.

Low, M. F. (1991). "Accounting for the Sakata Model." In *Elementary Particle Theory in Japan, 1930–1960*, eds. L. M. Brown, R. Kawabe, M. Konuma, and Z. Maki. *PTP Supplement* **105**: 216–25.

Maglic, B. C., L. W. Alvarez, et al. (1961). "Evidence for a $T = 0$ Three-Pion Resonance." *PRL* **7**: 178–82.

Maiani, L. (1976). "CP Violation in Purely Lefthanded Weak Interactions." *PL* **62B**: 183–6.

Maki, Z. (1964a). "The 'Fourth' Baryon, Sakata Model and Modified *B-L* Symmetry. I." *PTP* **31**: 331–2.

(1964b). "The 'Fourth' Baryon, Sakata Model and Modified *B-L* Symmetry. II." *PTP* **31**: 333–4.

(1989). "The Development of Elementary Particle Theory in Japan – Methodological Aspects of the Formation of the Sakata and Nagoya Models." *Historia Scientiarum* **36**: 83–95.

(1991). "The Composite Model." In *Elementary Particle Theory in Japan, 1930–1960*, eds. L. M. Brown, R. Kawabe, M. Konuma, and Z. Maki. *PTP Supplement* **105**: 204–10.

Maki, Z., and T. Maskawa (1971). "Fundamental Quartets and Chiral $U(4) \otimes U(4)$." *PTP* **46**: 1647–9.

Maki, Z., T. Maskawa, and I. Umemura (1972). "Quartet Scheme of Hadrons in Chiral $U(4) \otimes U(4)$." *PTP* **47**: 1682–703.

Maki, Z., M. Nakagawa, et al. (1960). "A Unified Model for Elementary Particles." *PTP* **23**: 1174–80.

Maki, Z., M. Nakagawa, and S. Sakata (1962). "Remarks on the Unified Model of Elementary Particles." *PTP* **28**: 870–80.

Maki, Z., and Y. Ohnuki (1964). "Quartet Scheme for Elementary Particles." *PTP* **32**: 144–57.

Maki, Z., Y. Ohnuki, and S. Sakata (1966). "Remarks on a New Concept of Elementary Particles and the Method of the Composite Model." In *Proceedings of the International Conference on Elementary Particles: In Commemoration of the Thirtieth Anniversary of Meson Theory*, ed. Y. Tanikawa. Kyoto, Publication Office, Progress of Theoretical Physics, 109–23.

Marchesini, G., and B. R. Webber (1988). "Monte Carlo Simulation of General Hard Processes with Coherent QCD Radiation." *Nuclear Physics* **B310**: 461–526.

Marshak, R. E., and H. A. Bethe (1947). "On the Two-Meson Hypothesis." *Physical Review* **72**: 506–9.

Mayo, D. (1983). "An Objective Theory of Statistical Testing." *Synthese* **57**: 297–340.

(1985). "Behavioristic, Evidentialist, and Learning Models of Statistical Testing." *Philosophy of Science* **52**: 493–516.

(1991). "Novel Evidence and Severe Tests." *Philosophy of Science* **58**: 523–52.

(1996). *Error and the Growth of Experimental Knowledge*. Chicago, University of Chicago Press.

Mayo, D., and M. Kruse (2001). "Principles of Inference and Their Consequences." In *Foundations of Bayesianism*, eds. D. Cornfield and J. Williamson. Dordrecht, Kluwer Academic Publishers, 381–403.

Medawar, P. (1964). "Is the Scientific Paper Fraudulent?" *Saturday Review*, August 1, 1964, 43–4.

(1969). *The Art of the Soluble.* New York, Barnes and Noble.

Merton, R. K. (1973). *The Sociology of Science: Theoretical and Empirical Investigations.* Chicago, University of Chicago Press.

Metropolis, N., and S. M. Ulam (1949). "The Monte Carlo Method." *Journal of the American Statistical Association* **44**: 335–41.

Muller, F., R. W. Birge, et al. (1960). "Regeneration and Mass Difference of Neutral K Mesons." *PRL* **4**: 418–21.

Musgrave, A. (1974). "Logical versus Historical Theories of Confirmation." *British Journal for the Philosophy of Science* **25**: 1–23.

Nakagawa, M. (1991). "A Note on the Sakata Model before the Full-Symmetry." In *Elementary Particle Theory in Japan, 1930–1960*, eds. L. M. Brown, R. Kawabe, M. Konuma, and Z. Maki. *PTP Supplement* **105**: 211–13.

Nakagawa, M., and H. Nitto (1973). "Further Comments on Quartet Models on Fundamental Particles with Fractional Charge." *PTP* **49**: 1322–39.

Narain, M. (1995). Oral History Interview by K. Staley. Tape Recording. October 19, 1995. Fermilab.

Ne'eman, Y. (1961). "Derivation of Strong Interactions from a Gauge Invariance." *Nuclear Physics* **26**: 222–9.

(1974). "Concrete versus Abstract Theoretical Models." In *The Interaction Between Science and Philosophy*, ed. Y. Elkana. Atlantic Highlands, New Jersey, Humanities Press, 1–25.

Neyman, J., and E. S. Pearson (1928). "On the Use and Interpretation of Certain Test Criteria for Purposes of Statistical Inference. Part I." *Biometrika* **20(A)**: 175–240.

(1933). "On the Problem of the Most Efficient Tests of Statistical Hypotheses." *Philosophical Transactions of the Royal Society (A)* **1933**: 289–337.

Nickles, T. (1992). "Good Science as Bad History: From Order of Knowing to Order of Being." In *The Social Dimensions of Science*, ed. E. McMullin. Notre Dame, Indiana, University of Notre Dame Press, 85–129.

Niu, K., E. Mikumo, and Y. Maeda (1971). "A Possible Decay in Flight of a New Type Particle." *PTP* **46**: 1644–6.

Ogawa, S. (1985). "The Sakata Model and Its Succeeding Development toward the Age of New Flavours." *PTP Supplement* **85**: 52–60.

Pais, A., and O. Piccioni (1955). "Note on the Decay and Absorption of the θ." *Physical Review* **100**: 1487–9.

Pakvasa, S., W. A. Simmons, and S. F. Tuan (1975). "Weak Current in Harari's Heavy-Quark Model." *PRL* **35**: 702–4.

Pakvasa, S., and H. Sugawara (1976). "*CP* Violation in the Six-Quark Model." *PRD* **14**: 305–8.

Parker, M. A. (1989). "The Search for Top in UA2." In *The Fourth Family of Quarks and Leptons: Second International Symposium*, eds. D. B. Cline and A. Soni. New York, New York Academy of Sciences, 10–28.

Partridge, R. (1995). Oral History Interview by K. Staley. Tape Recording. October 18, 1995. Fermilab.

Paschos, E. (1977). "Diagonal Neutral Currents." *PRD* **15**: 1966–71.

Peirce, C. S. (1931–1958). *Collected Papers of Charles Sanders Peirce.* 8 vols., ed. C. Hartshorne and P. Weiss. Cambridge, Massachusetts, Harvard University Press.

(1982–). *Writings of Charles S. Peirce: A Chronological Edition,* ed. Max H. Fisch. Bloomington, Indiana, Indiana University Press.

Pellegrini, C., and A. M. Sessler, eds. (1995). *The Development of Colliders.* New York, American Institute of Physics.

Perl, M. (1997). "The Discovery of the Tau Lepton." In *The Rise of the Standard Model: Particle Physics in the 1960s and 1970s,* eds. L. Hoddeson, L. Brown, M. Riordan, and M. Dresden. New York, Cambridge University Press, 79–100.

Perl, M. L., G. S. Abrams, et al. (1975). "Evidence for Anomalous Lepton Production in e^+e^- Annihilation." *PRL* **35**: 1489–92.

Peruzzi, I., M. Piccolo, et al. (1976). "Observation of a Narrow Charged State at 1876 MeV/c^2 Decaying to an Exotic Combination of $K\pi\pi$." *PRL* **37**: 569–71.

Pevsner, A., R. Kraemer, et al. (1961). "Evidence for a Three-pion Resonance Near 550 MeV." *PRL* **7**: 421–3.

Pickering, A. (1984). *Constructing Quarks: A Sociological History of Particle Physics.* Chicago, University of Chicago Press.

Pitt, J. (1999). *Thinking About Technology: Foundations of the Philosophy of Technology.* New York, Seven Bridges Press.

Prescott, C. Y. (1997). "Weak-Electromagnetic Interference in Polarized *e-D* Scattering." In *The Rise of the Standard Model: Particle Physics in the 1960s and 1970s,* eds. L. Hoddeson, L. Brown, M. Riordan, and M. Dresden. New York, Cambridge University Press, 459–77.

Prescott, C. Y., W. B. Atwood, et al. (1978). "Parity Non-conservation in Inelastic Electron Scattering." *PL* **77B**: 347–52.

(1979). "Further Measurements of Parity Non-conservation in Inelastic Electron Scattering." *PL* **84B**: 524–8.

Rapidis, P. A., B. Gobbi, et al. (1977). "Observation of a Resonance in e^+e^- Annihilation Just above Charm Threshold." *PRL* **39**: 526–9.

Resnik, D. B. (1994). "Hacking's Experimental Realism." *Canadian Journal of Philosophy* **24**: 395–412.

Richter, B. (1997). "The Rise of Colliding Beams." In *The Rise of the Standard Model: Particle Physics in the 1960s and 1970s,* eds. L. Hoddeson, L. Brown, M. Riordan, and M. Dresden. New York, Cambridge University Press, 261–84.

Riordan, M. (1987). *The Hunting of the Quark.* New York, Simon and Schuster.

Rochester, G. D., and C. C. Butler (1947). "Evidence for the Existence of New Unstable Elementary Particles." *Nature* **4077**: 855–7.

Rosenkrantz, R. (1977). *Inference, Method, and Decision: Towards a Bayesian Philosophy of Science.* Dordrecht, Reidel.

Rubbia, C., P. McIntyre, and D. Cline (1977). "Producing Massive Neutral Intermediate Vector Bosons with Existing Accelerators." In *Proceedings of the International Neutrino Conference, Aachen, 1976,* eds. H. Faissner, H. Reithler, and P. Zerwas. Braunschweig, Vieweg, 683–7.

Sakata, S. (1956). "On a Composite Model for New Particles." *PTP* **16**: 686–8.

    (1959). "Ryoshi rikigaku no kaishaku o megutte (On the Interpretation of Quan-
tum Mechanics)." In *Butsuri-gaku to hoho: Ronshu 1 (Physics and Its Method: Collected
Papers Vol. 1)*. Tokyo, Iwanami Shoten, 51–70.

    (1971a). "My Classics – Engels' 'Dialektik der Natur'." *PTP Supplement* **50**: 1–8.

    (1971b). "Theory of Elementary Particles and Philosophy." *PTP Supplement* **50**:
199–207.

    (1971c). "Concluding Remarks at the Meeting." *PTP Supplement* **50**: 208–9.

Sakata, S., and T. Inoue (1946). "On the Correlations between the Meson and the
Yukawa Particle." *PTP* **1**: 143–50.

Salam, A. (1968). "Weak and Electromagnetic Interactions." In *Elementary Particle
Theory*, ed. N. Svartholm. New York, Wiley, 367–77.

Salmon, W. (1984). *Scientific Explanation and the Causal Structure of the World*. Princeton,
New Jersey, Princeton University Press.

Sanda, A. I. (1998). "20 Years of Beauty Physics and 50 Years of Search for Discoveries." In *Twenty Beautiful Years of Bottom Physics: Proceedings of the b20 Symposium*,
eds. R. A. Burnstein, D. M. Kaplan, and H. A. Rubin. Woodbury, New York,
American Institute of Physics, 367–78.

Sarton, G. (1937). *The History of Science and the New Humanism*. Cambridge,
Massachusetts, Harvard University Press.

Sato, S., and S. Nakamura (1972). "An Interpretation of Niu's Event in the Theory
of the $O(4)$ Symmetry." *PTP* **47**: 1991–7.

Savage, L. J., ed. (1962). *The Foundations of Statistical Inference: A Discussion*. London,
Methuen.

Sawayanagi, K. (1979). "New-particle Search in Very-high-energy Nuclear Interactions of Cosmic Rays." *PRD* **20**: 1037–51.

Schwarzschild, B. (1985). "First Proton-Antiproton Collisions at Tevatron." *Physics
Today*, December 1985, 23.

Schwitters, R. (1997). "Development of Large Detectors for Colliding-Beam Experi-
ments." In *The Rise of the Standard Model: Particle Physics in the 1960s and 1970s*, eds.
L. Hoddeson, L. Brown, M. Riordan, and M. Dresden. New York, Cambridge
University Press, 299–307.

Shochet, M. (1995). Oral History Interview by K. Staley. Tape Recording. October
17, 1995. University of Chicago.

    (1998). Oral History Interview by K. Staley. Tape Recording. July 15, 1998. Uni-
versity of Chicago.

Sikivie, P. (1976). "Gauge Theoretical Realisation of the Superweak Model of *CP*
Violation." *PL* **65B**: 141–4.

Sliwa, K. (1992). Letter to the Editor. *Science*, October 2, 1992, 13–14.

    (2002). "Top Quark Physics at the Tevatron: Results and Prospects." *Acta Physica
Polonica B* **33**: 3861–7.

Sliwa, K., G. R. Goldstein, and R. Dalitz (1992). "Search for $t\bar{t}$ Events in the 'Semi-
leptonic' Mode, or, On Observation of Top Quark Production at the Tevatron"
(CDF-1751). Batavia, Illinois, Fermilab, May 15, 1992, CDF/HEAVYFLAVOUR/
ANA/CDFR/1751.

Snyder, L. (1994). "Is Evidence Historical?" In *Scientific Methods: Conceptual and
Historical Problems*, eds. P. Achinstein and L. Snyder. Malabar, Florida, Krieger
Publishing Company, 95–117.

Staley, K. W. (1999a). "Golden Events and Statistics: What's Wrong with Galison's Image/Logic Distinction?" *Perspectives on Science* **7**: 196–230.

(1999b). "Logic, Liberty, and Anarchy: Mill and Feyerabend on Scientific Method." *The Social Science Journal* **36**: 603–14.

(2003a). "Agency and Objectivity in the Search for the Top Quark." Paper presented at The Johns Hopkins University Conference on Scientific Evidence, Baltimore, Maryland, April 12, 2003.

(2003b). "Robustness and Security." Paper presented at the Twelfth International Congress of Logic, Methodology, and Philosophy of Science, Oviedo, Spain, August 2003.

Steel, D. (2003). "A Bayesian Way to Make Stopping Rules Matter." *Erkenntnis* **58**: 213–27.

Steinberger, J., W. K. H. Panofsky, et al. (1950). "Evidence for the Production of Neutral Mesons by Photons." *Physical Review* **78**: 802–5.

Suppes, P. (1962). "Models of Data." In *Logic, Methodology and Philosophy of Science: Proceedings of the 1960 International Congress*, eds. E. Nagel, P. Suppes, and A. Tarski. Stanford, California, Stanford University Press, 252–61.

Taketani, M. (1966). "On the Meson Theory of Nuclear Forces." In *Proceedings of the International Conference on Elementary Particles: In Commemoration of the Thirtieth Anniversary of Meson Theory*, ed. Y. Tanikawa. Kyoto, Publication Office, Progress of Theoretical Physics, 170–80.

(1991). "Physics and Philosophy." In *Elementary Particle Theory in Japan, 1930–1960*, eds. L. M. Brown, R. Kawabe, M. Konuma, and Z. Maki. *PTP Supplement* **105**: 86–98.

Tanikawa, Y., ed. (1966). *Proceedings of the International Conference on Elementary Particles: In Commemoration of the Thirtieth Anniversary of Meson Theory*. Kyoto, Publication Office, Progress of Theoretical Physics.

Tarjanne, P., and V. L. Teplitz (1963). "*SU*(4) Assignments for the Vector Resonances." *PRL* **11**: 447–8.

Tasaka, S., and Y. Yamamoto (1973). "On the Life Time of X-particles Found by Niu et al." *PTP* **50**: 1879.

Taubes, G. (1986). *Nobel Dreams*. New York, Random House.

Terazawa, H. (1980). "*t*-quark Mass Predicted from a Sum Rule for Lepton and Quark Masses." *PRD* **22**: 2921.

Timko, M. (1998). Oral History Interview by K. Staley. Tape Recording. July 16, 1998. Fermilab.

Tipton, P. (1994). "The Top Search at CDF." In *Lepton and Photon Interactions*, eds. P. Drell and D. Rubin. New York, AIP Press, 464–78.

(1995). Oral History Interview by K. Staley. Tape Recording. October 19, 1995. Fermilab.

Tollefson, K. (1998). "Top Quark Production and Decay Measurements from CDF." Batavia, Illinois, Fermilab, July 23–30, 1998, FERMILAB-CONF-98/389-E. Conference paper presented at ICHEP 98, Vancouver, British Columbia.

Tollestrup, A. (1995). Oral History Interview by K. Staley. Tape Recording. October 11, 1995. Fermilab.

(1996). "The Tevatron Hadron Collider: A Short History." In *History of Original Ideas and Basic Discoveries in Particle Physics*, eds. H. B. Newman and T. Ypsilantis. New York, Plenum Press, 499–523.

(1998). Oral History Interview by K. Staley. Tape Recording. July 10, 1998. Fermilab.

Tomozawa, Y., and S. K. Yun (1975). "Incorporation of CP Violation with a Unified Renormalizable Gauge Theory." *PRD* **11**: 3018–25.

Top Working Group. Minutes of the CDF Top/Heavy Flavors Working Group. Fermilab, Batavia, Illinois.

Treitel, J. (1987). "Confirmation with Technology: The Discovery of the Tau Lepton." *Centaurus* **30**: 140–80.

Tsai, P. (1971). "Decay Correlations of Heavy Leptons in $e^+e^- \to l^+l^-$." *PRD* **4**: 2821–37.

Ulam, S. M., and J. von Neumann (1947). "On the Combination of Stochastic and Deterministic Processes: Preliminary Report." *Bulletin of the American Mathematical Society* **53**: 1120.

Vaitaitis, A., R. B. Drucker, et al. [NuTeV] (1999). "Search for Neutral Heavy Leptons in a High-Energy Neutrino Beam." *PRL* **83**: 4943.

van der Meer, S. (1972). "Stochastic Damping of Betatron Oscillations in the ISR." CERN, August, 1972, CERN/ISR-PO/72-31. Rpt. in *The Development of Colliders*, eds. C. Pellegrini and A. M. Sessler. New York, American Institute of Physics, 1995, 261–9.

von Mises, R. (1972). *Wahrscheinlichkeit Statistik und Wahrheit*, 4th ed. Library of Exact Philosophy, ed. M. Bunge. Vienna, Springer-Verlag.

Walker, J. K., ed. (1977). *Colliding Beam Physics at Fermilab*. 2 vols. Batavia, Illinois, Fermilab.

Watkins, P. (1986). *Story of the W and Z*. New York, Cambridge University Press.

Weinberg, S. (1967). "A Model of Leptons." *PRL* **19**: 1264–6.

(1976). "Gauge Theory of CP Nonconservation." *PRL* **37**: 657–61.

Whewell, W. (1989). *Theory of Scientific Method*. Ed. R. E. Butts. Indianapolis, Hackett Publishing Company.

Wilczek, F., A. Zee, et al. (1975). "Weak-interaction Models with New Quarks and Right-handed Currents." *PRD* **12**: 2768–80.

Wilson, R. R. (1977). "The Tevatron." *Physics Today*, October 1977, 23–30.

Wilson, R. R., and A. Kolb (1997). "Building Fermilab: A User's Paradise." In *The Rise of the Standard Model: Particle Physics in the 1960s and 1970s*, eds. L. Hoddeson, L. Brown, M. Riordan, and M. Dresden. New York, Cambridge University Press, 338–63.

Wimsatt, W. (1980). "Randomness and Perceived-Randomness in Evolutionary Biology." *Synthese* **43**: 287–329.

(1981). "Robustness, Reliability, and Overdetermination." In *Scientific Inquiry and the Social Sciences*, eds. M. Brewer and B. Collins. San Francisco, Jossey-Bass, 124–63.

(1987). "False Models as Means to Truer Theories." In *Neutral Models in Biology*, eds. M. Nitecki and A. Hoffman. New York, Oxford University Press, 23–55.

Woodger, J. H. (1962). "Abstraction in Natural Science." In *Logic, Methodology, and Philosophy of Science*, eds. E. Nagel, P. Suppes, and A. Tarski. Stanford, California, Stanford University Press, 293–302.

Worrall, J. (1985). "Scientific Discovery and Theory-Confirmation." In *Change and Progress in Modern Science*, ed. J. Pitt. Dordrecht, Reidel, 301–32.

(1989). "Fresnel, Poisson and the White Spot: The Role of Successful Predictions in the Acceptance of Scientific Theories." In *The Uses of Experiment: Studies in the Natural Sciences*, eds. D. Gooding, T. Pinch, and S. Schaffer. New York, Cambridge University Press, 135–57.

Wray, K. B. (2000). "Invisible Hands and the Success of Science." *Philosophy of Science* **67**: 163–75.

Yanagida, T. (1979). "Horizontal Symmetry and the Mass of the t Quark." *PRD* **20**: 2986–8.

Yao, W. (1995). Oral History Interview by K. Staley. Tape Recording. October 18, 1995. Fermilab.

Yao, W., F. Bedeschi, et al. (1995). "Top Search in Lepton + Jets with SECVTX" (CDF-2989). Batavia, Illinois, Fermilab, March 3, 1995. CDF/ANAL/TOP/CDFR/2989 version 2.1.

Yeh, G. P. (1989). "Top Search Progress in CDF." In *The Fourth Family of Quarks and Leptons: Second International Symposium*, eds. D. B. Cline and A. Soni. New York, New York Academy of Sciences, 1–19.

(1995). Oral History Interview by K. Staley. Tape Recording. October 12, 1995. Fermilab.

(1998a). Oral History Interview by K. Staley. Tape Recording. July 8, 1998. Fermilab.

(1998b). Oral History Interview by K. Staley. Tape Recording. July 16, 1998. Fermilab.

Yoh, J. (1998a). Oral History Interview by K. Staley. Tape Recording. July 14, 1998. Fermilab.

(1998b). "The Discovery of the b Quark at Fermilab in 1977: The Experiment Co-ordinator's Story." In *Twenty Beautiful Years of Bottom Physics*, eds. R. A. Burnstein, D. M. Kaplan, and H. A. Rubin. Woodbury, New York, American Institute of Physics, 29–42.

Zahar, E. (1973). "Why Did Einstein's Programme Supercede Lorentz's?" *British Journal for the Philosophy of Science* **24**: 96–123, 223–62.

Zweig, G. (1964a). "An *SU* (3) Model for Strong Interaction Symmetry and its Break-ing I." CERN preprint 8182-TH-401. Rpt. in *Developments in the Quark Theory of Hadrons, A Reprint Collection. Vol. I: 1964–1978*, eds. D. Lichtenberg and S. Rosen, Nonantum, Massachusetts, Hadronic Press, 1980.

(1964b). "An *SU*(3) Model for Strong Interaction Symmetry and Its Breaking II." CERN preprint 8419-TH-412. Rpt. in *Developments in the Quark Theory of Hadrons, A Reprint Collection. Vol. I: 1964–1978*, eds. D. Lichtenberg and S. Rosen, Nonantum, Massachusetts, Hadronic Press, 1980.

Subject Index

ADONE accelerator, 26, 299n15
aplanarity, 164, 167

b-matter. *See* Nagoya Model
background estimate, 80, 128, 155, 169,
  272; Method 1, 183–4, 190, 217,
  306n5; Method 2, 184, 190, 306n5
Bayes's Theorem, 279, 310n17, 311n24
Bayesian inference, 249–50, 265, 279,
  285, 287, 307n8, 309n5, 311n24
beam cooling: electron, 49–50, 74;
  stochastic, 50, 74
Bernoulli process, 215
bias, 6, 142, 146–8, 155, 170–1, 182, 227,
  243, 264, 272, 281, 287–9, 308n3, 313;
  detection of, 193–200; prevention of,
  through predesignation, 6, 146–7,
  248–51, 262; prevention of, through
  pre-trial planning, 272–4. *See also*
  cuts, and tuning on the signal
big science, 1–3
blind box technique, 273–4
bosons, 8, 24, 300n21; intermediate
  vector (*W, Z*), 46, 50–1, 65, 70, 73–6,
  78, 84, 87, 89, 97, 104–7, 124, 142,
  144, 250–2, 255, 300n21
bottom (*b*) quark, 3, 8, 11, 15, 34–5, 37,
  65, 73–4, 76, 85, 96–9, 115, 124–8,
  142–4, 152, 154, 178–9, 219–21,
  251–3, 302n1, 304n17. *See also*
  upsilon

Brookhaven National Laboratory
  (BNL), 16, 26, 67, 79; Isabelle, 67,
  70–2, 79

Cabibbo matrix, 28–9
Cabibbo-Kobayashi-Maskawa (CKM)
  matrix, 29, 301n35
calorimeters, 74, 99–100, 163–4, 303n2;
  electromagnetic, 55, 58, 60, 62;
  hadron, 55, 58, 60–2, 65, 77, 203;
  segmentation of, 60–2, 65, 76, 85, 99,
  163
CDF detector, 4, 47, 52–102, 109–10,
  115, 130–1, 176–7, 219–20, 302n6; as
  allgemeine detector, 62, 85; as hybrid
  detector, 63
charm, 8, 14, 18–19, 299n17, 300n27,
  302n1; *X* particle as, 24–6,
  299nn17,18. *See also J/ψ* particle
citations, 8–11, 23, 25, 30, 34, 36, 41–4,
  173; as fulfillment of social
  obligation, 43; statistics on, at
  SLAC-SPIRES website, 297n1,
  301n37
Collider Detector at Fermilab (CDF)
  collaboration, 1–6, 45–7, 52–3,
  60–110, 120–1, 126, 130–2, 135–9,
  150–1, 156–7, 163–4, 166–7, 169–72,
  176–87, 202–3, 241–2, 273, 286–9,
  293, 302nn6,9, 303n3, 304n1, 306n5,
  309n12

spokesperson elections, 103
standard model, 1, 3, 8–10, 14–15, 34,
36–7, 45–7, 52, 87, 162, 220, 295,
299n18, 300n21
Stanford Linear Accelerator Laboratory
(SLAC), 26, 30, 36, 51, 95; Mark I
detector, 63, 73, 85; Positron Electron
Project (PEP), 46; Stanford
Positron-Electron Asymmetric Ring
(SPEAR), 30, 33, 73, 243
stopping rules, 221, 249–50, 279, 284–7,
309nn5,11, 311n22. *See also* bias,
prevention of through pre-trial
planning
SU(3) symmetry, 14–17, 19, 21, 31
success: epistemic, 240–3, 255, 262–4,
268–9, 308nn2,3; evidential, 240–2,
262, 264, 268–9, 308n2, 309n10;
experimenters', 241–2, 264–5, 269,
273, 291, 296, 308n3
Supersymmetry (SUSY), 116, 273, 295
symmetric (vector) models, 32, 36–7

τ lepton, 15, 31, 36, 106; discovery of,
31, 33, 36
τ neutrino, 8, 31, 36
tagged *Z* events, excess of, 140–3, 155,
162, 191
technicolor, 73, 76, 295
Texas towers, 90
Tevatron, 47, 50, 70–1, 76, 78–9, 81, 84,
88, 104, 295, 304n15. *See also* Fermi
National Accelerator Laboratory
(Fermilab)
themata, 38–40, 301n33
three-stage epistemology, 20–2, 37
top quark, 50, 65, 76, 80, 85, 87,
114–19, 133–5, 142, 163–4, 175–6,
187–8, 190–1, 215–20, 241–2, 275,
293–5, 304n15, 308n18, 309n12;
decay modes, 73–4, 97, 104–7,

111–12, 115, 118–19, 124, 144, 149,
160, 162, 164, 187–8, 219, 250–3, 255,
293–4; hypothesized, 3, 9–11, 29–32,
44; mass, 45–6, 104, 107–8, 144–7,
160, 166–7, 183–5, 187–8, 191, 216,
233, 242, 251–9, 276, 293–4; pair
production cross section, 97, 153,
160, 191, 199, 257, 293–5
tracking chambers, 55, 64, 81, 83–4, 89,
97, 99, 164, 178, 309n6
transverse mass analysis, 107–8
triggering, 61–3, 90
Tristan accelerator, 72
tuning on the signal. *See* cuts, and
tuning on the signal
two-meson theory, 18–20, 40

U.S. Department of Energy (DOE), 50,
67, 70–2, 75–7, 79, 82–3
Underground Area 1 (UA1)
collaboration, 85, 104; detector as
precursor to CDF's, 164; discovery of
W, Z bosons, 78; pseudodiscovery of
top quark, 79–80
Underground Area 2 (UA2)
collaboration, 78–80, 104, 107, 177,
181; detector as precursor to
D-zero's, 164
upsilon, discovery of, 34–6, 45, 51;
"oops-leon," 35
urbaryons, 17–19, 21–2, 24, 26, 38,
297n7, 298n12. *See also* Nagoya
model, revised

vector models. *See* symmetric models

weak interactions, 10–14, 16, 25, 27–8,
33, 36, 299n17

X particle, 23–7, 39, 299nn17,18. *See also*
charm, *X* particle as

Name Index

Gell-Mann, Murray, 12–16, 19, 21, 32, 36–7
Gerdes, Dave, 120, 123, 126, 130, 138, 157, 178–81
Giere, Ron, 249
Giromini, Paolo, 77
Glashow, Sheldon, 10, 13–14, 18–19, 24–5, 27, 30, 32, 36, 297n8, 299n17
Goldstein, Gary, 114, 116–23, 132, 183
Gooding, David, 202–3, 209, 212–13, 307n5
Grannis, Paul, 167
Grassmann, Hans, 114–15
Greenlee, Herb, 189

Haber, Carl, 95, 177
Hacking, Ian, 194, 237, 308n18
Halzen, F., 299n17
Harari, Haim, 31–3, 300n27
Hitlin, Dave, 51
Holton, Gerald, 38, 301n33
Howson, Colin, 224–5, 228, 234, 249, 285–6
Hughes, R. I. G., 204–6
Hughes, Richard, 108, 126
Hull, David, 42–3, 305n6
Huth, John, 135, 138, 151, 153

Incandela, Joe, 177–84
Innes, Walter, 35
Inoue, Takesi, 40

Jensen, Hans, 76–8, 111–12, 130, 135, 152
Jovanovic, Drasko, 112, 123, 130, 313–14

Kadane, Joseph, 265
Kadel, Richard, 82–3, 97
Kestenbaum, David, 185, 198
Keynes, John Maynard, 279–80, 282–4, 310n18
Kim, Shin-Hong, 117
Kitcher, Philip, 9, 173
Kleinfelder, Stewart, 95
Klima, Boaz, 164–5, 167, 187–8

Kobayashi, Makoto, 3, 8–11, 13–15, 24–30, 32–7, 39, 41, 43, 51, 297n2, 300nn22,24, 301nn28,35,36
Kondo, Kuni, 24–5, 71–2, 114–18, 132, 304n11
Kruse, Michael, 250, 309n5

Lederman, Leon, 15, 35, 70–1, 73–4, 81, 100, 103, 187, 301n29
Lee, Benjamin, 26
Lee, T. D., 301n28
Lenin, V. I., 19, 298n10
Leone, S., 115
Limon, Peter, 65, 68
Liss, Tony, 102, 112, 130, 135, 137–40, 157, 161–2, 169–70, 182, 241

Maiani, Luciano, 10, 13–14, 26–7, 32–4, 36
Maki, Ziro, 16–26, 38, 298n12, 301n35
Marshak, Robert, 40, 297n6
Maskawa, Toshihide, 3, 8–11, 13–15, 24–5, 27–9, 32–7, 39, 41, 43, 51, 297n2, 300n24, 301nn28,35,36
Mayo, Deborah, 3, 204, 206–7, 209, 211, 213, 222–6, 229, 237, 249–50, 262, 267–8, 275–6, 284–5, 290, 307nn11,13, 309n5, 311nn20,21
McIntyre, Peter, 50–1
Menzione, Aldo, 92–3, 95, 97–100, 130–1, 213
Merton, Robert, 42–4, 117
Metropolis, Nicholas, 307n9
Mill, John Stuart, 311n25

Nakagawa, Masami, 16–17, 22, 24–5, 301n35
Nanopoulos, Dimitrius, 34, 36, 43, 45
Narain, Meenakshi, 188–9
Ne'eman, Yuval, 14–16, 21, 37–40, 301n35
Newman-Holmes, Cathy, 144
Neyman, Jerzy, 223, 236–7
Nickes, Thomas, 5, 172–3
Niu, Kiyoshi, 23–6, 299n18